CAMBRIDGE GEOGRAPHICAL STUDIES

15 EMPIRICISM AND GEOGRAPHICAL THOUGHT

From Francis Bacon to Alexander von Humboldt

CAMBRIDGE GEOGRAPHICAL STUDIES

The frontispiece shows Alexander von Humboldt (1769–1859) with Mount Chimborazo in the background. Painted by Julius Schrader in 1859, it is the last portrait of Humboldt before his death.

EMPIRICISM AND GEOGRAPHICAL THOUGHT

From Francis Bacon to
Alexander von Humboldt

MARGARITA BOWEN

CAMBRIDGE UNIVERSITY PRESS

CAMBRIDGE
LONDON   NEW YORK   NEW ROCHELLE
MELBOURNE   SYDNEY

CAMBRIDGE UNIVERSITY PRESS
Cambridge, New York, Melbourne, Madrid, Cape Town, Singapore, São Paulo, Delhi

Cambridge University Press
The Edinburgh Building, Cambridge CB2 8RU, UK

Published in the United States of America by Cambridge University Press, New York

www.cambridge.org
Information on this title: www.cambridge.org/9780521105590

First published 1981
This digitally printed version 2009

A catalogue record for this publication is available from the British Library

ISBN 978-0-521-23653-9 hardback
ISBN 978-0-521-10559-0 paperback

TO JAMES
AND OUR FAMILY

CONTENTS

ILLUSTRATIONS

Alexander von Humboldt by Julius Schrader (1859), showing Mount Chimborazo in the background: the last portrait during Humboldt's lifetime. Reproduced by courtesy of the Schiller-Nationalmuseum, Marbach, West Germany. *frontispiece*

Illustrations 1, 3 and 5–7: Cambridge University Library. Illustration 2: the British Library. Illustration 4: the Oxford School of Geography.

ACKNOWLEDGMENTS

In addition to the reliance on a large volume of publications this study owes much to numerous institutions and the personal help of scholars in many parts of the world. At the University of Oxford valuable assistance was received during my visits in 1969 and 1973 to the School of Geography, and my special thanks are extended to Professor Jean Gottmann, to Miss Elspeth Buxton, librarian, and to Dr James Houston, formerly University Lecturer in Geography at Oxford and now Chancellor, Regent College, Vancouver. More recently I have been grateful for the excellent advice of Dr E. A. Wrigley and the editorial advisers of Cambridge University Press. I wish to express appreciation also for suggestions offered by Professor David Lowenthal of University College London, Dean George Tatham of York University, Ontario, Professor Joseph May of the University of Toronto, and Professor Clarence Glacken of the University of California, Berkeley; I am indebted too for the Humboldt researches of Dr Anne Macpherson at Berkeley. In Germany I am pleased to acknowledge the generous aid of Professor Hanno Beck and the late Professor Carl Troll of the University of Bonn, and in Switzerland of Dr Heinz Balmer who kindly made available a private collection of Humboldt's letters.

Of the many libraries that made this research possible, particular recognition is due to the Bodleian Library, the British Library, Cambridge University Library and the University Libraries of Bonn and Göttingen, as well as the Deutsches Literaturarchiv Marbach, where access was granted to Humboldt archives. Equally I must thank the librarians of the Australian National University and the Universities of Sydney, Melbourne and New England, the Library of New South Wales and the National Library, Canberra.

In Australia this project was commenced with the aid of a Commonwealth Post-Graduate Research Award from 1970 to 1974; my thanks are due also to Dr Evan Burge, formerly of the Australian National University, and Dr Joseph Powell of Monash University, for their comments. At the University of New England the stimulus for this work in the Geography Department owed a great deal to the foundation Professor, Gilbert Butland, for his interest in the history of geographical thought; and finally as an inter-disciplinary study this has benefited from the advice of Richard Franklin,

Professor of Philosophy, Alan Treloar, Reader in Comparative Philology, and members of the German Department at New England. I wish to express my gratitude to them, and above all to my husband James, for the challenge offered by his own ideas and for his constant help and encouragement.

INTRODUCTION

As part of a widespread reassessment of the positivist movement in science, this study considers its impact on geography during the period from 1600 to 1860, commencing with the sense-empiricism of Francis Bacon and concluding with the work of Alexander von Humboldt. A founder of modern geography, Humboldt was convinced of the need to change the established model of scientific procedure in his day: the formulation of that model during the preceding two centuries is examined here, along with concurrent developments in geographical thought, as shown in the leading texts of the time. That period, largely one of decline in geography, has received little attention from geographers themselves, who furthermore have tended to consider the history of the subject in isolation rather than in relation to contemporary developments in philosophy and science; in the absence also of any serious attention to the position of geography in recent works on the history and philosophy of science, the present study aims to provide a framework for further research and criticism. It is suggested that the theories of Humboldt, advanced in the same period as the positivism of Comte and Mill, or the historical materialism of Marx and Engels, offer a meaningful alternative for science today, leading to a new form of social empiricism. For geography, currently engaged in an effort to reconcile the radical–activist movement of the 1970s with the positivism of the mid-century quantitative revolution, the ferment of debate in the early nineteenth century shows a remarkable similarity with present conflicts in the sciences.

In the face of mounting world problems the entire scientific movement in the late twentieth century is undergoing a major revision of its aims, its methods, and its impact on society and the earth as a whole. Demands for social responsibility in science and a growing scepticism about the so-called scientific method itself are leading to a massive reappraisal of the whole philosophy of science that has been dominant for three centuries, and many observers see this as a new scientific revolution. One of the strongest pressures for change is a deepening concern that scientists will prove incapable of responding in time to solve the complex social and environmental problems now seen to be threatening the global ecosystem – problems that science frequently helped to create. Under the established framework of theory, technology and organizational structure in the sciences, most effort in the past was directed towards the solution of more specialized problems through relatively isolated programmes of research. Those areas of inquiry which were not considered amenable to standard scientific methods, or did not pursue the same kinds of questions, were excluded from the sciences, and those like geography which aspired to join their ranks were obliged to adapt the questions they asked and accept the kind of limitations imposed by the traditional model of 'exact science'. In a number of studies, including economics, psychology, education and geography, which after the Second World War attempted to identify with that model, the outcome was apparent: a trend towards specialization and restriction of inquiry, a preoccupation with data easily adapted to quantitative analysis, a rejection of wider moral issues and a reorientation of the entire discipline towards the accomplishment of precisely formulated research projects, a neglect of history and philosophy, and a concomitant loss of general theory. With the gradual recognition that earlier promises of rapid progress were not to be fulfilled, this approach has come under increasing criticism for its conceptual poverty and ineffectuality in dealing with contemporary problems. Drawing attention to the failure of the demigod scientist to cope with these issues, Wilbur Zelinsky, as a leading economic geographer, joined those Western critics who for some years had called for a revolutionary change in outlook, and in his 1974 Presidential Address to the conservative Association of American Geographers he emphasized 'the need for a profoundly different approach and perhaps even a totally novel philosophy of science'.[1]

The construction of a coherent alternative framework presents one of the major challenges of the next decade, and although Zelinsky's own Address offered little guidance on what form the new philosophy should take, it is clear that one of the first tasks of the growing revisionist movement is a more searching reassessment of the previous framework, in particular its prescriptions for scientific method and the theory of knowledge on which these were based.

Compounded of many ideas and subject to constant modification after it was formulated in the seventeenth century, the empirical method, as it came to be known, has played a significant part in directing the course of modern science. While it proved highly effective in organizing inquiry in numerous specialized sciences during the last three centuries, that method appears to have been largely inimical to the development of geography and so the position of this subject is of particular interest in the current debate. Until recently geographers showed a marked reluctance to engage in serious criticism of the established theory of science and in consequence allowed their study to remain methodologically at a disadvantage in relation to sciences like physics and chemistry for which the empirical method was expressly formulated. A large measure of the strength and apparent impregnability of the scientific movement since the seventeenth century has been derived from the theoretical justification it received in the form of a theory of knowledge developed and refined by a long line of philosophers, commencing with the British empiricists and continuing through to the modern positivists. In order to change that theory of scientific procedure it is necessary, as Humboldt saw over a century ago, to propose an alternative procedure based on a more appropriate theory of knowledge.

Science and empiricism

In general, science from its original connotation of 'knowledge' (Latin *scientia*) has come to imply the constant extension of knowledge about the world in accordance with an approved method for conducting inquiry. The first step in any programme of constructive change is then to consider what is meant by knowledge in this context and to examine the traditional claims for the validity and universality of the scientific or empirical method. Although the term 'empirical' is used frequently in connection with scientific inquiry, its definition and its implications are rarely examined. David Harvey for instance, in 1969 directed his major methodological study *Explanation in Geography* to the provision of 'guidelines for the conduct of empirical research in geography', without discussing the term 'empirical' itself or including it in his index.[2] Not only among practising scientists but even among reputable historians or philosophers this term has been used as if its meaning were self-evident, with empirical research commonly assumed to involve in the first instance some sort of direct contact through observation or experiment with the objects of study, a form of contract in which ideas or theory supposedly play no part at all. The historian Sir George Clark, for example, in considering the legacy of seventeenth-century empiricism, repeated that view without question: 'This immediate contact with the facts of nature, this attention to what is perceived by the senses, is a characteristic of modern science.'[3] The source of such a view can be traced to what is called the *sense-empiricism* of Francis Bacon. Founder of the British empiricist school of philosophy, Bacon gave expression to some of the most significant elements of later scientific method, and an important task in the search for a new theory is a critical re-examination of his views.

An essential characteristic of empiricism, which takes its name from the Greek word for experience (*empeiria*), is its commitment to the position that all knowledge is dependent upon experience, and this remains a widely accepted tenet of contemporary science and education: the main difficulty comes from the interpretation of experience. Where this is identified purely with some form of first-hand sense-experience, as in the Baconian tradition, the empiricist position can become restricted to a kind of *sensationism* which regards knowledge as a series of sense-impressions, or even to the sort of *objectivism* mentioned earlier which suggests that objects themselves, directly impinging on the senses, can provide accurate knowledge as long as the mind is not permitted to intervene. Prominent in scientific empiricism during the last three centuries, sensationist and objectivist views have been linked closely with the doctrine of *materialism* that led scientists to consider the world in terms of the mechanical actions of matter. Together, those views provided the foundation for what became the standard scientific method, as one based not on theories but on 'facts' obtained by strict observation and measurement, with the aim of securing exact knowledge through 'induction', by a procedure of dispassionate mathematical and logical analysis leading to the discovery of general laws. Science itself was extolled as a means of increasing man's power over the material world. In extending that method to the new study of sociology in the early nineteenth century, Auguste Comte gave the name of *positivism* to the movement which came to dominate the emerging social sciences as well as the physical sciences in the following years.

Positivism, as an extension of Bacon's form of sense-empiricism, gained its main strength throughout the nineteenth century by promising a sure means of progress for Western man in the use of knowledge for the control of nature, and its success was linked with the rise of industrial capitalism. A practical science, freed from abstractions and moral issues, was seen as the key to technological progress, and beyond this positivists regarded the scientific method of induction from facts or natural phenomena as the only acceptable and valid means of obtaining knowledge: all speculation on questions that could not be tested by experiment or were beyond the range of sense-experience was rejected, for it was claimed that the new positive science would eliminate traditional religion, idealist philosophy, and metaphysics, replacing these with ordered, objective and reliable knowledge of both man and nature. Social reformers, like Comte in France and the British utilitarians James Mill and his son John Stuart Mill, argued further that scientific progress through positivism would lead to a perfect form of society.

Few of its followers questioned the foundations on which the positivist programme rested, for one of its fundamental prescriptions was to rule out any discussion on the general assumptions underlying its formula for scientific method. Positivist philosophers themselves avoided such theorizing as 'metaphysical' and unscientific: John Stuart Mill, for example, concentrated on explaining the principles of logic involved in induction, while the school of logical positivists in the first half of the twentieth century went on to examine the limits of scientific knowledge and the problem of verifiability, using techniques of logical and linguistic analysis as a means of following the positivist dictum that scientists should restrict their attention to observable phenomena which can be measured, predicted or controlled. Practising scientists meanwhile readily adopted the precepts of Lavoisier, the French founder of modern chemistry, who in 1789 advised his followers to keep to the method of Bacon, working only from direct observations and modern experiments, and dispensing with the study of both philosophy and history, as impediments to progress in science. This anti-philosophical and anti-historical approach contributed to the rigid character of later positivism and proved an obstacle to its reform.

Today, as traditional positivism comes under increasing attack and it becomes clear that future progress depends on a reorganization of science to deal with world problems, a renewed significance is given to research on the history and philosophy of science in the task of re-evaluating traditions and restructuring the entire framework, or – to use an effective term – the whole *paradigm* of shared concepts, methods, goals and rewards, within which scientists operate.[4] The current emergence of a new pattern can be traced in a number of sciences, and indeed the high degree of concurrence evident in trends occurring in a wide variety of disciplines, in different parts of the world, lends support to the view that what Thomas Kuhn called a paradigm shift is now taking place.[5]

Geography and positivism

In geographical thought the reaction against positivism followed rather rapidly upon its serious introduction into the subject in the mid twentieth century. Geography entered its positivist phase much later than most other sciences, for until the 1950s it was generally considered that the scientific method was not strictly applicable in geography, with its emphasis on regional studies, lack of experimental research and inability to discover precise natural laws. In 1953 Fred Schaefer in the United States strongly criticized this 'exceptionalist' position for geography and called for geographers to use the same method as the other sciences but to differentiate their own study by limiting themselves to a concern with spatial relations: 'Spatial relations are the ones that matter in geography and no others.'[6] Schaefer's work had a tremendous impact in establishing the school of spatial analysis, and over the next decade a great deal of research in both physical and human geography was adapted to the current scientific model. That model, although Schaefer did not use the term, was clearly a positivist one, and the conflicts and changes that ensued in geography were associated with this.

It has become customary to describe the major changes as revolutions: the quantitative revolution of the fifties and sixties when a strong effort was made to introduce the standard scientific method, especially mathematical analysis, into geography, and then the radical revolution of the seventies, when social reformers and conservationists joined forces to call for new goals, new methods and a new focus for research. The present effort to reconcile those two movements in some kind of alternative framework represents, as I see it, a third revolution, both conceptual and social in its implications, and one that is liable to change the direction of geography for the rest of the century.

Signs of a growing criticism of positivist views began to emerge in geography during the 1960s. Throughout the previous decade, few geographers had followed John Kirtland Wright's tentative exploration of a new field in his 1946 Presidential Address to the Association of American Geographers on *the place of imagination in geography*. Wright defended the place of aesthetic subjectivity, the intuitive, and the methods of the humanities, alongside what he evidently accepted as the 'impersonal objectivity' of scientific geography. His position remained defensive and was scarcely strengthened by his rather coy references to imagination as 'hearkening to the Sirens': he attempted to moderate, but not to deny, 'the lofty observation point of the objective'. Sensitive and far-seeing, nonetheless, Wright recommended attention to the geography of knowledge and especially to what he called *geosophy*, described by him as an aspect of the sociology of knowledge concerned with the study of geographical ideas, both past and present, and including 'the relations of scientific geography to the historical and cultural conditions of which it is a product'. Constantly, he entered a plea for the history of geography, viewed more comprehensively than the traditional record of knowledge acquired in 'the core area of scientific geography'.[7]

In 1961 David Lowenthal went on to explore Wright's field of geosophy, with a paper 'Geography, experience, and imagination' which incorporated recent research in psychology, sociology and the philosophy of science, and dealt with important questions of epistemology or theory of knowledge (Greek *epistēmē*, knowledge).[8] In particular it drew attention to both the limitations and the complexity of human perception. Lowenthal's argument was a challenging one: 'sensing, thinking, feeling, and believing are simultaneous, interdependent processes'. Perception, he pointed out, is not a matter of the senses alone: seeing is conceptual as well as visual, involving information acquired prior to the present experience; thus all interpretations of stimuli are culturally conditioned. Acknowledging the separateness of individual experiences and the subjectivity of the private milieu, he emphasized at the same time the significance of the shared world view. Science as well as common sense, he argued, depends basically on a measure of concurrence regarding the nature of things, and for this he found support in Michael Polanyi's substantial work, *Personal Knowledge* (1958), with its criticism of naïve objectivism in science.[9]

On the whole, however, the positivist view of science remained unshaken in geography during the 1960s and Lowenthal's moves towards an alternative epistemology received no consistent development. Following the work of William Kirk in 1963, developing his earlier concept of the behavioural environment,[10] research in that field increased and perception studies multiplied, yet these were concerned invariably with the perceptions, not of geographers themselves, but of other groups such as the Chimbu or the Great Plains farmers. Although Harold Brookfield, in his 1969 article, 'On the environment as perceived', claimed that the growth of the humanistic view in science had dislodged the nineteenth-century idea of *absolute rationality*,[11] the objectivity of the scientific observer was rarely questioned: while it was recognized that the Chimbu are not always rational in their view of the world, it was assumed that scientists – and hence scientific geographers – would be rational and objective in their research. One of the strongest attacks on such objectivism was made by Preston James in 1967:

the direct observation of things and events on the face of the earth is so clearly a function of the mental images in the mind of the observer that the whole idea of reality must be reconsidered.

Obviously accepting the positivist definition of the term 'empirical', however, he went on virtually to reject empiricism altogether:

Empirical means wholly dependent on experience and observation without any reference to theory. But the discussion of the relationship between percept and concept suggests that anyone who claims to approach the observation of things and events with a blank mind is either fooling himself, or is a candidate for an insane asylum. Truly empirical study . . . does not exist.[12]

James raised a further important and related issue by questioning the notion of a purely inductive scientific method, proposing instead an inductive–deductive approach to explain the way the process of inquiry actually operates, and in a later work he continued his own contribution with his *All Possible Worlds* (1972), a lengthy study in the history of geographical ideas.[13] The confusion evident in his 1967 statement on empiricism, however, points to the need to clarify the distinction between the particular form of sense-empiricism known as positivism, and a wider interpretation of empiricism based on a less narrow concept of experience – one which acknowledges both the intellectual processes involved in experience and the dynamic social–environmental context in which it occurs.

An alternative to positivism: social empiricism

All science at present is faced with the need to formulate a new kind of empiricism, one that can offer an effective alternative to the positivist methodology, while retaining the useful guidelines it provided for research and the process of ordered inquiry. A number of trends are already evident, in geography as in other sciences, and the present question is whether these can be given co-ordination and coherence to form as it were a new paradigm. The first of these, a more rationalist approach stimulated by the work of Karl Popper and the neo-positivists, began to appear in geography during the sixties, in the form of a greater attention to theory as an essential part of the empirical method, a recognition that facts are always interpreted in terms of a theoretical framework, and a growing interest in the general concepts and propositions by which each discipline is identified. Wayne Davies in 1972 described these as amounting to a conceptual revolution in geography, and he noted the simultaneous emergence of general systems theory, stemming from Bertalanffy's earlier application of that theory in biology, as an indication of a return to integrating concepts and a concern for synthesis, in contrast to the kind of compartmentalization associated with the analytical method of a naïve empiricism.[14]

These developments are reminiscent of views being advanced in opposition to Comte's positivism in the early nineteenth century, when Alexander von Humboldt, as an acknowledged leader in applying the latest empirical methods in a number of sciences, was one of the first to call for a more *rational empiricism.* That period also was noted for its concern for organic and holistic views of man in nature, views that were castigated by the positivists of the time as idealism or romanticism, more suitable for poets than for practical scientists and with implications obviously unwelcome in the current economic and political climate. Given the context of Western industrialization with its destructive impact on both the landscape and the working classes, along with the world-wide extension of colonialism in the search for raw materials and markets, it is not surprising that the established view was a comfortable dualism, the model of man as master of nature.

After a long period when the mechanist and dualist views associated with the scientific revolution of the seventeenth century effectively deprived geography of a central integrating concept of nature, the current revival of holistic views in ecosystem studies now offers a more coherent framework for a subject dealing with the earth and its inhabitants. It is significant, again, to note that Humboldt, who emerged as one of the architects of a modern scientific geography, was also a pioneer in ecology, with his research on living plant communities in the early nineteenth century – studies that helped found biogeography and inspired the life work of Charles Darwin. Although it was the German naturalist Haeckel who coined the term *ecology* in 1868 for the study of organisms in relation to their environment, the foundations were already laid in Humboldt's plant geography, while the more comprehensive ecosystem concept of recent decades, which considers organisms and their environment as a single interacting system, can be seen outlined by 1850 in his final work *Kosmos*, where Humboldt wrote of man's place in the dynamic community of nature and even extended this early concept of a global ecosystem to include human culture and ideas.[15]

The radical implications of the ecosystem concept, in challenging the basis of the positivist world view, became apparent a century later in the conservationist movement of the 1970s when a series of United Nations conferences on the environment provided a focus for a growing number of publications and pressure groups expressing concern over the fragility of life-systems and protesting that the continued use of Western technology in the reckless exploitation of land and people could no longer be tolerated on what was now aptly called spaceship earth.[16] In geography the growth of ecosystem studies was accompanied by the emergence of social activists in the radical revolution of the seventies, although many of the new socialists, including William Bunge, Michael Eliot Hurst and Richard Peet, adopted Marxist views involving a materialist position and a theory of knowledge more in line with that of the positivism they were attacking. Others like Annette Buttimer, using the approach of continental existentialism and phenomenology, with its emphasis on the lived-world of personal experience, criticized positivist attempts to separate values from facts, intentionality from behaviour; they joined with Marxists in affirming the sociology of knowledge, a theory of knowledge which emphasizes the influence of the social context on all thought and thus on all scientific research. It can be seen that such a theory can be interpreted as an ecological approach to knowledge, and some implications of this are explored later in this work.

So far, then, three main trends have been noted in recent attempts to modify the positivist model: the development of rationalist views, the rise of the ecoscience movement, and an increase in radical demands for social action. Along with these can be discerned a strengthening of the tradition of liberal humanism, now no longer content to identify with the arts or humanities but actively criticizing positivism itself and demanding that science should contribute more effectively, not merely to a materialist ideal of progress, but to the conservation of life-systems and the creation of a more humane society. Whether it is to be expected that all these movements should be linked in some way remains an interesting question; however, it seems fair to say that elements of them all will be incorporated in the new paradigm for science, and that far from rejecting the empirical method entirely this will involve replacing the positivist version, based on Bacon's kind of sense-empiricism, with one that gives due recognition to the intellectual and social aspects of knowledge, as well as to the clear importance, underlined by radicals and humanists alike, of stressing the social responsibility of scientists. To take account of these aspects the term *social empiricism* has been introduced here in referring to the new theory, for although this might be regarded simply as a reformed version of positivism, the extent of the changes involved seems to justify the use of a different name.

An important element in social empiricism is an appreciation of the historical continuity of knowledge, as this is communicated through concepts and symbol systems. To understand the present, it is essential to be aware of traditions inherited from the past. History, in the sense of what we know about the past, is thus an integral part of the current world of ideas, and indeed the positivist neglect of the history of ideas has been a reflection of the poverty of scientism. In the search for a new scientific paradigm, historical research has a valuable function to perform, for the complex of beliefs, procedures and concepts that provided the operational framework for modern science is exceptionally difficult to analyse if observation is limited to the twentieth century. Even at the present time, some of the significant elements remain tacit rather than explicit, and numerous inconsistencies baffle any logical analysis. A historical study can show how the scientific movement evolved, incorporating various ideas, responding to particular problems and then tending to absolutize its procedures.

In geography the widespread neglect of historical and philosophical research during the positivist period had unfortunate consequences. Indeed, at the height of that period geography seemed to be facing the danger of approaching the end of the twentieth century as a discipline without a history, for in many institutions a generation of students was introduced to research in the subject with practically no awareness of its earlier traditions or its potential significance in a new phase of science; in terms of scientific theory they were taught virtually to be followers, where their general theory was limited to a study of methodology over the last century, for at that time geography appeared constantly in a defensive position, more preoccupied with adapting to the accepted notion of scientific procedure than with directing energy towards restructuring that theory.

More recently the followers of Marx have begun to restore some historical perspective by returning to a study of his works; however, their studies tend to be limited to the nineteenth century and face the serious problem of dealing with the complexities and apparent contradictions in Marxist writings – a problem made more acute by the inexperience of almost all geographers in such inquiry. A longer perspective is useful at this point, to show the continuity of positivist theory from the time of Francis Bacon and draw attention to the alternative to the theories of both Comte and Marx that was being formulated during the same period in Humboldt's work. In the two centuries between Bacon and Humboldt, philosophers and practising scientists combined to formulate, in the empirical method, a powerful instrument for inquiry, and its successes – particularly after Newton – tended to mask its deficiencies. The impact of this method on geographical thought during that time has received little attention, however, from either historians of science or geographers themselves, the present study being evidently the first in English for the period under review.[17] Indeed the two centuries before Humboldt have been called 'a dark age in the history of academic geography'.[18]

Empiricism and geography, from Bacon to Humboldt

As far as geography was concerned, the initial encounter with scientific empiricism was close to disastrous. Following the introduction of the new method during the seventeenth century, geography soon became obviously outpaced, and even displaced, by the progress of the specialized empirical sciences, for, in contrast to these, geography had to deal not with a specific group of material objects or forces but with a complex subject-matter that included man. By the end of the seventeenth century, geography had declined in prestige, and apart from such isolated efforts as those of Carpenter at Oxford in 1625 and Varenius at Amsterdam in 1650, few attempts were made to maintain it as a serious intellectual discipline concerned with furthering inquiry. Instead, although progress in cartography and exploration continued, the study of geography itself became essentially derivative, neither taught consistently nor seriously researched in universities or academies, and sustained in a mimetic textbook tradition supplemented by travellers' tales, compendia and gazetteers; in most texts open support for the heliocentric theory was avoided for virtually two centuries after Copernicus. Even the growing popular demand for geographical descriptions of the world throughout the eighteenth century seems only to have increased the volume of such works without any major improvement in their quality, while a significant display of interest in the subject by leading thinkers such as Kant and Herder towards the end of the century did not appear to stimulate immediate advances in research.

Then came a remarkable development: as physics had its Newton in the seventeenth century and chemistry its Lavoisier in the eighteenth, so, early in the nineteenth century, geography attracted the attention of one of the intellectual giants of the time, the German naturalist Alexander von Humboldt. Owing in part to his efforts, geography rapidly achieved greater scientific distinction, as he applied the empirical method of research with astonishing effectiveness to such aspects as vegetation, climate, landforms and regional studies. Beyond this, his work opened the way for geography to play a leading role in changing the empirical tradition itself: he saw the contribution of geography, in studying complex functional relationships and providing a coherent world view, as an integral part of scientific inquiry, and he questioned a theory of science that failed to encourage this. For more than a century, however, his ideas remained largely inactive. Perhaps the analogy with Newton or Lavoisier is not entirely appropriate, for whereas in physics and chemistry the stimulus given by those men was acted on quite rapidly, geographers did not take full advantage of the lead provided by Humboldt. In that respect he might be compared with Mendel in genetics or Copernicus in astronomy, where a considerable time lag occurred before their ideas received widespread recognition and application.

It must be appreciated that the full implication of Humboldt's theory of geography involved not only the introduction into that subject of empirical methods relating to the collection and processing of data, it also called for a basic modification in the whole notion of scientific method as this had emerged over the preceding two centuries. In the process of scientific inquiry Humboldt transferred some of the emphasis away from the objects studied, to draw attention to the importance of the concepts in the mind that make such study meaningful. Relying to a large extent on Kant's teachings for this modification of the Baconian theory of knowledge, Humboldt moved beyond Kant in suggesting that general concepts are not innate, but are built up out of human experience in the course of history. For this reason, in *Kosmos* where he outlined the limits and methods of his new science, he also traced the concept of cosmos in its historical development and indicated the way such a concept should be extended to provide a focus for understanding the dynamic world-complex; it can thus be claimed that the modern conceptual revolution in geography was initiated by Humboldt, although his attempt was in large measure premature.

It was perhaps too much to expect, given the depressed condition of geography as an intellectual discipline in his day, that in one leap, as it were, the subject could not only rise to the standard of other natural sciences in terms of procedure but also transcend them with a wider view of scientific inquiry. The geographers of Humboldt's time in general were ready to appreciate only the more conventional aspects of his work, and his successors – men themselves trained largely in the specialized sciences – found his revolutionary theory of science virtually incomprehensible. If anything they moved closer to the positivism of Comte or Mill and dismissed some of Humboldt's most powerful insights as evidence of an unscientific romanticism. Even the anarchist geographer Kropotkin, who admired Humboldt's approach to geography, did not appear to grasp the significance of the alternative theory of science that it implied, and indeed Humboldt's holistic concept of nature stood in opposition to the materialism of both Kropotkin and Marx. In the mid twentieth century, Schaefer set the pattern for the new school of positivists in geography by denying the relevance of *Kosmos* either to geography or to the methodology of science, and dismissing Humboldt's 'yearning for a synoptic view of the universe' as 'a characteristic of romanticism'.[19]

That view was repeated by a number of geographers, including David Stoddart, who in 1967 rejected any attempt to introduce into geography 'an outdated Romanticism drawn from Goethe and Alexander von Humboldt'.[20] Like Schaefer's this was an unsupported statement and it is noteworthy that it occurred in a discussion on the ecosystem idea in geography which contained no reference to Humboldt's contributions to that concept. Such comments, however, draw attention to the absence from the geographical literature in English of any study dealing with the movement known as romanticism and its actual relation to the development of science. A closer look at this question is clearly necessary in any re-evaluation of Humboldt's work and is especially significant if, as David Hooson suggested, we are now experiencing an 'Ecological Romantic Revival' after the 'Theoretical Quantitative Enlightenment'.[21]

The present research in part grew out of an interest in this problem, as it became clear that a new perspective on Humboldt's position could be gained only from a better understanding of the conceptual framework within which he worked, and for this it would be necessary to examine more carefully the formulation, during the preceding two centuries, of the models for both science and geography that he attempted to change. A biographical study of Humboldt however is not intended here: the basic concern is the tradition of empiricism in science and its impact on geography, and in that context his response remains one of the most challenging in the history of modern geographical thought. Although a high degree of respect has been accorded many of his achievements, some of his most valuable ideas have been disregarded, evidently because they failed to accord with the growing positivism of science. In the light of the current revisionist movement it seems time for a fresh look at his efforts to provide an alternative to the narrow empiricism of his day, for he emerges not only as an advocate of a more thoughtful empiricism and a pioneer of ecology, but also as a scientist actively engaged in social reform, moving frequently beyond a detached humanism into confrontation with the establishment in his opposition to colonialism, slavery and exploitation of the poor.[22]

A major obstacle in carrying out this project has been the lack of research in the history of geography, as reflected in the scarcity of monographs and the absence of even a catalogue of geographical publications from 1650 to 1860; it has been necessary to go some way towards filling this gap in the literature by commencing a survey of general texts in geography for that period. Within the limits of this study, emphasis in the main is on British and German works, although the value of French and other contributions is recognized as a subject for further research. Primary sources have been consulted wherever possible and fairly extensive quotations are included to give the flavour of the original, often using new translations. Here again the size of the task has grown with the recognition of a need to provide more adequate translations, particularly in the case of the classical Greek writers, Varenius, and Humboldt himself. Where not otherwise acknowledged, these are by the author The translating of Humboldt's work is of critical importance, since for the most part only nineteenth-century versions are available in English and these rarely convey his theory of science accurately, while many of his writings have not yet been translated, including even the unfinished final volume of *Kosmos*. The initial sections of the Introduction to that volume, which represent in effect his concluding statement on geography and scientific method, are presented here for the first time in English, and indeed to a large extent the task of interpreting those ideas can be said to have given rise to this study.

Humboldt regarded science as a social process, involving a constant reappraisal and development of ideas. The Baconian tradition has tended to emphasize the innovatory achievements of scientists, but we must not forget that these are the outcome of what is essentially a historical process and an educative one. In that process no one commences with a *tabula rasa* and we owe a great deal to colleagues and predecessors, not only to those whose work is accepted and used, but also to those whose views are disputed or corrected – as our own will be in turn. Strabo has put it rather well:

And if on occasion I shall be compelled to contradict even those whom in all other respects I follow most closely, I ask to be pardoned. For it is not my purpose to contradict everyone . . .   *Geography*, 1.2.1

CHAPTER 1

FOUNDATIONS OF MODERN EMPIRICISM

THE SCIENTIFIC REVOLUTION

A new cosmology

The seventeenth century in Europe introduced the modern period in science with a remarkable consolidation of scientific ventures as well as an increasing concern with the new empirical method by which these were achieved, and the scientific revolution which was an outcome of this brought significant changes in knowledge of the world. As the implications of the Copernican system of 1543 were explored, many established views of the universe were questioned and the idea of a fixed earth, long accepted in the geocentric theory of Aristotle and Ptolemy, was displaced by the concept, exciting to some and disturbing to others, of man's own planet turning through the vastness of space. Eventually the comforting belief in the stability of the earth at the centre of a finite cosmos had to be discarded, in spite of Church opposition.

Copernicus (1473–1543) indeed had accepted Ptolemy's Aristotelian model of the cosmos as a system of planetary spheres, bounded by a circle of fixed stars (see Fig. 1): his major changes were to locate the sun as its centre, while postulating the diurnal rotation of the earth on a circular orbit around the motionless sun. The Italian scholar Giordano Bruno (1548–1600), however, went on to an even more revolutionary conception, rejecting Aristotle and arguing, in his *On the Infinite Universe and Worlds* (1584), for the existence of innumerable worlds, all composed of the same elements and subject to the same laws of nature, in a limitless universe of constant change, a cosmic unity of material and spiritual aspects, in which all individuals and events are manifestations of an immanent divine Intelligence, and no part can be completely at rest. An infinite universe, he asserted, 'has neither centre nor circumference, but . . . every point also may be regarded as part of a circumference in respect to some other central point'.[1] His vision of universal relativity extended to the limitations of human perception:

We cannot apprehend motion except by a certain comparison and relation with some fixed body . . . if I were on the sun, the moon, or any other star, I should always imagine myself to be at the centre of a motionless world around which would seem to revolve the whole surrounding universe [311].

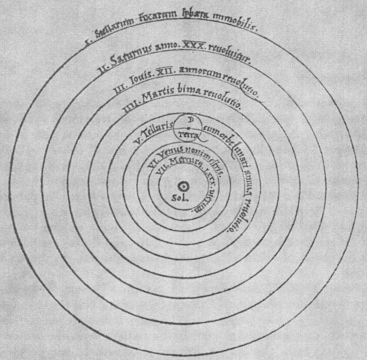

NICOLAI COPERNICI

net, in quo terram cum orbe lunari tanquam epicyclo contineri diximus. Quinto loco Venus nono mense reducitur. Sextum deniq; locum Mercurius tenet, octuaginta dierum spacio circũ currens. In medio uero omnium residet Sol. Quis enim in hoc

pulcherrimo templolampadem hanc in alio uel meliori loco po neret, quàm unde totum simul possit illuminare? Siquidem non inepte quidam lucernam mundi, alij mentem, alij rectorem uo= cant. Trimegistus uisibilem Deum, Sophoclis Electra intuentẽ omnia. Ita profecto tanquam in solio re gali Sol residens circum agentem gubernat Astrorum familiam. Tellus quocq; minime fraudatur lunari ministerio, sed ut Aristoteles de animalibus ait, maximã Luna cũ terra cognationẽ habet. Concipit interea à Sole terra, & impregnatur annuo partu. Inuenimus igitur sub hac

Fig. 1 A diagram of the Ptolemaic or Copernican universe, from Copernicus' *De revolutionibus orbium coelestium* (1543)

Along with this daring philosophy, Bruno's interest in magic, his admiration for England, where some of his major works were published, and his willingness to question Christian dogma, led finally to his imprisonment by the Inquisition, and he was burned at the stake in Rome as a heretic in 1600. His vision, however, remained too compelling to be denied. In the new cosmology, it became clear, the location of the earth in motion could be calculated only in terms of its relation to some other part of the system. This affirmation of a dynamic view was to prove a characteristic of the modern period, when the mathematical analysis of relationships became a feature of the emerging sciences.

In England Bruno's ideas were taken up by the queen's physician William Gilbert (1544–1603), who proposed a theory of the globe as a magnet in his influential Latin publication of 1600, *On the Magnet, Magnetic Bodies, and on the great magnet, the Earth. A new Physiology; demonstrated by many arguments and experiments.*[2] A significant feature of Gilbert's new science of nature was his attempt to increase the accuracy of his observations on magnetism by inventing an instrument for this purpose and conducting experiments with the loadstone. He was a pioneer, not only in studies of earth magnetism, but also in the experimental and empirical approach that was to bring about a virtual revolution in science. The world view of *De magnete* however was largely neglected during the next two centuries until, with a revival of interest in studies of earth magnetism and electricity, the contributions of Gilbert were acknowledged by Humboldt and other leading researchers. In his own time Gilbert's holistic concept of the earth did not gain wide acceptance when, in an age noted for its mechanics and the invention of clocks, efforts to replace the static medieval view of the earth as an ordered and final creation of providence led more often to explanations of the world as a mechanism, governed by natural laws. In astronomy, Galileo (1564–1642) became involved in applying mathematics to the task of studying the mechanics of moving bodies, although he was hampered by a lack of reliable data. Using the statistical records of Tycho Brahe in the same period, however, Kepler (1571–1630) was able after years of effort to construct a mechanical theory of the universe in his major work of 1609 which he called *The New Astronomy, based on Causation, or Physics of the Heavens derived from Investigations of the Motions of the Star Mars, Founded on Observations by Tycho Brahe.*[3] Suggested in this title are some important elements in the method of modern science, for Kepler's discovery of the first laws of planetary motion was based, as he indicated, on the ancient Greek belief in causality, but in the new astronomy those laws were to be treated as a problem in physics and given a mathematical expression, using data supplied by painstaking observation.

The general movement to scientific reform that took place in the seventeenth century can be seen as a necessary challenge to a rigid tradition, for many beliefs had been transmitted with little change from classical antiquity and the Church, which after the Thomistic synthesis of the thirteenth century had come in large part to support Aristotelian teachings, now acted to resist all contrary speculation. Already significant activity in Italy during the sixteenth century had marked a watershed in the history of Western science: not only was the Aristotelian cosmology, as interpreted in Ptolemy's geocentric system, effectively challenged by the Copernican theory but in addition the medieval interpretation of Aristotelian philosophy, relying as it did mainly on his logical treatises – known collectively as the *Organon* – was also subjected to criticism.

During the following century, with the centre of activity moving to Northern Europe, a more utilitarian and technologically-oriented science was fostered and criticism was extended. The work of Francis Bacon (1561–1626) early in the seventeenth century was designed specifically to replace the *Organon* of Aristotle with a method of scientific investigation based upon experience and incorporating the procedures of experiment and induction. Impatient at current preoccupation with the deductive logic of Aristotle and the conservative approach to knowledge inherited from the scholastics, Bacon called for a new birth of science, more closely allied with technology and directed towards the extension of knowledge as a means of increasing man's power over nature. Such knowledge in his view was to begin not with preconceived theories but with direct observation of nature itself, since he believed all generalizations must be drawn from particulars recorded by the senses. Apparently unaware of the extent to which Aristotelian thought was incorporated in his own theories he proclaimed the inductive method to be a 'new instrument', the *novum organum*, and in 1605 he gave his first major work on scientific procedure the challenging title *The Advancement of Learning*.

The reliance on experience that characterized the new kind of inquiry brought with it, to be sure, its own set of problems, so that a constant striving to control and interpret experience can be seen in the framing of experiments, the invention of more precise instruments, and the refinement of mathematics and logic for use in research. Nevertheless, as the century progressed, discoveries proliferated, confirming the growing confidence of those who believed that modern experimental science would succeed, where the Greeks had failed, in providing a physical explanation for practically everything in the world. There was a tendency to overlook the long process of investigation that had opened the way for their achievements.

THE HISTORICAL LEGACY

Although the emergence of scientific empiricism as a strong movement in Europe occurred during the seventeenth century, many of its tenets had been formulated in classical times. To a large extent the entire framework of argument in the emerging sciences, the vocabulary, the procedures of logic, the basic concepts and even the kind of questions asked, followed patterns established in ancient Greek and Hellenistic thought, and indeed the effort to break from the restrictions of that framework on some issues gave a revolutionary character to the new movement. A reaction from medieval scholasticism, with its reliance on book-learning and the authority of the ancients, was from the beginning a pronounced feature of the empiricist programme, and this contributed to an anti-historical bias that persisted well into the twentieth century.

In attempting to understand the ideas of empiricists since the seventeenth century it is important, then, to consider their response to the classical tradition, and this includes not only the various moves to reject aspects of Greek theory but also the important elements of the classical paradigm that were retained, often unrecognized, in modern empiricism. The way in which the classical authors have been translated and interpreted at different times must also be considered, of course, since even where the authenticity of surviving manuscript copies of their works is not in doubt, interpretations of language and meaning can vary widely, while scientists themselves in large part have relied more often on commentaries or translations which inevitably reflected the ideas of other scholars. With regard to the theory of induction, for instance, a distortion of the classical authors can be discerned, even in some reputable English editions, in that their work has been translated within a frame of reference derived from materialism and Baconian empiricism. An important task of the revisionist movement is to look more critically at translations now in use, for although any new interpretation will be conditioned by present ideas, this is preferable to using earlier versions now recognized as inappropriate.

The positivist injunction to ignore the classical inheritance has clearly had serious consequences in scientific thought, for in many cases it allowed the persistence of beliefs that might have been rejected had their authority been recognized, but which instead were retained as part of 'common knowledge', or even defended as self-evident truths. A case in point, of considerable importance in the Baconian view of scientific method, is the faculty theory of the mind, with its postulates of detached rationality and independent sense-perception as products of separate faculties which are linked in some way by a rather suspect faculty of imagination. The extent to which this Aristotelian model of the mind has been maintained, virtually unquestioned, in the course of modern empiricism is remarkable, to say the least.

The discussion of Greek thought, then, in the present context is by no means an antiquarian regression: what we are considering here are teachings that were sustained by empiricists of the modern period as an active part of their world of ideas. In geography also, the Hellenistic authors provided the standard model for practically all subsequent texts, virtually until the eighteenth century.

Greek philosophy: Plato and Aristotle

One of the many contributions to science made by the early Greeks was their concept of an inherent order in the cosmos, an order which they believed the human intellect to be capable of understanding, and over several centuries they were deeply concerned with the way such knowledge can be attained. Observation or experience formed one such way, and early efforts to establish an empirical method can be traced in their work, notably in the medical and geographical observations in *Airs, Waters, and Places* of the physician Hippocrates of Cos (born *c*. 460 B.C.), as well as in the biological researches of Aristotle (384–322 B.C.). Of equal significance to the development of empiricism were the mathematical investigations of Plato (427–347 B.C.), along with his teachings on the nature of reality and experience itself, although these have received less recognition among scientists.

The contribution made by the great philosophers of fourth-century B.C. Athens to the Western scientific tradition is one that continues to affect the very structure of modern thought on science. Drawing themselves upon the resources of the preceding centuries of vigorous discussion on the universe by the Ionic school and the Pythagoreans, they brought their own genius to bear on problems involved in all efforts to understand the world. For more than two thousand years their work exercised a strong influence in Europe, often through a variety of interpretations by their followers, sometimes more directly through their own writings – many of them lost for centuries and then recovered – and frequently through violent reactions from both. In this way Plato and Aristotle remained living forces, their works widely read and discussed throughout the period from Bacon to Humboldt, a period which included the classical revival of the eighteenth century, so that their ideas continued to form a significant part of the intellectual environment in which modern science developed. Plato and Aristotle as authors, then, must be considered from that point of view as virtual contemporaries of others whose works were being studied at that time.

Of the two, Aristotle has tended to be more highly regarded by empiricists because of his realist position and his concern for observation as the basis of an inductive method. Plato, on the other hand, with his doctrine of innate ideas and his emphasis on inner recollection as the way to knowledge, has more often been presented as an opponent of scientific observation, especially on account of his injunction to astronomers in the *Republic* to leave the starry heavens alone and instead pursue mathematics and philosophy. This attitude to Plato requires closer examination for it is tempting on the surface to accept a contrast between Aristotle as the practical man of science, occupied with studying the material world, and Plato as the vague philosopher who disregarded the evidence of the senses and gave priority to abstractions in asserting the reality of ideas or forms.

Arthur Koestler and other writers have been outspoken in their criticism of Plato as antagonistic to science.⁴ In the opinion of Dampier-Whetham, for instance, 'Plato was a great philosopher but in the history of experimental science he must be counted a disaster.'⁵ The distinguished Platonist, Paul Shorey, in the notes to his translation of the *Republic*, however, strongly denied this interpretation, suggesting instead that 'Plato as a whole is far nearer the point of view of recent science than Aristotle' and pointing out that the passages often taken as examples of 'Plato's hostility to science and the experimental method' must be considered in context.⁶

In the *Republic*, as Shorey indicated, Plato was criticizing the merely observational astronomy of his own day and proposing in its place a new intellectual study concerned with the geometry of bodies moving in ideal circles. Believing mathematics to be a necessary part of the training of an astronomer, in effect he anticipated the modern approach by urging attention to generalized problems on the model of geometry. He presented these arguments as the views of Socrates, his former teacher:

Socrates: 'It is by means of problems, then,' said I, 'as in the study of geometry, that we will pursue astronomy too, and we will let be the things in the heavens, if we are to have a part in the true science of astronomy and so convert to right use from usefulness that natural indwelling intelligence of the soul.'

In the dialogue the partner of Socrates replies wrily, 'That will multiply the labour of our present study of astronomy many times.'⁷

For Plato the true and the real are to be found not in individual material objects but in abstract mathematical relations. While this apparent paradox has troubled many empiricists, recent research supports Shorey's view that Plato actually was struggling with what has become a central problem of modern science, the study of motion. The movements of the stars 'in relation to one another', Socrates explains, 'can be apprehended only by reason and thought, but not by sight', and therefore 'we must use the blazonry of the heavens as patterns to aid in the study of those realities' [VII, 529 D–E]. Proposing motion in general as the subject of advanced studies in astronomy and harmonics, Plato criticized others for 'preferring their ears to their minds'. At the same time, in adopting the Pythagorean belief in the existence of an inherent harmony in numbers, Plato was led to make the doubtful assumption that no basis of experience is required for harmonics and that an *a priori* mathematics of acoustics is possible [VII, 531].

On issues such as this, Aristotle's concern with the world of daily experience, as shown in his own investigations of animals, recommended him as a better model for empirical science. At the same time, Aristotle's cosmology created serious problems for those who accepted his theories in succeeding centuries. Following the Pythagoreans he separated terrestrial from celestial mechanics and taught that the universe is geocentric and finite, with the earth, which 'must be at the centre and immovable' [*De Caelo*, 296b], occupying an inner sphere of change and decay, while the heavens were considered to be subject neither to change nor to the same laws that govern the earth: 'For order and definiteness are much more plainly manifest in the celestial bodies than in our own frame; while change and chance are characteristic of the perishable things of earth.'[8]

Plato, on the other hand, argued in the *Republic* [VII, 530 B] that since the stars are visible things and possess bodies it is absurd to expect that no deviation will occur in them. He therefore presented a more modern viewpoint, although the discoveries that proved him correct were delayed for two millennia. Rather than showing a constant antipathy to using the evidence of the senses, Plato can be seen here, as in the *Laws* and even in the *Timaeus*, striving to provide an adequate mathematical basis on which to interpret observations and make the phenomena amenable to science.

While the importance of Plato's mathematical contribution is now widely acknowledged, there has been a marked reluctance to accept other aspects of his philosophy as being relevant to empirical method. A rigorous examination of the nature of experience, however, is a fundamental task for empiricism, and in the *Republic* [VII, 514–17], with his parable of the cave, Plato provided one of the most dramatic illustrations in our literature of the problems associated with perception in human experience; indeed the captive audience he described there shows a rather uncomfortable resemblance to twentieth-century consumers of films and television. Prisoners in the cavern with their backs to the firelight see nothing but the shadows of objects projected on the wall; they must be freed and led out into the sunlight to recognize, gradually, the objects themselves. Socrates explains that in the interpretation of experience we ourselves are like the prisoners: 'we must', he said, 'liken the region revealed through sight to the habitation of the prison' [VII, 517 B].

The way to a clearer understanding of phenomena (Greek *phainein*, to appear) involves a process of learning: 'The release from bonds, and the turning from the shadows to the images that cast them . . . and the ascent from the subterranean cavern to the world above' is a 'progress of thought' which he called *dialectic*. The sciences in his view contribute to this process, for they are not to be concerned only with arguments over the shadows or appearances: in Plato's system of education, the sciences are intended to precede and lead up to the study of philosophy. He referred to 'all this study of reckoning and geometry and all preliminary studies that are indispensable preparation for dialectics' [VII, 536 D]. It is worth considering that there could be a distinct advantage in training modern philosophers in this way. At the same time it would be misleading to interpret the same passage as suggesting that the person who pursues one of the sciences does not need to study dialectic: it should not be assumed that Plato was recommending the modern anomaly where scientists occupied with specialized research may not be required even to examine the conceptual framework within which they operate, before being awarded a higher degree in philosophy. It is most unlikely that Plato would suggest that the education of a scientist should stop short of philosophy, since his primary concern was the defence of that study: 'I turned my eyes upon philosophy, and when I saw how she is undeservedly reviled, I was revolted' [VII, 536 C]. Evidently, the same kind of empiricism that resisted the application of mathematics to astronomy in his day also fostered a suspicion of philosophic questions.

Plato's philosophy, then, can be regarded as an energetic reaction against what he saw as a naïve empiricism. Observations without mathematical analysis are unlikely to reveal significant relationships, he argued, and above all he emphasized the importance of ideas in the quest for knowledge. The Platonic theory of ideas itself is still rather obscure and there is evidence that it was modified in the course of his writings. Of his proposed four levels of being – space, genesis, soul, and being – the position of soul (*psychē*) as intermediate between the immutable forms of being, and the phenomenal world of change or becoming, is in some sense transitional, and it may even be argued that Plato intended to bring his universals into meaningful relationship with events through the mind or *nous*. Whatever the intrinsic merits or deficiencies of his theory, the achievement of Greek thought in his century is marked by a constant effort to deal with the reality of ideas.

Following from Plato's efforts, Aristotle continued the attempt to make abstract concepts meaningful, using everyday examples. In the *Physics*, his treatise on the study of nature, he used the effective analogy of the builder who handles bricks and timber, yet must know 'the form of the house as well as the material'. A builder is inclined to regard himself as essentially a practical man, having little to do with ideas, yet before he constructs a house he must have a concept in his mind of the form it will take. In the same way, Aristotle suggested, nature (*physis*) has two aspects, of form and matter, so that it 'can neither be isolated from the material . . . nor is it constituted by the material'.[9]

The extent to which forms or ideas can be abstracted from the material, or remain associated with it, Aristotle set as a problem to be discussed more fully in a separate work which he called his *First Philosophy*. This was the work referred to by later editors, though not by Aristotle himself, as the *Metaphysics* since it was collated by them after (*meta*) the *Physics*. Noting that the concept of matter is relative, and that form can be linked with purpose, he went on to stress that it is the combination of matter and form which must concern the student of nature. Aristotle, like Plato, was critical of the materialist explanation of the world proposed by Democritus (*c.* 470–*c.* 400 B.C.) and other fifth-century Ionians: 'If we look at the ancients it would seem that physics is concerned only with the matter, for Empedocles and Democritus have little to say about forms.'[10]

In Aristotle's theory of matter and form the problem of the human mind and its relation to nature emerged as a crucial one, and he dealt with this question in his treatise *On the Psyche*.[11] There he argued that in man, mind is form or actuality, while body is matter or potentiality: the material aspect, in his view, remains potential until it is united with form in some object of nature. While this kind of logical distinction may not seem particularly satisfactory today, it represented at the time a strong attempt by Aristotle to link Plato's theory of forms with the world of change that they called nature. At the same time, in agreement with Platonic doctrine, Aristotle continued to maintain that mind is eternal and not subject to decay in the same way as the body: 'Mind (*nous*) seems to be something independent engendered in us, and to be imperishable.' Even when the individual perishes, he wrote, 'presumably the mind is more divine and is unaffected'.[12]

To that extent, then, Aristotle appeared to set mind apart from nature. In contrast to Plato he denied, further, that the mind has any organ comparable with the sense organs to provide it with a particular location in the body [III, 4: 429a]. Here it must be appreciated that in what is called Aristotle's faculty theory, the mind or *nous,* as the faculty or capacity (*dynamis*) in man for cognition and logical reasoning, was considered to be separate from the sensitive faculty. This distinction itself was in accordance with his division of all existence into contraries: 'What exists is either sensible or intelligible'; the senses, in his view, are concerned with what can be perceived, the cognitive faculty with what can be thought [III, 7: 431a]. From his work came a distinction between the senses as passive receptors of sensation, and the active intellect as the part of the *psyche* capable of ordering perceptions to produce knowledge of universals.

While maintaining that the mind is to a large extent independent, Aristotle nevertheless argued that the actual activity of thinking involves an association with the body: in all thinking, the intellect must be working on evidence provided by the senses, even in the case of 'the so-called abstractions of mathematics' [III, 8: 432a]. On that basis he justified his proposal to include the soul, as the principle of life, in a study of nature: 'This at once makes it the business of the student of nature (*physikon*) to inquire into the soul' [I, 1: 403a–403b]. In the process of thinking, he assigned a key role to the imagination, as the link which connects the functioning mind with the sense-faculty: 'imagination (*phantasia*) is different from both sensation and thought; imagination always implies sensation, and is itself implied by judgment' [III, 3: 427b]. However, he admitted that 'the question of imagination is obscure' and it seemed to resist clear tabulation in his list of aspects of the *psyche*, which in man he described as 'the faculties for nourishment, for appetite, for sensation, for movement in terms of place, and for thought'. These divisions themselves were related to his general classification of all living nature into vegetative, animal and rational aspects: plants have the nutritive faculty only, he stated, 'but other living things have the capacity for sensation too' [II, 3: 414a–b]. Differences between animals arise, according to Aristotle, because some have all these faculties and all senses, and others do not, while the capacity for thought or judgment (*dianoetikon*) is the prerogative of man alone. He made a clear distinction between the *psyche*, 'existing in all living things', and the *nous* which for him is the distinctively human part of the soul, as 'a faculty concerned with truth' and 'responsible for what is right and correct'. Thus he criticized Democritus, who 'actually identified psyche and mind; for he believed that truth is the appearance' [I, 2: 404a–b].

Aristotle's concept of the mind actively engaged in the search for truth is an inspiring one, implying the obligation for man to strive to realize his own potential humanity. This however was combined with an inflexible concern on his part to differentiate the nobility of man from the rest of nature. The scale of nature, as he presented it, is a static one – and it leads downwards from man. Writing prior to any concept of progressive evolution, Aristotle was not inclined to emphasize the similarities between man and the lower orders of living things.[13] His assertion that the inferior animals are incapable of reasoning was made without any reference to evidence and is one of the more regrettable of his teachings. However, it had far-reaching effects on the attitude to both perception and imagination that he transmitted to modern science.

The sense-faculty, Aristotle recognized, exists in the lower animals too, and since he acknowledged that some animals also show a capacity for both imagining and feeling, he assigned an inferior status to imagination and the emotions, in contrast to the rational faculty. That view was adopted later by Francis Bacon with scarcely any attempt to question the faculty theory itself. Similarly the assumption of the seventeenth-century empiricists, that perception occurs directly through the senses without the participation of the intellect, seems to be derived from Aristotle, who argued that since perceiving is a function of animals as well as man it cannot involve the mind, while that association of imagination with error which later became conspicuous in the methodology of modern science is also persistent in his work:

All sensations are true, but most imaginations are false . . . Nor is imagination one of the faculties which are always right, such as knowledge (*episteme*) or intelligence (*nous*) [III, 3: 428a].

Two uses of the imagination were noted by Aristotle: the first involving interpretation in the presence of the object, and the second in connection with memory in the absence of an object.

Imagination must be a movement produced by sensation actively operating. Since sight is the chief sense, the name *phantasia* (imagination) is derived from *phaos* (light), because without light it is impossible to see. Again, because imaginations persist in us and resemble sensations, living creatures frequently act in accordance with them, some, namely the brutes, because they have no mind, and some, namely men, because the mind is temporarily clouded over by emotion, or disease, or sleep [III, 3: 428b–429a].

Subsequently Aristotle found it necessary to ask how the lower animals can be said to have imagination at all, if, as he claimed, 'one regards imagination as some sort of thinking process' [III, 10: 433a]. His response was to posit two kinds of imagination: 'Imagination in the form of sense is found, as we have said, in all animals, but deliberative imagination only in the logical.' The mind however remains paramount in his view, except 'when the subject lacks self-control', for 'in nature the upper sphere always controls and moves the lower', therefore, 'the cognitive faculty is not moved but remains still' [III, 11: 434a].

In spite of what he considered to be its manifest limitations, then, Aristotle acknowledged that imagination is necessary for thought: 'Now for the thinking soul images take the place of sense-perceptions . . . Hence the soul never thinks without a mental image' [III, 8: 431b]. This became for him indeed an extension of the empirical principle: 'as no one could ever learn or understand anything without the exercise of perception, so even when we theorise we must have some mental image of which to think; for mental images are similar to objects perceived except that they are without matter' [III, 8: 432a]. The implication that sense-perceptions are still somehow '*with* matter' was left rather vague in Aristotle's epistemology; it was to emerge more strongly in modern scientific materialism. Even the Kantian *critique* achieved little immediate change in this tenet of empirical science, and the belief persisted that impressions of material objects are intrinsically more 'solid' and reliable than other mental images. Furthermore Aristotle argued that there can be no error in 'the thinking of indivisible objects of thought . . . where truth and falsehood are possible there is implied a compounding of thought into a fresh unity' [III, 6: 430a]. This view was to persist in the empiricism of Bacon and Locke, with their belief that the proper method of obtaining knowledge is to begin with 'particulars' or simple elements. They seem to have overlooked Aristotle's earlier statement in the *Physics* that in any investigation what we perceive initially will not be particulars but some kind of rather undifferentiated whole. It can be argued indeed that Aristotle showed himself here in favour of a gestalt theory of perception:

Now what to us at first are evident and obvious are [also] more confused; later the elements and principles of these become known by analysis.[14] So we must advance from total complexes (*katholou*) to constituent factors, for the whole is best known to the senses, a 'whole' being what the complex is: for many parts are included in the total complex [I: 184a].

The legacy of Aristotle's teachings on epistemology, as they were passed on in Baconian theory, can be seen, then, as firstly a kind of objectivism with regard to perception, since sensation is believed to be caused by the object independently of the mind. Secondly there is a twofold interpretation of imagination, so that it tends to be associated with fantasy and error, despite a tentative recognition of the essential part played by mental images in all thinking. Finally the separation of the reason from what is called a 'sense-faculty' is justified by a theory of nature in which rationality is seen as the special prerogative of man, as distinct from those aspects of the *psychē* which are common to the lower animals. Matter and mind are presented as contraries, and even imagination as the link between the two is deprecated, along with the emotions, as part of the sense-faculty of inferior organisms.

Here then was the argument prepared for the seventeenth-century rationalists who claimed that knowledge is accessible only by means of the cool dry light of the detached reason. Their opposition of mind to nature was reinforced by Christian teachings, and they shared with the empiricists of the time a suspicion of the emotions and the imaginative process as extraneous to scientific inquiry. This was an attitude that later helped precipitate the eighteenth-century romantic reaction with its defence of imagination and the feelings as having an essential function in all attempts to comprehend the world.

It is perhaps astonishing that the persistence of Aristotle's faculty theory in modern empiricism received so little recognition from Bacon's time onwards. Part of the difficulty seems to lie with the way in which the teachings of Aristotle were transmitted and interpreted in the centuries after his death, in the modified Aristotelianism of his followers. The Romans and the later Church fathers showed little interest in his works on natural or social science, and until the twelfth century only part of his writings on logic was directly available in Western Europe. With the arrival in the West of the first Greek texts of all his surviving works during the thirteenth century, Aristotle provided the stimulus for innovators such as Thomas Aquinas (*c.* 1224–74), who advocated a new realism and expansion of inquiry. By the sixteenth century, however, the continued dominance of Aristotle's syllogistic logic and his cosmology brought a strong reaction against the Aristotelianism that persisted in Church teaching, with its restriction of science to a search for cause or essence in a finite, ordered world. In breaking from that tradition, Bacon and other empiricists after him tended to reject Aristotle altogether, relying on commentaries or poor translations for their assessment of his work, and so failing to recognize, through a reading of the Greek texts themselves, the extent to which their own theories were conditioned by his ideas.

So pervasive indeed was the Greek influence in general, persisting through Hellenistic, Roman, Arabic, and Western culture, that almost all the new sciences, even after Bacon's time, commenced from the standpoint of classical inquiry. This was the case also in geography where in particular the model provided by the substantial texts of Strabo and Ptolemy remained prominent throughout the seventeenth century.

The classical formulation of geography: Strabo and Ptolemy

Geography is one of the oldest sciences. The term itself, meaning in its Greek derivation, drawing of writing (*graphē*) about the earth (*gē*), has been in use for over two thousand years, forming the titles of works in Greek by Eratosthenes (276–194 B.C), Strabo (*c.* 64 B.C.–A.D. 21) and Ptolemy (*c.* A.D. 90–168). The *Geographica* of Eratosthenes is known only from references, notably in Strabo; however, Strabo's *Geography* in seventeen books and the lengthy *Guide to Geography* of Claudius Ptolemaeus survived in manuscript copies and were widely reproduced. Strabo and Ptolemy diverged considerably in their notions of the scope and purpose of geography, and both devoted substantial introductory sections in their works to the task of defining the limits and methods of the discipline, especially in relation to the already established studies of mathematics, astronomy, geometry (earth-measurement) and physics. Strabo as a historian, and Ptolemy from his outstanding work in mathematics and astronomy, brought a wide experience to this task, reinforced by their familiarity with the Greek philosophical tradition, and together their writings provide a more thorough discussion of the theory of geography than any to be found before the time of Humboldt.

Strabo in his *Geography,* written evidently between 8 B.C. and A.D. 18, described in the last fifteen books the various parts of the known world, from Spain and Britain in the west to India in the east and finally Ethiopia and Libya in the south, after having previously in his first two books established the concept of the *ecumene* or inhabited world within which these lands occur, and discussed not only the mapping of the known world but also the general limits and functions of geography as he saw the subject in relation to other sciences.[15] Describing Eratosthenes as more of a mathematician than a geographer, Strabo claimed that the geographer must begin by accepting certain hypotheses regarding the general spherical shape of the earth and its relation to the heavens, without attempting to establish these for himself: 'the person who attempts to describe the parts of the earth (*chorographein*) must take many of the statements of physics and mathematics as hypotheses'. Since it is necessary to accept these 'as fundamental principles of his science, the geographer must rely upon measurements of the earth as a whole provided by the geometricians, who in turn rely on the astronomers, and they themselves on the physicists' [2.5.1]. Having himself mastered these principles, and assuming likewise in his students a knowledge of the mathematical sciences, the geographer, Strabo advised, should then

explain, in the first instance, our inhabited world (*oikoumenen*) – its size, shape and nature, and its relations to the earth as a whole; for this is the special task of the geographer. Next he must deal with the various parts of the earth and of the sea as well, giving an appropriate account of them [2.5.4].

For Strabo a fitting or appropriate account in geography is one that uses the methods of geometry and astronomy without going deeply into the question of causes, for this 'belongs to the student of philosophy alone'. As a Stoic himself, Strabo rejected the Aristotelian search for causes in studying the earth, although he followed Aristotle in accepting experience as a basis for knowledge. He relied on both 'sense-perception' (*aisthesis*) and what he called 'common conceptions' (*koinai ennoiai*) to justify the propositions he adopted. On the question of 'whether the earth is sphere-shaped', for example, he referred for support not only to a commonly accepted law on the nature of the universe, but also to

the phenomena observed at sea and in the heavens; for our sense-perception can bear testimony there as well as the common conception (*koinē ennoia*). For instance, it is obviously the curvature of the sea that prevents sailors from seeing distant lights at an elevation equal to that of the eye.[16]

He emphasized that observations should be as accurate as possible and that they should be made with a consideration for their utility: in selecting detail, 'that which is useful and more trustworthy should always be given precedence' over what is merely famous or charming, for 'utility above all is our standard in empirical matters (*empeirias*) of this kind' [1.1.16–19]. It was his intention that the *Geography* 'should be generally useful' to educated men – 'to the statesman and to the public at large – as was my work on *History*'. At the same time he stressed that 'the present work is a serious one, and one worthy of a philosopher', a work dealing, not with petty or insignificant matters, but 'with great things only, and with wholes' [1.1.22–3].

Although Strabo is usually credited with establishing the tradition of geography as merely regional description of a rather encyclopedic kind, it must be remembered, then, that he preceded that part of his work with a quite lengthy examination of the history and theory of geography as a whole. He gave particular attention also to the problem of mapping the known world, recording the work done in this respect by Eratosthenes and his followers with their mathematical researches on the earth. As he pointed out, these had led by the time of Hipparchus (*c.* 150 B.C.) to the establishment of propositions concerning the circumference of the earth and the location of known landmasses in relation to various 'circles' – those forming zones or *climata* parallel to the equator, and others, called meridians, passing through the poles.

It was this task of mapping the *ecumene*, or inhabited earth, that Ptolemy some years later in his *Geography* advanced as the chief concern of the subject: 'to record (*katagrapsai*) our inhabited world (*oikoumenen*) as much as possible in accordance with truth'.[17] Like Eratosthenes, three centuries before him at Alexandria, Ptolemy emphasized a more mathematical approach within geography itself, maintaining a close relationship with astronomy, and relegating the regional descriptions of Strabo in large part to a separate study of chorography. For Ptolemy the purpose of geography is to provide 'a view of the whole, analogous to the drawing of a whole head', whereas 'chorography has the purpose of describing the parts, as if one were to draw only an ear or an eye'. Geography is still to be concerned with the main regions, for these are 'the principal features of the ecumene and most readily fitted in on the proper scale'. As in a drawing, however, the first task is to establish the correct proportions and fill in the principal parts, 'in order that the representation may be perceived as a whole'. Therefore 'geography must first consider both the form and size of the whole earth as well as its position in relation to its surroundings'; then the main task is to show 'locations and general outlines' of known lands by establishing mathematically 'the correct proportion of distances' and 'determining relative positions'. After a discussion on map projection, the majority of his own *Guide to Geography* consists of tables of co-ordinates, giving latitude and longitude for hundreds of places to provide the basis for constructing maps of the known world.

In Ptolemy's attempt to separate a mathematical geography from a descriptive chorography can be seen the beginning of a long series of efforts to solve the methodological problems of the subject by effecting some kind of division in it. This was of course the process by which various specific sciences were defined and created from the period of early Greek philosophy onwards. Geography, however, seems to have resisted the kind of dismemberment suggested by Ptolemy. There is the practical difficulty that in order to describe the whole earth, one must know a good deal about individual regions: this is a telling instance of the problem in human knowledge that occupied the Greek philosophers for centuries before Ptolemy, that of relating universals to particulars, generalizations to individual instances. In geography the epistemological issue has obvious practical outcomes.

Ptolemy justified his exclusion of chorography from geography – and with it, incidentally, most of Strabo's concern with human activities – on the grounds that they differ not only in subject matter and scope, but also in method. Assuming that chorography in describing localities does not require the mathematical method, he argued that geography deals with quantities rather than the nature of things and so has more in common with astronomy in this respect. Claiming that knowledge of astronomy is more reliable, Ptolemy went on to restate Aristotle's geocentric view:

The heavens, rotating about us, are revealed to man's understanding through mathematics . . . while the earth is represented through a model (*eikonos*) since . . . it cannot be traversed in its entirety . . . by the same man [I.1.9].

Like Strabo before him, Ptolemy recognized the problem of the empirical method in geography and he noted that in constructing such a map or model of the earth, 'the descriptions given by travellers are of the highest importance'. The value of these reports, he added, depends upon the 'theoretic understanding' of geography that the observers have, and also on the method adopted, the use of instruments such as astrolabes and sun dials giving a distinct advantage in accuracy, as he saw in his own day, to observations of stars as compared with estimations of ground distances in *stadia* by travellers. He therefore recommended that geometry or earth measurement should be supplemented by astronomical observations, so that the direction of distances traversed can be calculated, as well as the latitude and relation of any distance to the circumference of the earth [1.2.1–3].

His determination to make geography a more precise study, concerned chiefly with providing an accurate map or model of the earth, brought Ptolemy into conflict with the more descriptive approach of Strabo, who gave much greater attention to human activities, believing that 'geography is directed mostly towards the needs of states'.[18] Conflicting interpretations with regard to the place of man in geography and the role of quantification are thus already evident in their work.

Meanwhile, after Ptolemy, although Arab scholars maintained some vigour in the study, there was a notable decline in geography as a whole until the revival of cartography in the fifteenth century during the age of European discovery. To some extent the earlier traditions of astronomy and geography were preserved in the Christian cosmographies, most of them in Latin, that continued to appear until well into the seventeenth century, although in the tradition of the rather strange work by Cosmas of Alexandria, *Topographia Christiana*,[19] such books tended to show a preoccupation with maintaining a study of the universe as a support to religion, reinforcing at once the Platonic idea of cosmic design and the biblical teachings regarding the history of the world. In this, geography shared in the general decline of learning that followed the collapse of the Roman empire, when the remnants of classical scholarship were transmitted uncritically, under the aegis of the Church.

Medieval and renaissance developments

Signs of a renewed concern with empiricism appeared as early as the twelfth century in Europe, with the translation into Latin of Aristotle's logic and the geometry of Euclid (*c.* 330–*c.* 260 B.C.) which itself represented a development of Platonic thought. At that time a number of scholars repeated Aristotle's dictum that there is nothing in the intellect that was not first in the senses, while Robert Grosseteste (*c.* 1170–1253) and Roger Bacon (1214–94) stressed the dependence of natural science on sense-experience.[20] They proposed the use of inductive procedures in the rejection of postulates, while acknowledging the element of probability in all affirmative statements, some 700 years before Karl Popper's argument that the falsity of a hypothesis can be proved, even though it is not possible to prove that any so-called universal law will always operate as predicted. That element of uncertainty did not prevent the thirteenth-century thinkers from promoting a new concept of science, one which accepted the neo-Platonic postulate of a basic mathematical harmony in the universe, while setting as the immediate objective a search for mathematical laws of nature that can be used to explain and predict events or effects in the world. This was opposed to the view, promoted by Aristotelianism, that explanation consists in defining innate 'essences'.

Throughout the fourteenth century there was increased attention to those matters that can be observed and expressed in terms of quantities. As a keystone of scientific method this was to stimulate effective research, although in its extreme form it favoured a positivistic and materialistic approach of dealing only with what can be measured, predicted or controlled. An early association of materialism with empiricism is suggested in the work of William of Ockham (*c.* 1284–1349), who recommended studying the behaviour of moving objects rather than searching for some reason for motion and claimed indeed that nothing exists except individual things accessible to observation.

A dramatic extension to the European awareness of the world as a whole occurred during the fifteenth century with the expansion of trade and successful voyages of exploration: the discovery of the American continent by Columbus in 1492 was the culmination of a long period of observation and inquiry. Along with the development of navigation and the invention of printing, the mental horizons of the renaissance were extended also by a renewed contact with Greek literature. The renaissance of fourteenth-century Italy had depended largely on a revival of Latin literature and the humanist movement of that time with its reverence for the classical past seems to have given little encouragement to scientific inquiry or to technical advances associated with the crafts; in the universities the humanist studies (*studia humanitatis*) recommended for the cultured citizen included history, letters, and eloquence, and these came to be referred to colloquially as the *humanities*.[21] Thus a separation between the humanities and sciences might be traced to renaissance Italy. Nevertheless, after the teaching of Greek had been introduced successfully to the university of Florence by Emmanuel Chrysoloras (*c.* 1350–1415) at the beginning of the fifteenth century, a stimulus was given to natural science as the great works of ancient Greece and Alexandria again became available in the West.

Among the first works translated into Latin in the time of Chrysoloras was the *Geography* of Ptolemy.[22] Its lasting influence is indicated by the number of editions that followed – over forty in the next 200 years – first in manuscript, and then after the development of printing in 1450, as printed editions.[23] Maps were re-drawn according to Ptolemy's system of parallels and meridians, often using new projections, but generally following the co-ordinates provided in his text – although it is not clear whether any map by Ptolemy himself actually survived.[24] Commencing with some of the earliest editions, modern maps and contemporary discoveries were included; in the 1508 Rome edition the New World was added to the Ptolemaic world map – an interesting change, for Ptolemy's serious underestimation of the earth's circumference had encouraged Columbus to sail westwards from Spain to Cathay. Despite its errors, and sometimes because of them – indeed the search for Ptolemy's great southern continent or *Terra incognita* persisted into the eighteenth century – Ptolemy's *Geography* provided a focus for renaissance attempts to explore and conceptualize the world. Along with the *Geography* of Strabo, translated in Italy around 1440, it offered a stimulus to the growing revival of geography. This gathered momentum during the sixteenth century in northern Europe, when an interest in local surveys and topographies showed a marked increase. The current cosmographies, however, still relied heavily on Ptolemy in dealing with the world, some of them being little more than descriptive texts accompanying a collection of maps.

One of these, an *Introduction to Cosmography* of 1507, became famous for its proposal that the New World be named America after Amerigo Vespucci, a Florentine sea-captain who was believed to have visited the South American continent prior to Columbus.[25] Its authorship remained in some doubt and subsequently Humboldt, from the researches for his *History of the Geography of the New Continent*,[26] suggested that the name *Hylacomylus* on the edition he consulted must refer to Martin Waldseemüller (1470–1518), since the Greek name, like the German, means 'miller of the woods'. The geographer Waldseemüller, with Matthias Ringmann (*c.* 1482–1511), professor of cosmography in Basle, worked from 1505 to produce the important 1513 Strasburg edition of Ptolemy, and Humboldt referred to the close association of these two men in summarizing his interpretation for the 1847 volume of *Kosmos*.[27] For his own part Humboldt was concerned to defend Amerigo from accusations of connivance in the affair; at the same time he was clearly proud to assign a German source for the name America in giving credit to Waldseemüller, and his explanation has been widely accepted.[28]

Recent German research, however, has argued that Ringmann, already known to have written the descriptive text for Waldseemüller's 1511 map of Europe, was actually the author of the *Cosmographiae Introductio*, which was probably intended to accompany Waldseemüller's 1507 *Universalis Cosmographia* – a world map unknown in Humboldt's time, which was rediscovered only in 1900. According to Franz Laubenberger it was at Ringmann's instigation that Waldseemüller in 1507 used the name America for the first time on both the world map and a globe (recovered in 1890); Waldseemüller himself apparently became sceptical about Amerigo's priority as a discoverer and did not use the name again – it does not appear in the

Ptolemy published after Ringmann's death – although it was adopted by others and perpetuated.[29] The effective introduction of the name in this way is, however, an interesting indication of the significance of cartography in the sixteenth century, and Humboldt's point was well taken that it was not in Spain that the name given to the New World originated, but in Lorraine: the centre of geographical research had moved northwards.

In 1540 a new edition of Ptolemy was produced at Basle by Sebastian Münster (1489–1552) whose *Cosmographiae universalis* of 1544, reproduced in many editions, was to remain a standard text for over a century. Münster's *Cosmographia* contained numerous maps – many from his Ptolemy edition – along with a large number of detailed illustrations and a rambling Latin text devoted mainly to Europe, with a shorter section on Asia in which some of the tales of Pliny (A.D. 23–79) were repeated. His regional descriptions of Europe, based on political units, suggest the foundations of a vigorous human geography in his discussion of agricultural, industrial and city life, yet his commentary with its lack of organization and uncritical use of sources is in strong contrast to the relative order and control of information on his maps. The descriptive geography of Münster is the kind that brought disparaging comments from Varenius a century later; nevertheless, along with the more mathematical and astronomical cosmography of Peter Apian (Apianus or Bienewitz, 1495–1552), it exercised a strong influence on geographical thought for more than a hundred years. Use of the term *cosmography* at that time, it should be noted, could indicate either a religious emphasis, as in Münster's case, or the inclusion of astronomy, as in the case of Apian, although it was often used interchangeably with *geography*.[30]

During this period an important contribution to empiricism was made as the age of exploration became the great age of cartography in Italy and later in northern Europe. In the Netherlands a century of leadership in that field commenced with the impressive *Theatrum Orbis Terrarum* (Theatre of the Whole World) of Abraham Ortelius (1527–98), and the *Atlas sive Cosmographicae Meditationes* of Gerard Mercator (1512–95). With this work Mercator, who himself prepared maps for the *Ptolemy* edition of 1578, introduced the term *atlas* to describe the modern systematic and standardized collection of maps that soon superseded those of Ptolemy. Sixteenth-century cartography represented a joining of two medieval traditions: the diagrammatic *mappae mundi,* or T–O maps, intended as plans to illustrate church teachings about the world, and the nautical charts or *portolans* drawn for navigators, from the fourteenth century on, with greater attention to accuracy. One significant feature was the extent to which modern cartographers were prepared to leave gaps where there was a lack of data, indicating the need for further exploration: this is a characteristic of an effective scientific method, to locate areas for further research, rather than to obfuscate by glossing over ignorance. The new maps were based increasingly on local observations and instrumental measurement of data, interpreted mathematically in terms of projection and scale; the symbolic representation of this information in maps relies to a very small extent on language, so that the map can be considered one of the oldest techniques for the non-verbal recording of empirical inquiry in the natural sciences. Like all attempts to fix present knowledge in a world of change, maps can be deceptive of course if interpreted as an exact record of what exists: at its best a map indicates rather effectively the limits of knowledge, and indeed what geographers have been able to accomplish through a consistent use of scale and projection is to exercise some control over the degree of distortion involved. Every map, however, remains a generalization – even the modern map constructed with data collected mechanically to minimize human error – and it is just as absurd to assume complete objectivity in map-making as in any other area of science.[31] The cartographer faces the dilemma of all empiricists: the map remains a symbolic representation of experience.

The notion of experience began to receive closer attention from intellectuals during the fifteenth century. Leonardo da Vinci (1452–1519) vigorously supported the empiricist point of view, although most of his notes on science remained unread for almost three centuries. Commenting on this, Humboldt argued that the development of natural science would have been considerably advanced, had Leonardo's proposals been put into effect in his own time:

Like Francis Bacon, and a full century before him, he held *induction* to be the only reliable method in natural science: *dobbiamo cominciare dall' esperienza, e per mezzo di questa scoprirne la ragione* [we must commence with experience, and by means of this discover the cause].[32]

During the time of Leonardo the relation between observations and theory became of increasing concern, and the attempt to draw a strict logical distinction between *induction* as a procedure of working from investigation of particulars towards theory or generalization, and *deduction* as the drawing of inferences from general statements, dates from this period. The term 'induction' itself was derived from the Latin *inducere*, to lead, after Cicero's use of *inductio* to translate Aristotle's *epagogē*,[33] a term which signified a *method* (*meta hodos*, after the right path) by which investigation can lead to understanding. Greek literature was still being passed through a filter of Latin scholarship at that time; otherwise the method of the new empiricism might have been called, not induction, but *epagogy*, in the same way that *pedagogy*, the leading of children, became accepted along with its Latin equivalent, education. This did not happen, however, and the new interpretations of inductive method showed a marked change of emphasis from that of Aristotle.

In his *Physics* Aristotle's use of *epagogē* suggested a procedure of reasoning by taking account of individual instances, as contrasted with the alternate technique in debating of bringing forward logical arguments in proof.[34] Aristotle himself seems to have attached no certainty to inductive reasoning, indeed he evidently believed that in the imperfect sublunary sphere of change and decay the best that could be achieved was some kind of plausible speculation concerning material things. Among Italians of Leonardo's time, however, it was argued that a science of nature based on *esperienza* (experience and experiment) and following a proper method could arrive at complete truth. Reason, in the form of logical or mathematical demonstration, formed an important part of this method as it was outlined in the early sixteenth century by Pietro Pomponazzi (1462–1525) and Jacopo Zabarella (1533–89) at the University of Padua.

Opposing that tradition of demonstration the Spanish humanist Juan Luis Vives (1492–1540) pointed out that universal principles cannot be obtained by induction since it is impossible to enumerate all the particulars on which the universal must be established. Vives himself, in discussing the practical arts, with which he was especially concerned, argued that these arise from particular experiments reported by the senses, with rules being 'derived by the work of the mind'.[35] In scientific inquiry Vives appealed to direct experience of nature, although he was, as Blake has noted, 'extremely vague . . . as to the method of such an empirical science'.[36]

This vagueness as to procedure was not altogether dispelled by later proponents of the new scientific method; nevertheless, their confidence in achieving certitude in the study of nature remained virtually unshaken until the mid eighteenth century. Throughout that period one legacy of Aristotelian thought was retained unquestioned. While the school of Baconian sense-empiricists argued that certainty is to be attained by means of experiment and the evidence of the senses, the rationalist school relied instead on the operation of the reason, and especially on the application of mathematics and deductive logic: what both groups had in common, however, was an acceptance of the Aristotelian antithesis between sense and reason, a distinction that was to have profound consequences for the empirical sciences.

FRANCIS BACON: THE *NOVUM ORGANUM*

By the turn of the seventeenth century the gathering impetus of technological change and scientific inquiry began to make itself felt in Europe, although the universities in general were still occupied with scholastic debates and a derivative study of the classics. Francis Bacon (1561–1626), Baron Verulam, Viscount St Albans, and for a time Lord Chancellor of England, gave public expression to a growing popular belief when he declared that a complete renewal of knowledge was required. Through an ambitious series of works that he planned under the title *Instauratio Magna* (great renewal) Bacon hoped to rejuvenate learning by developing both a new scientific method and a new system of sciences. Many of Bacon's most influential statements were not of course original to him, but can be traced to renaissance humanists, medieval nominalists and classical authors. He was exceptionally significant, however, as a popularizer of such ideas in Europe during a time of change, his achievement being the combination of these into a practical programme, with an emphasis on the utility of knowledge and incorporation of the technical tradition.

Bacon's first important book, *The Advancement of Learning*, was published in English in 1605, more than half a century after the great work of Copernicus and several years before the publication of the discoveries of Galileo and Kepler. His major work on scientific method, the *Novum Organum* of 1620, was written in Latin and soon created wide interest in Europe with its plan for the systematic development of science.

Bacon's theory of induction

Bacon described his new instrument as 'a method of intellectual operation' to assist the human mind in 'finding out new works' through vigorous sciences. Rejecting for this purpose 'the logic now in use', derived from Aristotle, he argued that it does not 'help us in finding out new sciences', but tends instead to fix errors:

The syllogism consists of propositions, propositions consist of words, words are symbols of notions. Therefore if the notions themselves . . . are confused and over-heavily abstracted from the facts, there can be no firmness in the superstructure. Our only hope therefore lies in a true induction.[37]

Published with *Novum Organum* was his introduction to *The Great Instauration* – for which comparatively few of the projected works were ever actually completed by him. Here Bacon included a strong complaint on the previous neglect of inductive method by the logicians; only Plato, in his view, had previously attempted the use of the kind of induction now proposed. Rejecting earlier forms of disputation and demonstration, Bacon outlined his own theory of knowledge: nature provides 'particulars', and experience is derived from these through the senses. Next the understanding produces the propositions and 'middle terms' or axioms that themselves lead to 'the most general propositions' or principles of nature, and from these again new particulars are derived: 'Now my plan', he wrote, 'is to proceed regularly and gradually from one axiom to another, so that the most general are not reached until last.' This process, which he called *Interpretation of Nature*, is 'a kind of logic' but one which deals with the problem of experience: 'what the sciences stand in need of is a form of induction which shall analyse experience and take it to pieces, and by a due process of exclusion and rejection lead to an inevitable conclusion'. In explaining how this should be done, Bacon did not go much beyond the thirteenth-century technique of falsification. Nevertheless, he claimed that such a form of judgment 'which is extracted not merely out of the depths of the mind, but out of the very bowels of nature' would lead not to arguments but to works and enable man 'to command nature in action'.[38]

Bacon's project was a grand one in its scope. The reform and development of learning depended, he saw, on an effective system of inquiry, and the outlook of modern science is suggested in his work. A faith in man's progress towards control over nature, a vigorous assertion of the utility of knowledge and a commitment to study the world of change recorded by experience, are accompanied by an urgency to demonstrate the certainty of the knowledge and the immediacy of the experience on which all these are seen to depend. Many of the contradictions implicit in later scientific theory are evident in Bacon's thought, for in his confidence that his method would bring about a new era of discovery he tended to overlook the inconsistencies in his own theory. To begin with, the effectiveness of his inductive procedure at each stage was not examined closely, for he was preoccupied with establishing the contrast between his method and that of the ancients.

In *The Advancement of Learning*, which contains an early statement of all the preceding arguments, Bacon opposed deduction from theory without reference to experience. Referring to this as a faulty kind of induction, he condemned

the Induction which the Logicians speak of, and which seemeth familiar with Plato, (whereby the Principles of Sciences may be pretended to be invented, and so the middle propositions from the Principles); their form of induction, I say, is utterly vicious and incompetent . . . They hasted to their theories and dogmaticals, and were . . . scornful towards particulars.[39]

Bacon himself was never scornful towards particulars, for his outlook was decidedly materialist and he was if anything inclined to be rather cavalier with theories. He failed to consider the possibility, already raised in Aristotle's work, that 'particulars' may not be the starting point of perception at all but are rather a product of subsequent analysis of complex impressions. The assumption of direct contact with the facts of nature was as important to Bacon's theory as to the method of later positivists.

At the same time he acknowledged the argument of the Sceptics who 'denied any certainty of knowledge or comprehension' and argued that 'the knowledge of man extended only to appearances and probabilities' [126]. Bacon's answer to the sceptics was that the senses can provide access to the truth, given proper assistance:

Here was their chief error: they charged the deceit upon the senses; which in my judgement, notwithstanding all their vacillations, are very sufficient to certify and report truth, though not always immediately, yet by comparison, by help of instrument, and by producing and urging such things as are too subtile for the sense to some effect comprehensible by the sense, and other like assistance.

These were the aids to perception that scientists were to take up most enthusiastically, the conversion of data to observable forms with the use of instruments being especially effective during Bacon's own lifetime in the case of air pressure, temperature, and magnetism. Admitting that, with all this, serious obstacles remained in the way of certainty in knowledge, Bacon was prepared to lay the blame on the intellect as distinct from the senses, and continued in the same passage:

But they ought to have charged the deceit upon *the weakness of the intellectual powers, and upon the manner of collecting and concluding upon the reports of the senses* [126–7; my italics].

With the use of the right method, Bacon was confident that truth can be reached by the human mind and according to his plan, additional sciences were to be invented and organized to provide further help to the intellect. Those sciences, however, when they emerged were to inherit both his urge to reach certainty and his suspicion of intellect. In separating senses and intellect he was clearly adopting the faculty theory of the mind, advanced by Aristotle and reinforced in the teachings of Thomas Aquinas, although he apparently never questioned the source of this theory, or even recognized it as such; and indeed it should be noted that Aristotle himself referred to the divisions of the *psychē* not as faculties but in terms of a potential capacity for action: he spoke of 'the capacity for sensation' rather than 'the senses'. Taking his own metaphor from optics, Bacon in *The Great Instauration* described the intellect acting as a mirror to reflect 'the genuine light of nature' directed to it by the aided senses: 'All depends', he claimed, 'on keeping the eye steadily fixed upon the facts of nature and so receiving their images simply as they are' [32]. There would be no problem at all, in his view, 'if the human intellect were even, and like a fair sheet of paper with no writing on it' [24].

Our minds, however, according to Bacon are possessed by false notions that can be cleared away only by true induction. In the *Novum Organum* of 1620 [I, 38–44] he developed this into the famous theory of the four 'idols', using the term in the original sense of the Greek word, as an illusion. First, he listed the *Idols of the Tribe* as those common to 'the tribe or race of men'. Showing some loss of confidence in the reliability of the senses, he stated that

All perceptions as well of the sense as the mind are according to the measure of the individual and not according to the measure of the universe. And the human understanding is like a false mirror, which, receiving rays irregularly, distorts and discolours the nature of things by mingling its own nature with it.

Secondly, the *Idols of the Cave* he declared, recalling Plato, are those 'of the individual man. For every one . . . has a cave or den of his own, which refracts and discolours the light of nature.' Then there are the *Idols of the*

Market-place, arising from discourse and a confusing use of words. Finally, Bacon condemned the false notions derived from 'the various dogmas or philosophies, and also from wrong laws of demonstration', calling these *Idols of the Theatre*, 'because in my judgement all the received systems are but so many stage-plays, representing worlds of their own creation after an unreal and scenic fashion'.

Bacon's own philosophy, emphasizing sense-experience as the source of scientific knowledge, attracted many followers and provided the foundations for the movement later known as empiricism; in his own writings, however, the term 'empiric' is used in most cases in a derogatory way, reminiscent of Plato's attitude to uncritical observation. In *The Advancement of Learning*, for example, Bacon quoted the medical treatise of Celsus (*c*. A.D. 30) concerning 'the Empirical and dogmatical sects of physicians', who found medicines and cures first and then discussed the reasons and causes [123]. Comparing the cures of 'learned physicians' with those of 'empirics and old women', he concluded 'they be the best physicians, which being learned incline to the traditions of experience, or being empirics incline to the methods of learning' [116]. Most previous empirics in his view were to be condemned for conducting random experiments without systematic organization, or trusting to chance experience instead of attempting to exercise control over their discoveries. In the *Novum Organum* he even included William Gilbert in his attack on empiricists who construct complete systems on the basis merely of a few experiments:

The Empirical school of philosophy gives birth to dogmas more deformed and monstrous than the Sophistical or Rational school. For it has its foundations not in the light of common notions . . . but in the narrowness and darkness of a few experiments. To those therefore who are daily busied with these experiments . . . such a philosophy seems probable and all but certain; to all men else incredible and vain. Of this there is a notable instance in the alchemists and their dogmas; though it is hardly to be found elsewhere in these times, except perhaps in the philosophy of Gilbert. Nevertheless . . . I foresee that if ever men are roused by my admonitions to betake themselves seriously to experiment and bid farewell to sophistical doctrines, then indeed through the premature hurry of the understanding to leap or fly to universals and principles of things, great danger may be apprehended from philosophies of this kind [I, 64].

A new enthusiasm for experiment soon spread as Bacon predicted, to the extent that modern science in its early stages was called the experimental philosophy. Whether this was initiated chiefly by Bacon's admonitions, however, is doubtful, since the experiments of Galileo and William Gilbert predated 1620. The degree of Bacon's antagonism to William Gilbert indeed has aroused much comment among his critics, as Gilbert in the same period appears to have been practising rather effectively the kind of scientific method advocated by Bacon. In retrospect, Humboldt for example put this down to Bacon's 'very limited knowledge, even for his time, in mathematics and physics', noting that he rejected the Copernican doctrine which Gilbert himself adopted.[40] Gilbert's theory of magnetism, of course, stood in opposition to the kind of mechanistic view favoured by Bacon and his followers.

The cult of science

Bacon's own attitude to experiment was not altogether consistent. In his 1605 treatise he seemed to place little weight on experiment in his methodology, but in *The Great Instauration* the experiment is presented as the chief means for the correction of the senses:

> This I endeavour to accomplish not so much by instruments as by experiments . . . such experiments, I mean, as are skilfully and artificially devised for the express purpose of determining the point in question. To the immediate and proper perception of the sense therefore I do not give much weight; but I contrive that the office of the sense shall be only to judge of the experiment, and that the experiment itself shall judge of the thing. And thus I conceive that I perform the office of a true priest of the sense (from which all knowledge in nature must be sought, unless man mean to go mad).[41]

From the point of view of empiricism this is one of the most significant passages in Bacon's writings. All knowledge in nature, he claimed, must be obtained through the senses, and here he not only gave authority to the *sensationist* philosophy of the following century but tended also to dogmatize some of the results of his own battle with scepticism. The senses, he admitted, are subject to errors, omissions and self-reference; nevertheless, he had already claimed in *The Advancement of Learning* that his method, though it may 'begin with doubts . . . shall end in certainties' [34]. Therefore in his desire to establish certainty at some point he turned to the controlled experiment to provide direct contact with reality. The function of the sense, he promised, is only to judge the result, *the experiment itself shall judge of the thing.*

Here is perhaps the most dangerous part of Bacon's doctrine, for it promoted the objectivism that became implicit in the work of later scientists – the belief that from direct confrontation with facts or objects themselves, and from instruments or test situations, knowledge is obtainable with a minimum of interference from the mind. In his attitude to the experimental method, Bacon provided a classic example of the objectivist fallacy, for he failed to take into account not only the interpretation of results but also the extent to which the framing of an experiment already determines the kind of results it will show. His method proved extremely productive in encouraging active, ordered research, but in the hands of people even less experienced in philosophy than Bacon himself and other intellectuals of his time, it could degenerate into what was almost a cult of science, with the experiment providing a kind of secret link with the unknown. The dedicated pursuit of impersonal exactitude, an urge to secure absolutes, and a reforming zeal based on the conviction of direct access to truth, contributed to the priestly character of modern science after Bacon and it is not surprising to find that he referred to himself as *a true priest of the sense*. This may be an instance of a reaction taking on the very characteristics of that which it is rejecting, for Bacon set out in 1605 to separate religion from science in his theory.

In *The Advancement of Learning* Bacon accepted divine revelation as a source of knowledge in the scriptures but argued that divinity is concerned with 'the book of God's word', while philosophy deals with nature or 'the book of God's works'. Denying that 'inquiry into these sensible and material things' would encourage atheism, Bacon claimed it would instead promote a natural theology – the understanding of God as 'the first cause' through his works [6–8]. Some of Bacon's statements of course may have been tactical, for this treatise was addressed to King James I, head of the Church of England.

Bacon's attempt to free scientific inquiry from religious dogmatism was to have far-reaching effects and his novel admonition in the *Novum Organum* to 'give to faith that only which is faith's' became a guideline in the scientific movement that followed. His own concern was to overcome the traditional preoccupation with 'abstract forms and final causes and first causes', and he advised his followers to be guided by utility in avoiding the two extremes, neither 'abstracting nature till they come to potential and uninformed matter, nor on the other hand . . . dissecting nature till they reach the atom; things which, even if true, can do but little for the welfare of mankind' [I, 66].

Utility is a constant theme in Bacon's writings, contributing to the powerful effect of this theory, for his emphasis on what is active, productive and useful in knowledge provided a link between fourteenth-century empiricism, and modern scientific materialism and pragmatism. Directing attention to dynamics, he condemned in the *Novum Organum* the search for primary forms and elements by those who 'make the quiescent principles, *wherefrom*, and not the moving principles, *whereby*, things are produced, the object of their contemplation and inquiry' [I, 66]. Evident here, although not acknowledged, is the teaching of William of Ockham, whose materialism is suggested also in Bacon's affirmation that 'in nature nothing really exists beside individual bodies, performing individual acts according to a fixed law'. Bacon expressed a distinctly modern viewpoint, however, in advocating the application of science to the control of natural processes, as he argued that 'the investigation, discovery, and explanation' of universal laws would provide the foundation of both knowledge and action. Science would extend man's power to effect changes in nature:

Although the roads to human power and to human knowledge lie close together, and are nearly the same, nevertheless on account of the pernicious and inveterate habit of dwelling on abstractions, it is safer to begin and raise the sciences from those foundations which have relation to practice, and to let the active part itself be as the seal which prints and determines the contemplative counterpart [II, 2].

Bacon's tendency to separate the abstract from the useful was to set an unfortunate precedent in scientific thought, although in the *Novum Organum* he himself insisted that active and contemplative are to be closely linked; indeed, with a statement that anticipated the 'truth is what works' dictum of nineteenth-century pragmatism, he declared, 'what in operation is most useful, that in knowledge is most true' [II, 4]. Like the pragmatists after him, then, Bacon stressed utility in scientific inquiry, insisted that the deductive logic of the syllogism cannot produce new knowledge, and directed effort towards the analysis of experience. The pragmatic assertion that the mind selects from a continuum of experience was not part of Bacon's theory, however, for he always considered particulars to be the grounds of experience.

The concept of experience is a crucial one in Bacon's philosophy, yet his consideration of it is neither thorough nor consistent. His chief concern was to distinguish random experience from controlled or experimental experience, and here he made a useful contribution by urging the recording of available evidence, to provide what he called *literate* experience, for as he asserted in the *Novum Organum* – referring probably to the craft tradition – 'experience has not yet learned her letters' [I, 101–3]. At the same time the function of stored knowledge was not integrated effectively into his theory, which was marked by a strong antipathy towards book-learning in science: in *The Advancement of Learning* he condemned 'the overmuch credit that hath been given unto authors in sciences, in making them dictators' as being 'the principal cause that hath kept them low at a stay without growth or advancement' with the result that the sciences contrasted poorly with the steady refinement of the 'arts mechanical . . . sailing, printing and the like'. Bacon's support for the linking of science with technology and the development of experiment in place of excessive reliance on tradition was well justified in his day; however, in addition he tended to denigrate altogether the process of learning from others as a way of acquiring experience. For instance, noting that man's knowledge is derived either from divine revelation or from the light of nature', he added that 'The light of nature consisteth in the notions of the mind and the reports of the senses: for as for knowledge which man receiveth by teaching, it is cumulative and not original' [85].

Vicarious experience – the experience of others, communicated through symbols – presents a major problem for any empiricist who is intent on asserting the primacy of direct personal experience, and Bacon, like many pragmatists up to the twentieth century, found difficulty in integrating this wider concept of experience into his theory. Instead he tended to limit experience to the perceptions and resulting mental notions of the individual

concerned; in addition he placed a premium on originality, thus emphasizing the subjectivity rather than the sociology of knowledge, in a way that was
to become characteristic of the whole empirical movement in the following
century.

Bacon's low regard for the value of learning from others stood in evident
contradiction to his own general argument that progress in science depends
on organized co-operative efforts – such co-operation itself implying effective communication of ideas. His *Advancement of Learning* actually commenced with the argument that the understanding of nature had been
hampered in the past, less by 'the capacity of the mind' than by such
impediments as 'shortness of life, ill conjunction of labours, ill tradition of
knowledge over from hand to hand' [5]. He even went on to ask the King's
support in providing institutions more encouraging to science than the
established colleges, since 'this excellent liquor of knowledge, whether it
descend from divine inspiration, or spring from human sense, would soon
perish and vanish to oblivion, if it were not preserved in books, traditions,
conferences, and places appointed, as universities, colleges, and schools, for
the receipt and comforting of the same' [62]. Requesting further assistance
and reward of those inquiring into the sciences, he proposed the establishment of 'a fraternity in learning' among the universities of Europe [67].

Primarily of course Bacon's intention in *The Advancement of Learning*
was to create a change of direction away from the conservation of past
learning and towards new inventions, and he contrasted his own approach
with those who 'tend rather to augment the mass of learning in the multitude
of learned men than to rectify or raise the sciences themselves' [62]. Later, in
the preface to his *Novum Organum*, he became even more explicit in
separating these two approaches. Denying that he objected altogether to the
use of the received philosophy, he remarked that it remained quite suitable
'for supplying matter for disputations or ornaments for discourse – for the
professor's lecture', and so forth. Indeed, he added, 'I declare openly that
for these uses the philosophy which I bring forward will not be much
available . . . It does not flatter the understanding by conformity with
preconceived notions. Nor will it come down to the apprehension of the
vulgar except by its utility and effects.' He continued by outlining what is
probably the first direct proposal for the separate development of a scientific
elite:

Let there be therefore . . . two streams and two dispensations of knowledge; and in
like manner two tribes or kindreds of students in philosophy – tribes not hostile or
alien to each other, but bound together by mutual services; let there in short be one
method for the cultivation, another for the invention, of knowledge [58].

Although Bacon himself went on in his preface to recommend co-operation
between the two groups, it is clear that he regarded the new method of
discovery as being essentially of a different kind and hence able to function

independently, having as it were a monopoly on invention. Furthermore, he promised that unlike the other it leads to certain knowledge and therefore should be followed by all those who joined with him in aspiring 'to overcome, not an adversary in argument, but nature in action; to seek, not petty and probable conjectures, but certain and demonstrable knowledge' [59]. That division was to take effect in later centuries with the growing separation between the sciences and humanities.

Bacon's approach to the study of nature was taken up with much enthusiasm, not only by those who like himself were tired of overmuch disputation and classical philosophy, but also by the new and growing class of skilled men from the mechanical tradition who welcomed the chance to participate in a scientific inquiry which demanded no philosophical background at all. To some extent the cause inevitably suffered from the excesses of the converts; nevertheless, much of the problem stemmed from Bacon's insistence on the complete independence of the scientific method. In order to achieve progress in the sciences, Bacon seemed prepared by 1620 to repudiate his view of learning as a social process, and, with the confidence of a revolutionary, called in the *Novum Organum* for drastic measures: 'there is no hope except in a new birth of science; that is, in raising it regularly up from experience and building it afresh; which no one (I think) will say has yet been done or thought of' [I, 97].

In view of the previous experiments and observations of Kepler, Galileo, Gilbert and others it is doubtful that Bacon was as much the solitary pioneer in this as he liked to see himself. No one else, however, seems to have approached the rashness of his proposition for founding a scientific method on the revolutionary myth of the fresh start. What must be done, he believed, in order to create a pure natural philosophy, was 'to sweep away all theories and common notions, and to apply the understanding, thus made fair and even, to a fresh examination of particulars'. Given the prevalence of error and confusion in the thought of his own day, the degree of his impatience with the past can of course be appreciated when he added, 'human knowledge, as we have it, is a mere medley and ill-digested mass, made up of much credulity and much accident'. The important point is Bacon's evident conviction that scientific invention can be advanced by first eliminating ideas and theories altogether. This was a theory of knowledge that was to have a lasting impact on science over the next three centuries. Determined to free himself from prejudice and superstition, he did not hesitate to advocate the most extreme objectivism, although it must surely be assumed there was at least some element of exaggeration in his assertion that the ideal researcher, in order to 'apply himself anew to experience and particulars', should be a person of 'unimpaired senses, and *well-purged mind*'.[42]

Even in that age of utopias it seems difficult to believe that such a proposal could be taken seriously, since it contradicted Bacon's own assumption that notions and axioms form an essential part of a continuing process of inquiry. Deduction from what he called 'Learned Experience' was accepted by Bacon elsewhere in the *Novum Organum* as a corollary of his method of induction [I, 110]. Moreover, he argued that new particulars and experiments would be derived from the axioms established by induction. It seems, however, that his belief in the priority of particulars led him to the objectivist view that the scientific method can begin at some point with a blank mind. Indeed, to make sure that the intellect starts from facts and not from general principles which are 'notional and abstract and without solidity' Bacon insisted that in the sciences 'the understanding must not therefore be supplied with wings, but rather hung with weights, to keep it from leaping and flying' [I, 103–4].

The rather severe puritanism of Bacon's outlook added to the anti-intellectual tendency evident in his view of the scientific method, and in the preface to *The Great Instauration* his objectivism is again apparent. Condemning the method of those who have merely 'cast a glance or two upon facts and examples and experience, and straightway proceeded, as if invention were nothing more than an exercise of thought', he contrasted this with what he considered to be his own approach:

I, on the contrary, dwelling purely and constantly among the facts of nature, withdraw my intellect from them no further than may suffice to let the images and rays of natural objects meet in a point, as they do in the sense of vision; whence it follows that the strength and excellency of the wit has but little to do in the matter [14].

In spite of his determination to sweep away all theories, withdraw his intellect, and generally encourage his followers to approach the study of nature with well-purged minds, Bacon nevertheless did not himself claim to separate empiricism from rationalism. Rather, by presenting his method of inquiry so that his errors might be corrected and his work continued, he hoped 'to have established for ever a true and lawful marriage between the empirical and the rational faculty, the unkind and ill-starred divorce and separation of which has thrown into confusion all the affairs of the human family' [15].

Similarly in the *Novum Organum* Bacon claimed to combine for the first time the experimental and the rational approach in science. Previously, he wrote,

those who have handled sciences have been either men of experiment or men of dogmas. The men of experiment are like the ant; they only collect and use; the reasoners resemble spiders, who make cobwebs out of their own substance. But the bee takes a middle course; it gathers its material from the flowers of the garden and of the field, but transforms and digests it by a power of its own.

Clearly, this analogy could lead to difficulties if pursued too far; neverthe-less, Bacon believed he had made his point in favour of the bee: 'The true business of philosophy', he stated, is not to rely 'solely or chiefly on the powers of the mind', but to 'take the matter which it gathers from natural history and mechanical experiments', and store it, not 'in the memory whole, as it finds it', but 'in the understanding altered and digested' [I, 95].

One of the serious deficiencies of Bacon's theory of knowledge, then, was his tendency to underestimate the active role of the mind in experience, since the operation of the intellect, in his view, commences only with the consideration of axioms and causes, not with perception, and can even be suspended at will. Apart from the influence of Aristotle here, it is likely that the prevalence of rote learning in his day encouraged him to assume that perception itself precedes thought and can therefore occur without any action of the mind, simply by objects impinging directly on the senses. Thus the kind of empiricism he inspired was itself strongly inclined towards objectivism, tending by the nineteenth century to exhibit, along with increased specialization, a strong antagonism to general philosophy. Yet this does not seem to be the kind of development that Bacon himself intended at all, for elsewhere in his writings he stressed the continuity of knowledge and the function of a universal science in sustaining this.

Bacon offered no clear directives for the construction of a universal science. Repeatedly, however, he insisted on the need to regard the various sciences as parts of a total complex of knowledge – or, in his phrase, as regions in a continent. For instance, in recommending in *The Advancement of Learning* the development, alongside *divine philosophy* and *natural philosophy*, of new studies in *human philosophy* or *humanity*, he anticipated some of the studies that later were to become the social sciences, pointing out, in a reference to Plato, that 'the *knowledge of ourselves*' deserves 'more accurate handling'. He warned at the same time that these studies should not be cut off from natural philosophy as a whole: *Humanity*, he stated,

is but a portion of natural philosophy in the continent of nature: and generally let this be a rule . . . that the continuance and entireness of knowledge be preserved. For the contrary hereof hath made particular sciences to become barren, shallow, and erroneous, while they have not been nourished and maintained from the common fountain [105–6].

This was an eloquent statement of a position that Alexander von Humboldt was to defend throughout his life; it is therefore ironical that the later partitioning of the natural sciences and their separation from philosophy, against which Humboldt in *Kosmos* protested in terms as strong as Bacon's own, should have been based on a notion of objective, experimental science derived from Bacon himself. Meanwhile in 1605 Bacon argued that 'philosophy and universality' are not 'idle studies', but provide the basis of action in 'all professions' [63].

In the *Novum Organum* Bacon was even more outspoken on the danger of separating individual sciences from the general study of natural philosophy, calling it the 'great mother of the sciences' and declaring that it should no longer be neglected:

Meanwhile let no man look for much progress in the sciences – especially in the practical part of them – unless natural philosophy be carried on and applied to particular sciences, and particular sciences be carried back again to natural philosophy. For want of this, astronomy, optics, music, a number of mechanical arts, medicine itself, – nay, what one might more wonder at, moral and political philosophy, and the logical sciences, – altogether lack profoundness, and merely glide along the surface and variety of things; because after these particular sciences have been once distributed and established, they are no more nourished by natural philosophy [I, 80].

He argued that natural philosophy provides a source of stimulation through the study of common problems, and among these he included not only 'motions, rays, sounds, texture and configuration of bodies', but also 'intellectual perceptions' [I, 80].

Bacon's own consideration of the function in science of perceptions or concepts was by no means extensive and the relationship of ideas to experience was to be explored with more care by his followers. While 'common notions' are incorporated in his theory of knowledge and therefore seem to form in his view an essential part of a science, they are not examined with any thoroughness except in the context of his discussion of *idols* or errors. The tendency to associate mental notions with error, rather than assigning to them an integral function in the process of acquiring experience, is therefore implied, if not intended, in Bacon's writing. Part of the difficulty stemmed from his insistence that inquiry begins directly with particulars and preferably with matter itself, although in *The Advancement of Learning* he had already recognized, at least to some extent, the importance of concepts in science:

It is the duty and virtue of all knowledge to abridge the infinity of individual experience, as much as the conception of truth will permit . . . by uniting the notions and conceptions of sciences [95].

In suggesting how this could be achieved it is significant that Bacon proposed, not a modification of the scientific method itself, but a series of sciences with different functions and at different levels of generalization.

Bacon's classification of learning

In *The Advancement of Learning* Bacon classified all human learning as *philosophy, history,* or *poesy,* each being related to one of what he saw as the three parts of the understanding: *philosophy* to the reason, *history* to the memory, and *poesy* to the imagination [69]. Poesy, which he considered to be linked with music and concerned with 'affections, passions, corruptions, customs' – expressed in poetry or prose – was listed as 'one of the principle portions of learning'; at the same time he assigned it no serious function in the progress of knowledge and indeed tended to denigrate poesy as mere '*feigned history*' in which 'the acts or events of *true history*' are distorted by the imagination: for whereas the reason 'doth buckle and bow the mind into the nature of things', the imagination, 'being not tied to the laws of matter, may at pleasure join that which nature hath severed, and sever that which nature hath joined' [82–3]. History also, according to his objectivist view, does not involve the use of reason, being limited to the recording of items as a function merely of the senses and memory: 'it is the true office of history to represent the events themselves . . . and to leave the observations and conclusions thereupon to the liberty and faculty of every man's judgment'. Using the term *history* in the wider sense of the Greek *historia*, inquiry or observation, he suggested four divisions: natural, civil, ecclesiastical, and literary history, describing their task specifically as the collection of particulars or facts.

Philosophy – divided into *natural philosophy, humanity*, and *divinity* or natural theology – was to be concerned with the search for explanation or causes, so that natural history would provide the basis for natural philosophy, of which natural science is a part. Natural philosophy in Bacon's scheme was itself divided into an operative or practical aspect called *natural prudence*, concerned with the invention of experiments and the production of effects or changes in nature, and a speculative or theoretical part, *natural science*, dealing with the investigation of causes in nature. Next he proposed two divisions of natural science itself, physical and metaphysical – *Physique and Metaphysique* – according to the particular causes each was to investigate. To a large extent Bacon was continuing to argue in an Aristotelian framework, the reference here being to the theory of 'the four causes' as they came to be called in medieval discussions, although Aristotle's term *aitia* is more accurately translated as 'condition'. In the *Physics* Aristotle argued that inquiry into all change in nature involves a search for principles or essential conditions: the material condition, such as bronze for a statue (material *aitia*); next 'the form and the model', conformity to which makes the thing what we say it is (formal *aitia*). Then there must be 'an agent of change' that produces the effect (efficient *aitia*); and lastly the *end* (*telos*), 'the purpose for which the process was initiated' (final *aitia*) [II, 3: 194b–195].

Already in the seventeenth century the direction of inquiry into nature was towards a study of effects and away from the scholastic preoccupation with causes and essences. It may be argued that the medieval interpretation of *essence* (from Latin *esse*, be) differed from that of Aristotle, who referred in the *Physics* simply to *being* or 'that which is' [II, 2: 194b 10] and in discussing being or essence indicated the complex nature of each entity: 'essence – the whole and the synthesis and the form' [II, 3: 195a 20]. Whatever the case, Bacon's contribution to the reaction against Aristotelianism in his own day was to allocate the search for causes to particular sciences, while planning to eliminate it altogether from natural history. As he explained in *The Advancement of Learning*:

Physique, taking it according to the derivation, and not according to our idiom for *medicine*, is situate in a middle term or distance between Natural History and Metaphysique. For natural history describeth the variety of things; physique, the causes, but variable and respective causes; and metaphysique, the fixed and constant causes [93].

According to Bacon the two functions of description and analysis can be separated completely, so that natural history was to supply simply enumeration or description while Physique was to go further, and bringing the reason to bear, investigate causes, although still subject to certain restrictions. Physique was limited to a consideration of the two conditions that he regarded as most closely related to matter – the material cause and the agent of change – with the implication that nature itself can be discussed adequately in materialist terms. In contrast Aristotle in his *Physics* had discussed all four *aitiai*, noting moreover that a combination of these may be required to produce a given effect in nature [II, 3: 195a].

Bacon himself did not reject entirely the classical search for forms. In assigning to Metaphysique 'the inquiry of formal and final causes' he dismissed what he called 'the received and inveterate opinion that the inquisition of man is not competent to find out essential Forms or true differences'; in *The Advancement of Learning* he argued in effect that the problem can be solved by a pragmatic and materialistic approach to the Platonic notion of forms:

It is manifest that Plato, in his opinion of Ideas, as one that had a wit of elevation situate as upon a cliff, did descry, *that Forms were the true object of knowledge*; but lost the real fruit of his opinion, by considering of Forms as absolutely abstracted from matter, and not confined and determined by matter [94].

This criticism of Plato was itself taken from Aristotle, who had insisted that form and substance must be considered together in nature; Bacon's own notion of the form however, became notably different from that of either Plato or Aristotle. In the *Novum Organum* he referred to forms as 'fundamental and universal laws' of nature, noting with an eye to utility that understanding of these will lead to control of natural processes, since 'any profound and radical operations on nature . . . depend entirely on the primary axioms'. [II, 5]. At the same time the form for Bacon is not an ideal abstraction in a Platonic realm of being, it is *the thing itself*, fluctuating in accordance with natural changes: 'the Form of the thing is the very thing itself, and the thing differs from the form no otherwise than as the apparent differs from the real' [II, 13].

Beyond such brief references, however, Bacon in the *Novum Organum* did not pursue the problem of appearance and reality, one of the great epistemological issues of the previous two millennia. He seemed to take it for granted that further investigation would justify a mechanical explanation of nature in which such problems would be of little importance. Inquiry into forms, in his scheme, was to begin with the study in *physics* (Physique) of the particles of which things consist, even though these may be too small for the senses to perceive: 'every natural action depends on things infinitely small, or at least too small to strike the sense' [II, 6]. Guided, in some way that he did not specify, by the primary axioms or forms of *metaphysics*, the investigator will be led, Bacon claimed, not to subtleties but in the direction of simplicity, to the 'real particles' of which things consist. He was quick to point out that it was not the theoretical atomism of Democritus that was being proposed and he followed Aristotle in rejecting not only the notion of Democritus that atoms are unalterable but also his theory of a void or vacuum:

Nor shall we thus be led to the doctrine of atoms, which implies the hypothesis of a vacuum and that of the unchangeableness of matter (both false assumptions); we shall be led only to real particles, such as really exist. Nor is there any reason to be alarmed at the subtlety of the investigations, . . . on the contrary, the nearer it approaches to simple natures, the easier and plainer will everything become; the business being transferred from the complicated to the simple; from the incommensurable to the commensurable; . . . from the infinite and vague to the finite and certain; . . . And inquiries into nature have the best result, when they begin with physics and end in mathematics [II, 8].

Some of the characteristic beliefs of modern empiricism are evident here in his emphasis on measurement and his assumption that what can be measured will be known, that certainty is to be found in what is finite. Acknowledging that the complex is vague and difficult to quantify, Bacon suggested a way of escape through analysis, towards what he saw as the greater simplicity of the particle. He could not be expected to foresee the complexity of quantum physics or the impact of the Heisenberg uncertainty principle. Meanwhile his own theories were to have lasting consequences in the kind of sciences they promoted.

The problem of geography

Bacon's entire programme was oriented towards action and it achieved its effect dramatically in encouraging systematic study in the natural sciences. A theory of science, however, which argued that the unit can be more meaningfully studied than the complex was likely to prove inhibiting to geographical work, and it is not surprising that there was little evidence in the following years of a successful effort to apply the new method to geography: from that point of view it provided an interesting testing-ground for the workability of his scientific theory. Bacon tended to see things in terms either of unity or variety, general or particular, and he had a strong aversion to vague generalities with no practical relevance. There is virtually no provision in his method for the kind of study of processes, distributions and complex groupings that later was to be associated with geography. Although he made the tentative proposal that a new part of Physique should give consideration to processes in nature, he did not link this in any way with cosmography, and it was only in the late eighteenth century that the French *géographie physique* seems to have become concerned with that possibility. Moreover, the difficulties inherent in Bacon's classification of learning are apparent in the case of geography, or cosmography as he usually termed it, which in his outline of sciences was divided between Metaphysique, where it was classed as a part of mixed mathematics, and history, where it was divided again between natural and civil aspects.

In his system Bacon regarded mathematics as a branch of Metaphysique within natural science, distinguishing in *The Advancement of Learning* between *pure* and *mixed* mathematics – the pure mathematics being geometry, 'handling quantity continued', and arithmetic, quantity 'dissevered'. Mixed mathematics, on the other hand, 'considereth *quantity determined*' and to this group he assigned cosmography, noting the importance of mathematics in the study and control of nature:

For many parts of nature can neither be invented . . . nor demonstrated nor accommodated unto use with sufficient dexterity, without the aid and intervening of mathematics; of which sorts are *perspective, music, astronomy, cosmography, architecture, enginery* . . .

Furthermore he predicted an increase in the number of such studies: 'And as for the Mixed Mathematics, I may only make this prediction, that there cannot fail to be more kinds of them, as nature grows further disclosed' [98–100]. The relation of geography to mathematics remained an important issue, and although Bacon himself did not elaborate further, the question was taken up by Varenius later in the century as a basic problem in his *Geographia generalis*. The major part of geography, however, was assigned by Bacon to the *history of cosmography*, this being, in his view,

compounded of natural history, in respect of the regions themselves; of history civil, in respect of the habitations, regiments, and manners of the people; and *mathematics*, in respect of the climates and configurations towards the heavens: which part of learning of all others in this latter time hath obtained most proficience [79].

It is significant that Bacon regarded the *history of cosmography* as a compound subject, with the collection of data relating to man being the concern of civil history and that relating to the 'regions themselves' a part of natural history. The grounds for a division between human and physical geography were already in evidence at this time and indeed the idea of an opposition between man and nature was encouraged in Bacon's work, with his emphasis on the extension of man's power over nature. In his *Preparative toward Natural and Experimental History* Bacon referred specifically to the natural history of geography as a study dealing with features of the earth and sea, and not with matters related to civil life, although he implied that geography as a whole is concerned with both aspects.[43] Bacon, however, gave no consistent attention to geography and did not attempt to resolve this problem. Nor did he explain what kind of relationship should be maintained, by those who might practise the subject, between *cosmography* as a part of *mixed mathematics*, that is, as a branch of Metaphysique concerned with essential forms or final causes, and the *history of cosmography* which presumably deals only with particulars. In some way, he apparently assumed, information collected in the *history* would provide an objective basis for subsequent reasoning. Bacon was by no means clear on these issues and he evidently allowed a place for mathematics in the *history of cosmography* as well, where it was required for the calculation of location and of 'climates' or zones based on latitude.

The chief application of mathematics to geography in Bacon's time was in connection with astronomy and navigation, and while he showed little regard for the astronomical theories of Copernicus, stating indeed that he expected to see them disproved by induction, Bacon was more enthusiastic about the progress of exploration by sea, using this frequently as a symbol for the extension of human knowledge in general:

And this proficience in navigation and discoveries may plant also an expectation of the further proficience and augmentation of all sciences; because it may seem they are ordained by God to be coevals, that is, to meet in one age . . . as if the openness and thorough passage of the world and the increase of knowledge were appointed to be in the same ages.[44]

Completion of both projects, Bacon claimed in the *Novum Organum*, would be merely the work of a few years, if men sought new knowledge with the same energy that they sought new lands [I, 112]. He urged the active exploration of the intellectual sphere:

And surely it would be disgraceful if, while the regions of the material globe, – that is, of the earth, of the sea, and of the stars, – have been in our times laid widely open and revealed, the intellectual globe should remain shut up within the narrow limit of old discoveries [I, 84].

This was an effective analogy in seventeenth-century England, and one especially appropriate for encouraging a monarch to support scientific inquiry as generously as voyages of discovery. At the same time, in a theory of science it remains to some extent deceptive to suggest that exploration involves simply sailing forth to a confrontation with new lands, or that regions once visited have been 'revealed'.

At one point in this discussion Bacon's use of the exploration analogy appeared to lead him closer to recognizing the importance of the conceptual element in research, as he realized that to stimulate an active enthusiasm for inquiry he must encourage confidence in the outcome, establishing in effect a prior concept of what is to be discovered:

I will proceed with my plan of preparing men's minds; of which preparation to give hope is no unimportant part . . . And therefore it is fit that I publish and set forth those conjectures of mine which make hope in this matter reasonable; just as Columbus did, before that wonderful voyage of his across the Atlantic, when he gave the reasons for his conviction that new lands and continents might be discovered . . . which reasons, though rejected at first, were afterwards made good by experience, and were the causes and beginnings of great events [I, 92].

Returning quickly to his emphasis on particulars, however, he went on to maintain that 'the strongest means of inspiring hope will be to bring men to particulars; especially to particulars digested and arranged in my Tables of Discovery . . . since this is not merely the promise of the thing but the thing itself' [I, 92]. Bacon's confident realism was to be challenged in the next decade by the Cartesian theory of doubt and in the following century the notion of *the thing in itself* in relation to human knowledge was to be examined more critically by Kant. In the meantime he proceeded to lay the foundations for the dedicated pursuit of natural history, the study that by the end of the century was to come close to supplanting traditional geography.

Natural history according to Bacon's plan should be more than the earlier 'histories and descriptions of metals, plants and fossils', including even Aristotle's history of animals. Their lack of quantification provided the main basis for his criticism: 'Nothing duly investigated, nothing counted, weighed or measured, is to be found in natural history.' The purpose of the new *Natural and Experimental History*, he asserted, is 'to supply the understanding with information for the building up of philosophy'. Therefore, to provide adequate grounds of experience he proposed 'to collect a store of particular observations', not only concerning 'the variety of natural species' but also 'experiments of the mechanical arts', for he rejected vigorously the tendency to dissociate learning from technical matters. Noting, however, that 'the mechanic, not troubling himself with the investigation of truth, confines his attention to those things which bear upon his particular work', he suggested the gathering together of a new 'variety of experiments, which are of no use in themselves, but simply serve to discover causes and axioms'.[45]

From the outset Bacon had stressed the need for an increase in mechanical aids for research. Referring in *The Advancement of Learning* to the alchemists 'who call upon men to sell their books and build furnaces', he pointed to the example already set by cosmographers and others:

Unto the deep, fruitful, and operative study of many sciences, especially Natural Philosophy and Physic, books be not the only instrumentals . . . spheres, globes, astrolabes, maps and the like, have been provided as appurtenances to astronomy and cosmography, as well as books: . . . likewise . . . some places instituted for physic have annexed the commodity of gardens for simples of all sorts, and . . . the use of dead bodies for anatomies.

Urging that such trends should be encouraged, he offered a prediction which in view of the expansion of laboratories and complex apparatus in the following centuries might be regarded as something of an understatement: 'In general, there will hardly be any main proficience in the disclosing of nature, except there be some allowance for expenses about experiments' [65].

Foreseeing, however, that the industrious collection of instances might itself lead to confusion, Bacon assured the readers of *Novum Organum*: 'let no man be alarmed at the multitude of particulars'; as a pioneer he intended to show others 'the true road . . . one in which the labours and industries of men (especially as regards the collecting of experience), may with the best effect be first distributed, and then combined' [I, 132]. The distribution of effort indeed was to become so pronounced a feature of scientific research after Bacon that by the end of the next century such leading thinkers as Kant and Humboldt pointed to the necessity of balancing this with increased attention to the combination of results. Meanwhile, concerned that men should not 'be kept for ever tossing on the waves of experience', Bacon stressed the importance of order in his method.[46] The particulars of natural history, collected by means of the senses, were to be recorded and arranged in tables as an aid to the memory, so that the intellect could then begin the work of extracting axioms from experience, as the basis for new experiments.

Throughout the *Novum Organum* Bacon made it clear that he wanted his method to proceed as mechanically as possible: 'For my way of discovering sciences goes far to level men's wits, and leaves but little to individual excellence; because it performs everything by the surest rules and demonstrations' [I, 112]. Without questioning his own ability to provide such guidance, he went on to claim a universal application for his method of induction, not only to natural philosophy but also to 'the other sciences, logic, ethics, and politics . . . For I form a history and tables of discovery for anger, fear, shame, and the like; for matters political; and again for the mental operations of memory, composition and division, judgement and the rest: not less than for heat and cold, or light, or vegetation, or the like' [I, 127]. Although he expressed an interest in studying the nature and functions of the mind, Bacon refused to admit any significant role for general concepts in the inductive method itself. Claiming that he did not intend 'to found a new sect in philosophy', he added:

nor do I think that it matters much to the fortunes of man what abstract notions one may entertain concerning nature and the principles of things . . . for my part I do not trouble myself with any such speculative, and withal unprofitable matters [I, 116].

Asserting constantly that knowledge is derived from experience, Bacon was nevertheless reluctant to extend the concept of experience beyond the immediate reports of the senses, and by regarding sense-perception as a passive function of reception rather than as an active process of interpreting or 'picturing' reality, he diverted attention from the mental processes of image construction and concept formation as important aspects of scientific thought. Having accepted the Aristotelian division between body and mind, senses and reason, he too relied on imagination to provide the active link between them, while maintaining the dual view of *phantasia* as the power to produce mental images, either as fanciful illusions, or else as faithful representations of reality, essential in all knowledge. He even suggested that through the communication of *human tradition* the imagination can be used to counteract the complete preoccupation with the present that the senses or affections otherwise tend to produce:

The affection beholdeth merely the present; reason beholdeth the future and sum of time. And therefore the present filling the imagination more, reason is commonly vanquished; but after that force of eloquence and persuasion hath made things future and remote appear as present, then upon the revolt of the imagination reason prevaileth.[47]

Through an intellectual process of imagining, he suggested, what is remote and general must be made to become part, as it were, of present experience. Altogether Bacon indicated an important function for the imagination in his theory of knowledge, yet in his method of induction he did not discuss the use of imagination, preferring to leave rather vague the way in which the mind is supposed to make a transition from 'particulars' through notions to general principles.

Bacon paid little attention also to the basis on which problems for research are selected; scientific thought, in his terms, involved simply the controlled intellect, working almost mechanically on materials provided by the senses themselves. The other aspects of the sense-faculty – will, appetite, the feelings – were to have nothing to do in the matter and apparently were to be eliminated, along with imagination, from the method of science. Significantly, these were the same aspects that in his Aristotelian division of the mind into rational and moral faculties were assigned responsibility for determining morality, aided by imagination in effecting action. This rejection of imagination and emotion remained characteristic of scientific empiricism after Bacon, in spite of a strong challenge to that view by the romantic school during the eighteenth and nineteenth centuries. It fostered also the idea of science as being not only impartial and objective in the search for knowledge, but even above morality altogether in the effort to gain control over nature. Almost without question, for more than three centuries, science along with technology became virtually synonymous with progress and Bacon received acclaim as the prophet of the new way, offering an escape from the past.

The extent of Bacon's originality and influence on his contemporaries is difficult to determine, but his major works on method, including the *Novum Organum* (1620), were published before Galileo's famous *Il Saggiatore* of

1623 and even earlier than the writings of Descartes, who referred frequently to Bacon's ideas. His *New Atlantis*, published posthumously in
1627, provided with its idea of a research institute the inspiration for the
Royal Society, the famous scientific group whose foundation members
publicly acclaimed the experimental philosophy of Lord Verulam and his
theory of heat as motion. His own attempts at experiments, however, seem to
have been rather ineffectual, with the possible exception of the final one with a
chicken packed in snow, which might have given an early stimulus to the
frozen food industry had it not led to his premature death from pneumonia.
Bacon himself seems to have held some reservations about publishing his own
natural and experimental history: according to the editor W. Rawley, who
produced the *Sylva Sylvarum* shortly after Bacon's death, the famous author
had frequently expressed concern that 'it may seem an Indigested Heap of
Particulars'.[48]

Whatever the limitations of his own work, Bacon's impact on scientific
thought was enormous and in large part his theory of induction provided the
foundation for modern sense-empiricism. Reinforcing the mechanistic view
of the universe, his theory of scientific method offered a strong encouragement to specific research while at the same time giving a pronounced
materialist and objectivist bias to the empirical movement. In assessing his
contribution it must be remembered that the deficiences of his theory were
rarely questioned by later scientists, despite his own insistence that his
errors should be found and corrected. The majority of his followers probably
wanted to believe that a foolproof method of procedure had been found;
tired of what they most likely saw as interminable philosophizing, they
responded to his urge to action, and the tendency for sciences to develop
separately from the philosophical and literary traditions became pronounced in the following centuries. Meanwhile, the development of
geography was if anything hampered by Bacon's kind of empiricism and
indeed it was more than a century before significant attempts were made to
assert its value as a study dealing with complex interrelationships and the
general concept of the world as a whole.

CHAPTER 2

SCIENCE AND GEOGRAPHY: THE SEVENTEENTH-CENTURY ENCOUNTER

SCIENTIFIC METHOD: THE RATIONALIST CONTRIBUTION

During the first half of the seventeenth century the scientific theories of Francis Bacon in England were rivalled on the Continent by those of the Italian, Galileo Galilei, and the French rationalist, René Descartes, and in the following decades the ideas of those three outstanding thinkers were combined to formulate a new programme for science. Bacon, outlining what came to be called the empiricist position, challenged the sciences to justify their position in relation to the progress of knowledge as a whole, and to conform to the demands of a particular scientific method, one based on experiment and direct sense-experience. The method of Continental rationalism, on the other hand, relied on the correct application of human reason, the use of mathematics and logical deduction from universal principles in science to achieve certainty. Throughout the century, scientific thought was marked by efforts to effect a reconciliation of those views, with the aim of providing an absolutely reliable guide to inquiry.

The search for a method or *ratio*, a prominent feature of European learning since the previous century, and one reflected in many texts of that period, represented an attempt to regulate and make more efficient the task in hand, whether that might be map-making, speed-writing, or the teaching of Latin grammar. Linked with this, however, was a constant desire to establish finality, to find the best method, the absolute truth, and in science the determination to acquire certain knowledge of the world gave a stimulus to some kinds of research, while acting as an impediment to others. Astronomy, optics and mechanics, for example, gained a distinct advantage from the new theories, but many traditional studies, geography among them, soon came into conflict with the growing demands for certainty and objectivity in science, while the mechanist world view favoured by the rationalist school provided an added complication for inquiries concerned with living organisms, and particularly with man. In many ways geography was seriously hampered by the new theoretical framework, and its subsequent development will be considered in the following chapters in terms of a response to the concept of scientific procedure that emerged in this period.

Galileo: science and mathematics

Galileo (1564–1642) gave a strong impetus to the view, central to the rationalist position, that science is primarily mathematical as well as experimental in method. Scientific inquiry, in his opinion, should be confined to the study of matter and forces, and this can be accomplished only through measurement and calculation. In dealing with the natural world he rejected the kind of questions asked by Aristotelians on essential natures, promoting instead a search for observed regularities and mechanical principles. To assist in the observation of measurable effects he constructed new instruments, producing the first thermometer by 1603 to enable the recording of heat in degrees rather than as a range of qualities. He helped develop also the compound microscope and the telescope to provide further extensions to the senses, opening up a new range of discoveries. Over many years, through a series of observations and experiments on falling bodies, commenced even before he became professor of mathematics at Pisa in 1589 and later at Padua, he proceeded to expose some of the errors in Aristotelian teachings presented to him as a student. Similarly the doctrine of the 'changeless heavens' was discredited with the observations of new stars reported by Tycho Brahe in 1603 and Galileo in 1604.

Supporting the Platonic view, Galileo argued that nature is mathematically constructed, asserting furthermore that its laws can be discovered with certainty and given a mathematical expression. The book of nature, he declared in his *Il Saggiatore* (The Assayer) of 1623, is written in the language of mathematics, so that the study of geometry provides the key to its comprehension; he extended the scope of geometry to deal with time and motion as well as with the traditional problems of lengths, areas, and volumes. In addition he made a significant distinction, one that was to persist in later science and philosophy, between what came to be called primary and secondary qualities. Identifying quantity, size, shape and motion as real and objective properties of matter, he recommended that science should be concerned with these, since mathematics can be applied only to what is measurable. Tastes, smells and sounds on the other hand, he claimed, depend on the sense organs of the being that perceives them, and have no independent existence. Regarded as non-measurable, these tended to be ignored in science. The materialist and mechanist approach of Galileo, along with his notion that science is measurement, found a widespread acceptance.

In 1632, despite Church warnings against heresy, Galileo published in Italian his *Dialogue Concerning the Two Chief World Systems*, presenting arguments in favour of the Copernican as well as the Ptolemaic theory. Any serious challenge to Aristotelian physics, as Galileo realized, was extremely hazardous at a time when the geocentric view was still approved by the Church in Rome. Ecclesiastical antagonism to heliocentric theories indeed had been intensified by a fear of their association with renewed interest in cults of sun-worship and the kind of magical religion of nature, revived in the Hermetic tradition of the renaissance and promulgated by Bruno, who saw the Copernican hypothesis as a portent of a new philosophy of man's place in the unity of an infinite universe sustained by a single cosmic motive force. Anxious to dissociate himself from all occult beliefs, Galileo advanced the empiricist position, defending conclusions drawn from observation rather than authority. At the same time he was even prepared to call on the authority of Aristotle to justify an alternative explanation of the solar system, arguing that Aristotle himself had declared 'that what sensible experience shows ought to be preferred over any argument', and therefore 'it is better Aristotelian philosophy to say "Heaven is alterable because my senses tell me so", than to say "Heaven is inalterable because Aristotle was so persuaded by reasoning."' With the telescope, Galileo pointed out, 'we possess a better basis for reasoning about celestial things' and he argued that Aristotle would not have objected to the replacement of his own theory of the universe in this way but indeed would have approved such an approach.[1]

Dismissing also Aristotle's division between the terrestrial and celestial spheres, Galileo followed Bruno in establishing the doctrine of uniformity in the working of the material universe as one of his major contributions to later scientific thought. Earth and the rest of the universe, according to this theory, share in a mechanical relationship and are controlled by uniform laws. The sciences that would discover these, in his opinion, must be based on mathematics, and on this account he offered some criticism of the magnetic philosophy of William Gilbert, while declaring himself to be its supporter: 'I have the highest praise, admiration, and envy for this author, who framed such a stupendous concept . . . What I might have wished for in Gilbert would be a little more of the mathematician, and especially a thorough grounding in geometry . . .'[2]

The dialogue form of Galileo's *Two Systems*, in which both sides of the argument were presented, did not prevent his being brought to trial by the Inquisition in 1633 and convicted of spreading heretical views. His punishment aroused widespread concern, and the desire to free natural science from Church control encouraged an enthusiastic response to the pronounced separation of natural from moral philosophy advocated in his final publication. This was the *Discourses and Demonstrations Concerning Two New Sciences* of 1638, a major contribution to modern physics and mechanics. In Galileo's view the natural world, mechanical in its function and measureable in its effects, must be distinguished from the moral or social world; philosophy therefore, which formerly dealt with knowledge in general, must be divided accordingly. For his followers, then, he was unable to replace the unitary medieval view of the world with a modern synthesis. Like Bruno and Gilbert, he accepted the idea of an infinite universe, but his doctrine of uniformity was not extended to all parts of it: the notion of a mechanical world, though transferred subsequently to the biological sciences with the idea of the plant or animal as a mechanism, was not applied to the social sphere. Galileo's division of natural from moral philosophy, along with his distinction between mathematical and subjective phenomena, became incorporated in the theory of exact and specialized science that remained paramount in the following centuries. For geography in particular it posed an obstacle to coherent development, since the old custom of dividing civil or political geography from the rest of the subject was now given a further theoretical justification. Eventually, as Humboldt came to realize, the successful establishment of a modern science dealing with the world as a whole, including man, must involve a reappraisal of the entire concept of scientific method inherited from Galileo and Francis Bacon.

To a certain extent of course the doctrines of later empiricism represented a change of emphasis from that intended by those men themselves, and Albert Einstein, in a preface to the dialogue on the *Two Systems*, offered a perceptive defence of Galileo on this account:

It has often been maintained that Galileo became the father of modern science by replacing the speculative, deductive method with the empirical, experimental method. I believe, however, that this interpretation would not stand close scrutiny. There is no empirical method without speculative concepts and systems; and there is no speculative thinking whose concepts do not reveal, on closer investigation, the empirical material from which they stem. To put into sharp contrast the empirical and the deductive attitude is misleading, and was entirely foreign to Galileo . . . The antithesis Empiricism *vs*. Rationalism does not appear as a controversial point in Galileo's work . . . His endeavours are not so much directed at 'factual knowledge' [Wissen] as at 'comprehension' [Begreifen]. But to comprehend is essentially to draw conclusions from an already accepted logical system.[3]

These incisive arguments from the outstanding physicist of the twentieth century were directed against the orthodox views of his day, for the belief that the empirical, experimental method is opposed to speculation had become deeply entrenched since the time of Galileo. Similarly the assumption that a clear distinction can be made between an inductive or empirical method and a deductive or rational approach had recurred as a persistent feature of scientific thought, after the formulation of modern rationalism in the works of Descartes. In effect, the contrasts between Cartesian 'rationalism', and the 'empiricism' of Galileo and the Baconian school, were stressed more than their similarities.

Descartes: *Le Traité du Monde*

Like Galileo in the same period, Descartes (1596–1650) set out to write a treatise on physics, to be based on the Copernican system of the world, and supported by observations and experiment. In letters written during 1630 to his friend, the friar, Marin Mersenne, he promised to have his *Treatise on the World* finished in three years. By April 1632, however, he was ready to admit the problem of bringing such a cosmology to a conclusion: 'after the general description of the stars, the heavens and the earth, I did not originally intend to give an account of particular bodies on the earth . . . In fact, I am now discussing in addition some of their substantial forms, and trying to show the way to discover them all in time by a combination of experiment and reasoning.' Evidently following Bacon's plan to discover *forms*, he reported 'making various experiments to discover the essential differences between oils, . . . waters, . . . etc.'.[4] At the same time, seeking a wider principle by which individual instances collected according to 'the Baconian method' might be explained, he proposed to continue his studies in astronomy, convinced that there existed a regular and 'natural order' among the stars:

The discovery of this order is the key and foundation of the highest and most perfect science of material things which men can ever attain. For if we possessed it we could discover *a priori* all the different forms and essences of terrestrial bodies, whereas without it we have to content ourselves with guessing them *a posteriori* from their effects [23–4].

With the news in 1633 however of the Inquisition's condemnation of Galileo's *Massimi Sistemi del Mondo*, Descartes announced his intention to withhold publication of his own *Traité du Monde* since it, too, 'included the doctrine of the movement of the earth'. Explaining to Mersenne his firm decision to avoid conflict with 'the authority of the Church' and 'to live in peace', he recalled the case of 'the Antipodes, which were similarly condemned long ago', and expressed the hope that 'in time my World may yet see the day' [26–7].

It is likely, of course, that the difficulty of completing the work itself played a part in his decision. As Humboldt, who shared the same problem, commented in the third volume of *Kosmos*, Descartes had attempted in his *Traité du Monde* 'to include the whole world of phenomena: the heavenly sphere and everything that he knew of animate and inanimate earthly nature . . . in the correspondence with Father Mersenne we find him frequently complaining about the slow progress of the work and the difficulty of ordering so many matters'. In his introduction to the unfinished final volume of *Kosmos*, Humboldt referred to the same matter again, this time in his own defence, noting that the complaints of Descartes would have been even more bitter had he worked in the expanding scientific world of the nineteenth century.[5]

Although Humboldt expressed regret that the Inquisition deprived the world of a great work in the *Traité du Monde*, it seems from certain comments of Descartes himself, and from parts of the *Treatise* subsequently published, that with respect to geography it would have provided no major contribution to the literature, and cannot in that sense be considered a forerunner of *Kosmos*. Sections from the *Traité du Monde* appeared in 1644 as Parts III and IV of his *Principia Philosophiae*. There, in Part III, on *The Visible World*, Descartes dealt with the earth in its relation to the sun, to the 'fixed stars', and the other planets, using a strategy that has since been criticized as 'fictionalism'. While rejecting the Ptolemaic system (Principal XVI), he explicitly denied that the earth moves (XIX), and claimed that he set forth the hypothesis of Copernicus and Tycho, which is agreed to be false, simply because 'true and certain' explanations of phenomena may still be deduced from it.[6]

In Part IV, on *The Earth*, what Descartes provided was not a geographical description but a set of principles regarding problems that later came to be assigned to modern physics and chemistry, on the nature of matter, heat, gravity, the elements, and so on. He admitted that from his original plan 'a Fifth and a Sixth Part, the fifth treating of living things, that is of animals and plants, and the sixth of man' were not completed, since he lacked the time and financial support to carry out the necessary experiments. It is clear that even by 1644 Descartes had not completed a thorough description of his *World*, as he called it. His book on the earth is more important as far as geography is concerned for the mechanical and mathematical kind of science that it promoted. The notion of the human body as a machine, one that provided a theoretical model for modern physiology, is introduced here, along with the general principle that 'all the knowledge man can have of nature' must be derived from the *figures, magnitudes* and *motions* of bodies, understood according to 'the principles of Geometry and Mechanics'.[7]

The final excerpts from the *Traité du Monde* were published posthumously in 1664, under the title *Le Monde, ou Traité de la Lumière*. Along with the *Treatise on Light*, later editions included the *Treatise on Man*, and his *Description of the Human Body* of 1648.[8] Elaborated here are the results of his researches on optics, as well as on anatomy and physiology; however, he showed no concern with developing a comparable science of geography. Like Galileo, Descartes interpreted the science of nature as the mathematical study of matter and motion, his mechanical hypothesis contributing a great deal to the founding of modern physics and physiology. In the history of geographical thought his chief importance is derived from those of his ideas that helped mould subsequent notions of scientific method. These were expressed in his first publication, the famous *Discours de la Méthode* of 1637.

Descartes: on method in science

Writing to Mersenne in March 1636, Descartes announced that he had four treatises, all in French, ready to be published anonymously under the general title: 'The Plan of a Universal Science to raise our Nature to its Highest Degree of Perfection . . .'[9] The work appeared however the next year as the *Discourse on the Method of rightly conducting the Reason and seeking for Truth in the Sciences*. In the *Discourse* the first precept of his logic was the famous principle of doubt: 'to accept nothing as true which I did not clearly recognise to be so'. The second was to divide problems into parts and the third to order his reflections, rising from the most simple objects 'by degrees, to knowledge of the most complex, assuming an order, even if a fictitious one', among them.[10]

Of all his conceptions, the first that he could accept as certain was the existence of the self: '*I think, therefore I am.*'[11] This, he added in his *Principia philosophiae* (1644), 'furnishes us with the distinction which exists between the soul and the body, or between that which thinks and that which is corporeal', since, while we are certain of the existence of our own thoughts, 'we still doubt whether there are any other things in the world'.[12] For a justification of the *existence of material things,* Descartes in his *Principia* appealed, somewhat surprisingly it seems now, to the reliability of God. Arguing that the necessity of God's existence is proved by the clear conception man has of such a perfect being, he declared that all ideas of corporeal objects which are clear and distinct must be produced by such external objects themselves, since God is no deceiver and would not provide man with 'the light of nature, or the faculty of knowledge' in order to produce false perceptions.[13]

In many ways Descartes was in sympathy with the empirical tradition of Francis Bacon, and indeed made his own distinctive contribution to it. Like Bacon and Galileo he was a realist, accepting the existence of an external world as revealed in experience, and like them he was a materialist, in that he conceived that world to be constructed of particles of matter. However he was even more explicit than Bacon in making a separation between *matter* as divisible and having extension, on the one hand, and *mind,* unextended and indivisible, on the other. Thought, the chief attribute of mind, included for Descartes all operations of the soul – sensation, imagination, willing and feeling, as well as reasoning. He defended the notion of free will in man, while making an effort to reconcile this with the Christian doctrine of divine pre-ordination.[14]

The Cartesian dualism was an attempt to overcome the inability of the mechanical hypothesis to account for everything in the world, by claiming that the human mind is itself outside the material system. Furthermore it enabled him to accept the determinism implicit in the mechanistic theory as applying to the material world only, and not to man, so justifying the free-will argument. An opposition between mind and matter was already a feature of the Church doctrine, found earlier in Aristotle, that man by his power of reason is set apart from the natural order and designed to dominate it. Descartes wrote in October 1637,

This theory involves such an enormous difference between the souls of animals and our own that it provides a better argument than any yet thought of to refute the atheists and establish that human minds cannot be drawn out of the potentiality of matter.[15]

The argument that was designed to refute the atheists and demonstrate the existence of God came to provide a central assumption for a good deal of subsequent empiricism. Even among atheistic scientists, the dualist approach tended to determine the form of discussions on nature and mind until the twentieth century.

In his epistemology Descartes went further than Bacon in exploring the operation of the intellect in perception. He therefore questioned more critically the extent to which certain knowledge of the external world is possible, asserting at the same time the reality of the mind and the importance of thought. Emphasis in his philosophy is on the effective use of reason, a subject more superficially treated by Bacon; however, his rationalism did not exclude for Descartes an empirical approach. Within his framework of doubt, he still accepted the reliability of most sensory perception, even if he saw this arising from divine influence; experiment and mathematical analysis were essential to his method of scientific inquiry. Descartes was always interested in studying the natural world, offering, for example, a mechanical theory of the universe, of light, of human physiology. While he was often in error – opposing Harvey, for instance, on the function of the heart, suggesting the pineal gland as the seat of the brain, or declaring that water particles are like eels – he also made lasting contributions to the natural sciences, particularly physics and physiology. The main question on which he differed from the empirical view of knowledge was in his assertion that some ideas, of simple notions such as existence or certainty, and of the laws laid down by God in nature, are *inborn in our minds.*[16] Empiricists came to take profound exception to this, arguing that all knowledge must be derived from experience, *a posteriori.* The debate as to whether *a priori* knowledge is possible continued through the following centuries, obscuring as many issues as it clarified. In particular it tended to force empiricists into the untenable position that all reliable knowledge is derived from immediate sensory experience.

With the polarization of opinions in the dispute over *a priori* ideas, few stopped to ask the question, 'prior to what?' In the religious context of the time, Descartes and his rationalist followers argued that such ideas are provided by God prior to all human experience. Rejecting that possibility, scientific empiricism moved to an extreme objectivist position, denying not only innate ideas, but even the importance of *a priori* knowledge in the form of concepts or expectations already in the individual's mind prior to present experience. An outcome of this was the tendency of scientists to underrate the significance to empirical inquiry of the historical development of ideas and their communication through education.

Impact on geographical thought

It was in this context that the natural sciences took shape during the next centuries, until Humboldt at the height of his career found himself opposed to almost the entire empirical tradition in his concern with examining and developing the ancient concept of cosmos in his own geographical work. Even in recent times many geographers evidently remained puzzled by that aspect of Humboldt's contribution. For a philosopher to bring arguments back to general ideas seemed to them understandable, but for a distinguished scientist to do so was incomprehensible, and this led to charges that Humboldt in his declining years must have succumbed to a less rigorous romanticism. Like Einstein a century later, however, Humboldt was an exceptionally competent thinker who came to question not only the mechanical theory of the universe propounded by the followers of Galileo and Descartes, but also the assumption that natural science is concerned only with measurable aspects of the world and therefore should be kept separate from speculative philosophy.

In spite of his efforts the division between scientists and philosophers became more pronounced in the century after Humboldt. Some consideration of theoretical problems, particularly those related to the search for a correct scientific procedure, continued to be accepted even by the strictest empiricists as justifiable, but the philosophy of science as they pursued it rarely questioned the basic assumptions of scientific empiricism itself on central issues of epistemology or ontology. Discussing the effect of attitudes to science and nature in the scientific revolution, Crombie argued convincingly that 'the 20th-century dichotomy between the philosophy of science of scientists and that of philosophers can be seen in embryo even in the 17th century. Each tending more and more to ignore the writings of the other, the division solidified in practically all European educational systems in the 19th century, to the increasing disadvantage of both sides.' Scientists, as he pointed out, were interested usually in solving problems arising from their scientific work, being preoccupied especially with the relationship between specific theories and 'the mechanical philosophy of nature in terms of which it was assumed that all explanations in physics must be given'.[17]

One corollary of the mechanist hypothesis that posed special problems for geography in the period after Descartes was the view of the human mind as being outside the natural order of the material universe. In physiology this had already raised the question of how mind and body can interact, and the suggestion put forward by Descartes that interaction occurs through the pineal gland did not resolve the difficulty. To a large extent the problem was ignored or evaded in most of the specialized sciences established during the following centuries, but in geography it was to prove a serious impediment to the development of the subject as a whole, and the place of man in the study became an increasingly troublesome issue. As the distinction between moral and natural philosophy was reinforced by Cartesian teachings on the dualism of mind and matter, the implication that the study of man has no place in natural science was to have a continuing impact on geography, where such a separation is patently difficult to sustain. Even in Humboldt's work, two centuries later, that problem was never resolved, and the position of human geography in his *physische Weltbeschreibung* was left in doubt as one of the intriguing questions of the unfinished *Kosmos*. Meanwhile, with the adoption of a strict division between moral philosophy and the natural sciences it began to seem a deficiency on the part of geography that it could not be assigned completely to either.

GEOGRAPHY 1599–1650: THE SEARCH FOR A SYSTEM

As the scientific revolution gathered momentum in the seventeenth century, then, geography from the beginning appeared to occupy a rather anomalous position and it attracted the serious attention of none of the leading thinkers. Faced with the question of how the traditional themes of geography should be related to the new theories of science, those who continued to teach the subject found no clear programme available, for neither the Baconian nor the Cartesian school offered a viable solution. To the rationalists, a descriptive work like Münster's *Cosmographia* in the tradition of Strabo could not be regarded as scientific, and while the mathematical, cartographic geography of Ptolemy was more acceptable to them, any attempt to move away from that model, by incorporating local descriptions or the old study of civil geography, led to immediate conflict with their theories of science. Meanwhile Bacon, it has been noted, gave geography no place at all in the part of natural science he called *physique,* but divided it between *mixed mathematics* as a part of metaphysics on the one hand, and *civil* and *natural history* on the other, each with its own method and with no clear indication of how any relation might be effected between them. In the circumstances a confident development of geography was scarcely to be expected.

Geographical texts in the early seventeenth century were in general slow to show the effects of the new movements in science, and while no detailed study seems to have been made of this process, it is evident that for a long time Strabo and Ptolemy remained the chief models. During that period the use of the title *cosmography*, with its association with a tradition of biblical emphasis, was discontinued after the 1605 *Cosmographiae generalis* by the Dutch writer Paulus Merula (1558–1607), although further editions of his book continued to appear until 1648.[18] The name *geography* was revived, and with it an explicit distinction between a mathematically oriented *general* geography, concerned with the whole sphere, and a *special*, or *topical* part, devoted to the description of particular regions and including a study of the population. This appears to have been an attempt to effect a compromise between the approaches of Strabo and Ptolemy, bringing back the *chorography* of Ptolemy, while respecting his attempt to distinguish it from general geography in terms of both scale and method. The kind of division that resulted was to provide a recurring theme in subsequent geographical thought.

Throughout the seventeenth century, progress was slow. Classical geography continued to receive considerable attention as a support for Latin and Greek history, but recent advances in cartography or exploration were rarely incorporated in the general texts of the time, while the Ptolemaic theory of a geocentric universe continued to be taught without serious argument until mid-century. Examination of a number of works from this period by English and German writers shows the early efforts at providing a modern theory of geography as being largely tentative and meeting with comparatively little response. In England the five authors considered, all of whom worked at Oxford, made a decisive break from Latin in their publications, yet even that change to the vernacular does not seem to have stimulated any significant improvement. Instead, all indications point to a concern to serve a market both highly conservative and uncritical in outlook, interested in curiosities and making few demands for innovations or intellectual quality.

Abbot: the arid compilation

A work which remained popular in England during this period, appearing in a number of editions until 1664, was *A Briefe Description of the Whole Worlde*, published in 1599 by an Anglican bishop, George Abbot (1562–1633), Master of University College, who lectured on geography at Oxford. With headings in Latin and text in English, this small book provides an assortment of information on the main countries of the world, from Spain through Europe, Asia, Africa, and various islands, to 'America or the New World', where Abbot enlarged on the vices of Spanish oppression. No maps are included and indeed there is no pretence at a systematic geography at all. After a single preliminary sentence, 'The globe of the earth doth eyther shew the sea, or land', he proceeded to naming the seas, following this with notes on the various kingdoms, including stories from their history and classical references at random.[19]

This is the kind of descriptive geography, based on political units, that Varenius was to criticize for its neglect of general principles. However, it provided a model that was often followed and E. G. R. Taylor, discussing Abbot's book in her history of English geography from 1583 to 1650, condemned the whole *genre*: 'this arid little compilation is for the most part a mere catalogue of place-names, forerunner of the dreary geographies which held the field in English schools right down to the twentieth century. To name is to know: this was the dictum of the unenlightened pedagogue.'[20]

These comments on the conceptual aridity of English geography are interesting and it is to be regretted that in addition to her useful bibliographical studies of the period Professor Taylor did not herself expand on that theme and initiate a positive statement of theory, along the lines suggested here.

Keckermann: a system of geography

In Germany a more scholarly approach to geography, and one offering a prototype for subsequent attempts to organize the subject matter into an effective system, is found in the Latin *Compendious System of all Mathematics: Geometry, Optics, Astronomy, and Geography* of Bartholomew Keckermann, published in 1617.[21] Based on lectures given in 1605, it included a brief *Commentary on Navigation* dating from 1603. Keckermann, a prolific writer of text-books, had already published in Latin a *System of Theology* (1602), of *Logic* (1603), of *Ethics* (1607), and a *System of Rhetoric, general and special* (1608), as well as the work reproduced in his 1617 compendium, a *System of Geography in two Books* (1611).[22]

The geography of Keckermann with its pronounced dependence on classical authors was on the whole conservative in approach; indeed, the 1661 edition, reproduced with few modifications at Oxford, continued to quote Aristotle's *de Caelo* as authority and teach that 'the earth lies immobile at the centre of the whole universe'.[23] One feature of interest however is Keckermann's attempt to place geography in a classification of sciences. The mathematical sciences, to which he assigned geography, were classified in his scheme as either *Primary Theoretical* or *Secondary Practical*, the practical sciences being geodesy and mechanics, in which architecture was included. Geography was given a place in the theoretical sciences, which themselves were divided into *Abstract* (geometry and arithmetic), and *Concrete* – these being either *Qualitative* (optics and music) or *Corporeal*. It was in this last group that he placed 'Astronomy which is mobile', and 'Geography which is immobile', thereby associating geography with a static and materialist approach that was to be retained for a long time [36].

Keckermann also made some effort in his section on geography to provide a theoretical introduction to the subject, commencing with its history. He referred in his preface to the contributions of Columbus, Amerigo Vespucci, and Francis Drake in the history of exploration, before presenting some standard arguments on the utility of geography for politics, philosophy, medicine, astrology, theology, and especially for mathematics and history. Elsewhere, although Ptolemy was his constant guide, he included some references to the recent historical background of the study – the cosmographies of Münster and Apian, the writings of the French author Jean Bodin (1529–96), and in cartography the work of Mercator, Ortelius, even Erasmus and Tycho Brahe.

In his introductory chapter *On the Nature and Divisions of Geography* Keckermann offered a definition relatively modern in its emphasis on differentiation and on a mathematical, scientific approach: 'Geography is the science of the measurement and differentiation of the earthly globe [Geographia est scientia de mensura et distinctione orbis terrarum].' Cosmography in his view is more general, including the whole natural science of geography and astronomy, while geography is narrower and special in comparison, describing the terrestrial globe, its lands and seas. Similarly, following a plan already suggested in the work of Münster and Apian, he divided geography itself into two parts, 'one general, the other special', recalling that Ptolemy in Book I of his *Geography,* Chapter 1, had distinguished between geography as the representation and description of the whole earth and chorography as the representation and description of the various regions in particular [351]. Thus general or universal geography deals with the entire globe while the special or particular part 'treats the parts of the earthly globe *in specie*' [445]. He stressed in principle the importance of mathematics in each section and gave evidence on the use of maps in both, although he provided none in his own special geography, while the world map in his 1617 edition is primitive even for his own time and more reminiscent of medieval T–O maps [530]. His special geography with its notes on Europe, Asia, Africa and the New World was brief and marked by numerous references to Greek authors. The general philosophy, though more substantial, was still heavily dependent on Ptolemy and showed no real commitment to any modern scientific movement. Nevertheless, the work was influential during the seventeenth century, evidently providing a source for Heylyn (1621), Carpenter (1625) and Pemble (1630) in England, as well as Gölnitz (1643) and Varenius (1650) in Europe. As Manfred Büttner noted in his study of general geography before Varenius, Keckermann was responsible for introducing the actual terms *general* and *special*, his aim evidently being to release geography from both its dependence on the classical tradition and its recent preoccupation with natural theology, the description of the earth as a product of divine providence; he thus deserves recognition as the modern founder of a 'neutral' or secular general geography, a study later developed more fully by Varenius.[24]

Heylyn's *Microcosmus*

Peter Heylyn (1599–1662) was little more than twenty when his first work was produced at Oxford, where he was engaged in giving the statutory lectures on cosmography during the long vacation at Magdalen College. This youthful effort, under the title *Microcosmus, or a Little Description of the Great World. A Treatise Historicall, Geographicall, Politicall, Theologicall* (1621), was evidently well received, reaching an eighth edition by 1639, although by that time even its author evidently became disturbed by the number of errors it contained. In this work Heylyn made an attempt to break out of the limits set for geography by Ptolemy, and he began by promising a definition less narrow in extent. When it appeared, however, his own suggestion proved to be little more than a return to Strabo and the later tradition of Münster: '*Geographie* is a description of the earth, by her parts and their limits, scituations, inhabitants, cities, rivers, fertilitie and observable matters, with all other things annexed thereunto.'[25] His brief *Generall Praecognita to Geographie*, with its marginal reference to Keckermann, offered only the traditional divisions into continents, islands, circles, etc., while his notes on various countries were equally undistinguished.

As the divisions or *Species* of geography, Heylyn listed *Hydrographie*, the study of the sea, along with *Topographie* and *Chorographie* as the description of the land, and in this again he was following Strabo. 'Hydrography' itself was the name used by the Cambridge geographers, Dee and Cunningham, in the sixteenth century for the study of seas, lakes and rivers, a study approached with increased enthusiasm at this time for its relevance to navigation, but one already outlined intelligently by Strabo.[26] Heylyn went on to praise the utility of geography, referring like Strabo to its value for the statesman and stressing also its use for navigation and trade, for studies such as astronomy, or medicine (*Physicke*), and especially for history. Above all Heylyn emphasized the close interdependence of geography and history, defining history as 'a memoriall or relation of all occurrents observable, hapning in a commonwealth' and declaring in a colourful statement that, 'As Geographie without Historie hath life and motion but at randome, and unstable: so Historie without Geographie like a dead carkasse hath neither life nor motion at all.'[27]

Cluverius: history and geography

The close connection of geography with history is evident also in the *Introductio in universam geographiam* of the German writer Philip Cluverius (1580–1622), written in Latin and published two years after the author's death by Elzevier of Holland in 1624. Produced clearly in the tradition of Münster's cosmography, this remained a standard work throughout the century, appearing in more than thirty Latin editions in several countries during the next hundred years, with translations into French by 1631, English by 1657 and German by 1678; the English edition of 1657, under the title *An Introduction into Geography both Ancient and Moderne*, was published at Oxford where Cluverius himself had spent some time in the early part of the century.

Apart from the use of the term geography in the title, indicating a concentration on the earth and the exclusion of astronomy, there is little to distinguish this text from the cosmographies of the previous century. Book I, dealing with the globe in general, is stereotyped in its approach, treating briefly of latitude, longitude and the ancient division of the earth into latitudinal zones or *climata*, together with notes on the measurement of distances and direction, on the winds, oceans and navigation, and finally the arrangement of land masses – summarized under the traditional headings of Continents, Peninsulas, Isthmuses and Islands. The other five books give descriptions of various countries, commencing as Strabo did with Spain, and showing a similar emphasis on history, since Cluverius also was primarily a historian. Even in the augmented edition of 1683, with its maps, diagrams, and tables, the whole work remains stolidly conservative in its approach.[28]

Noting the inferiority of the general section in the *Introductio*, R. E. Dickinson pointed out that the mathematical geography shows no advance on the work of Apian written a century earlier; indeed, 'Cluverius does not know the views of Copernicus; for him, the earth remains the centre of the universe.'[29] Whether Cluverius was actually ignorant of the Copernican hypothesis is of course open to question since, recalling the fate of Bruno in 1600 and the submission of both Galileo and Descartes in 1633 to Church suppression of Copernican doctrines, it must be allowed that Cluverius and Keckermann before him in Germany may have preferred to avoid trouble on such a controversial issue, even if they had known of the heliocentric theory.

Conflict over that question probably retarded the development of a modern cosmography in this period. Even Nathaniel Carpenter, who condemned a slavish following of Aristotle and praised both Copernicus and Gilbert in his *Geography*, published at Oxford in 1625, was reluctant to appear to contradict the Scriptures and finally expressed only a cautious support for the theory of the earth's daily rotation or 'Magneticall Revolution' on its axis, while continuing to state as a specific theorem: '*The Terrestrial Globe is the Center of the whole world'*.[30] Although investigation of that question could form an interesting subject for further research, it is outside the scope of the present inquiry, the main concern here being with the imposition on geography of a conservative viewpoint throughout the century.

Carpenter, the scholarly empiricist

Carpenter (1589–1628), a Fellow of Exeter College, provided in this period one of the most intelligent and mature attempts, within the limitations of thought in his day, to clarify the theoretical position of geography and relate the study both to established traditions and to the most recent developments in learning. This effort stands as an isolated one in his own country, and the failure of his work to elicit a strong response marks the retreat of English geographers from any serious engagement with the crucial ideas of the century. His treatise of 1625, *Geography delineated forth in two Bookes*, seems to have made a very small impact, reaching only to a second edition in

1635 and being rarely mentioned by later writers. Learned, thorough, and often rich in insights, it represents apparently the first and the last attempt in England during the seventeenth century to secure for geography some measure of academic esteem in conjunction with the sciences of the time.

Although in this work Carpenter avoided any confrontation with Church orthodoxy on the motion of the earth, he tackled directly from the beginning those other Ptolemaic and Aristotelian doctrines which in his view impeded development. In his first chapter he discussed a number of critical issues with regard to geography: its relation to cosmography, to chorography, to natural philosophy and the accepted divisions of scientific inquiry, and finally to ancient and modern theories on the nature of the earth itself; here he rejected Aristotle's idea of internal form in favour of Gilbert's theory of magnetism and his own rather dynamic concept of the whole earth as a 'naturall Harmony or order, arising from the parts working together' [I, 11].

Commencing with the problem of Ptolemy's separation of chorography from geography, Carpenter defined geography, in a manner similar to Keckermann, as 'a *Science* which teacheth the *Measure* and *Description* of the whole *Earth*', and then proceeded to make the inclusion of chorography in the study a central issue. Here he was not content merely to follow the practice already adopted by Keckermann and other writers, but endeavoured to provide a clear refutation of the Ptolemaic view. While acknowledging in some detail the ways in which the two studies were distinguished by 'our approved *Ptolomie*', especially with regard to the closer relationship of geography to the mathematical sciences, Carpenter went on to refute Ptolemy with a logical argument:

I see no great reason why *Chorographie* should not be referred to *Geographie*; as a part to the whole; forasmuch as the objects on which he hath grounded his distinction, differ only as a generall and a speciall; which being not opposite, but subordinate . . . cannot make two distinct Sciences [I, 1–4].

Until in the following years the promulgation of the Cartesian dualism accentuated the problem of including man in geography, Carpenter's effort seemed likely to resolve the issue of the place of chorography in the science. Geography is properly termed a *Science*, he stated, since it has 'no other end but knowledge', whereas arts are 'directed to some further work or action'; it is clear he differed from Bacon on the utility of science, although he seemed to follow the Baconian approach in classifying sciences. From the outset he classed geography as a *physical* science in respect of its subject which is 'the whole ·Globe of the Earth', and as 'a mixt Science', both *mathematical* and *historical*, 'in the manner of *Explication*', the mathematical aspect being applied chiefly to the figure, site and circles of the sphere, the historical aspect to the study of places [4–5].

In selecting names for the two divisions of geography Carpenter attempted an innovation: although aware that 'the common and received division of this *Science* amongst *Geographers*, is into the *Generall* or universall part; and the *Speciall*, he proposed instead the terms 'Sphaericall and Topicall' [I, 5]. His Topical section itself in Book II proved to be unlike the Special Geography of Keckermann, for instead of offering a description of various countries, it outlined a general approach to the study of individual regions – their measurement and mapping, location, differentiation, and finally their description, which should be organized, he suggested, according to the *Naturall* aspects, of water or land (Hydrography and Pedography), and *Civill* aspects 'which concern the inhabitants' [II, 203]. Carpenter specified that topography (the study of places) should deal with the *habitable* parts of the earth, implying a constant reference to man, and he devoted the last four chapters to an examination of civil aspects, discussing the diversity of nations with respect to their lands and their education.

A similar effort to break with tradition is evident in his first book, on *Sphaericall Geographie*, for although the last seven chapters present a conventional account of the earth in terms of circles, zones, and so on, the early chapters are devoted to a more searching examination of theoretical issues, incorporating in the third and fourth chapters what he referred to as a 'Magneticall Tract' [I, 76]. Here he discussed the work of his countryman Gilbert, 'who to his everlasting praise hath troden a new path to *Philosophie*', being the first to show 'that the Earth it selfe was a meere *Magneticall* body'. Carpenter distinguished geography from natural philosophy or physical sciences which deal with the properties of natural bodies, and therefore limited his own consideration of magnetism to 'some generall grounds . . . which I hold necessary for a *Geographer*', his first task being to show the poles, meridians and other circles to be 'not bare *Imaginary* lines . . . but to be *Really* grounded in the magneticall nature of the Earth' [I, 45–8], and then proceeding to consider the various motions of the Copernican hypothesis, especially the daily revolution of the earth, as magnetical motions, requiring no external agent.

This concept of nature as self-regulating must have conflicted with the approved religious doctrine of divine control, and indeed in defending the theory of the earth's motions, Carpenter was not only satirical of the 'learned *Ignorance*' that supported Aristotle's system of celestial spheres, he was also openly sceptical of those who would have 'the Angels to be like galley-slaves' commencing or regulating all motion in nature [I, 78–81]. Whatever the reception encountered by these proposals, at a time when the mechanistic views favoured by Bacon, Galileo and Descartes were gaining wider acceptance, it is significant that the next serious attempt to incorporate studies of magnetism in geography is found in the work of Humboldt, who himself articulated a concept of nature similar in many respects to Carpenter's view of the earth as a natural harmony arising from the functioning of its parts. While Humboldt in *Kosmos* made no reference to Carpenter and may not personally have been familiar with his book, nevertheless the parallels between the work of the two men are interesting, indicating at the least a precedent for some of Humboldt's ideas. Carpenter for instance extended his discussion not only to earth magnetism but also to the occurrence of earthquakes and particularly to the character of the equatorial regions, all themes central to Humboldt's research.

In terms of method, furthermore, Carpenter strongly favoured an empirical approach, his model in this case evidently being Gilbert rather than Bacon, and he stressed frequently the importance of experience as a source of proof, as well as the use of experiments and observation in correcting any *Philosophie* [I, 8, 45]. Urging the careful investigation of specific questions, such as the variations in temperature of places on the same latitude, or evidence of what now would be identified as altitude sickness on 'the ridge of the mountaines *Andi* in *America*', he encouraged 'curious industrie', and ridiculed the 'juggling Scholler which assignes the cause to be a sympathie, antipathie, or some occult qualitie' [II, 26]. At the same time he showed awareness of the problem involved in selection from the evidence of experience, although it was chiefly in topography that he acknowledged this difficulty [I, 4]. On that account indeed he decided to avoid altogether what he classified as the special part of topography, the description of small places, it being 'an infinite taske in the whole earth to descend to all particulars which come in our way' [II, 2]. Encountering here the problem that was to plague subsequent empiricists, including geographers, Carpenter seemed unable to apply in his topical geography the method he had already used effectively, if perhaps intuitively, in the spherical part – the method of selecting information in terms of a significant central concept. Two centuries were to elapse before that method was to be articulated more clearly for geography in Humboldt's work; nonetheless, Carpenter's attention to the concept of nature in conjunction with his empiricism is significant.

A final parallel with Humboldt in this regard is Carpenter's emphasis on change in nature: 'Every place of the Earth hath beene subject to much mutation in the process of time, as well in Nature of the *Soyle* as of the *Inhabitants*'. [II, 7]. For Carpenter, change in the land meant not only progress from imperfection to perfection under the guiding hand of man, but also degeneration through wars or neglect:

A good instance whereof we may finde in the land of *Palestine*: which in times past by God himselfe was called, *A land flowing with milke and hony*, for the admirable pleasantnesse and fertilitie of the Soile: yet at this day, if wee will credit travellers report, a most barren Region, devoid almost of all good commodity fit for the use of man [II, 10].

Carpenter emerged here as an early conservationist, warning of man as a destructive agent, yet on this, as on many of the central issues of theory raised in the first book of his *Geography*, his ideas seem not to have been sustained by his followers.[31]

The Civil War period in England

After Carpenter's work, a small book which evidently had a much greater popular appeal, probably as a school text, was Pemble's *A Briefe Introduction to Geography* of 1630. Written, as the title-page proclaimed, 'by that *learned* man, Mr William Pemble, Master of Arts, of Magdalen Hall in Oxford', and 'very necessary for young students in that science', it continued through a number of editions to 1685. Pemble defined geography as 'an art or science teaching us the generall description of the whole earth', in this way neatly avoiding the issue of its scientific status.[32] Author also of *A Summe of Morale Philosophy*, published at Oxford in 1632, Pemble was a textbook writer rather than a geographer and clearly was not concerned with advancing the theory of the subject. He divided geography into a general and a special part, after the manner of Keckermann, and dealt only with the general part in this text, following the traditional topics of circles, zones, horizons, etc. On the position of the earth he was quite specific, and using diagrams to support his point, refuted Copernicus with a firm statement: 'The earth resteth immovable in the very midst of the whole earth [*sic*]' [12]. Although it is clear from his next statements that he intended to write 'midst of the whole world', the error reinforces the impression of bumbling respectability associated with geography texts in England for the next century.

Bound with this edition of Pemble's *Introduction*, as a kind of special geography, is an equally mediocre work, its title advertising the contents as *A Geographicall and Anthologicall Description of all the Empires and Kingdomes, both of Continent and Ilands in this terrestriall Globe. Relating their Scituations, Manners, Customes, Provinces, and Governments*. Attributed to Robert Stafford who signed the Dedication, it probably represents the work of his tutor, mentioned here by Stafford and named as 'John Prideaux, Rector of Exeter College and since Bishop of Worcester' in a manuscript note which also gives the date of printing as 1618, compared with 1634 on the title page.[33] This text is probably indicative of the teaching of geography in Oxford in the early part of the century, immediately prior to the work of Heylyn and Carpenter; indeed, many parallels with Heylyn's book are evident, including the division of the subject into chorography, hydrography and topography, although this author would prefer to assign hydrography to 'the art of navigation' [1–3]. His own effort to 'descend unto particulars' in describing the various continents proves, however, with its lists of places and trite comments, rather dismal and the continued publication of such works is evidence of the low standard of geography in England at the time that both Baconian and Cartesian thought were making their impact in Europe.

In this period, of course, scholarly activity was interrupted in Europe under the impact of the Thirty Years War (1618–48) and England too was in turmoil during the Civil War of 1641–5 when a Parliamentary militia was raised at Cambridge and Oxford was occupied by Royalist troops. With the defeat of Charles I, England entered its period of Parliamentary rule under Cromwell and after the eventual execution of the King in January 1649 it continued as a republic until the restoration of Charles II in 1660. Immediately after the war a Parliamentary Commission replaced a number of Royalist supporters at Oxford with members of the so-called Invisible College, a group including Robert Boyle, Robert Hooke and Thomas Sprat who were interested in pursuing the new experimental philosophy and later joined in forming the Royal Society of 1660. If the model for that society was Bacon's *New Atlantis*, an additional stimulus was the visit to London in 1641 of Johannes Amos Comenius (1592–1670), the famous Moravian reformer who was invited by Samuel Hartlib to present to Parliament his plans for a pansophic 'College of Light' and the achievement of a utopian society through universal knowledge. The war, however, cut short that project and Comenius returned to Europe, settling finally in Amsterdam where he produced, as one of his most interesting works, the first illustrated school text, his famous *Orbis Sensualium Pictus*, a kind of encyclopedia in Latin with a vernacular parallel text; after the first edition of 1658 at Nuremberg in Latin and German a Latin–English version appeared the following year at London (see Fig. 2) with the title translated as *A World of Things Obvious to the Senses, Drawn in Pictures*, and it went on to world-wide popularity with translations in many languages. Its few pages on geographical topics were well illustrated with sketch-maps and diagrams, although it continued to affirm Aristotle's cosmology: 'The Heaven is wheeled about and encompasseth the Earth, standing in the middle.'[34] Meanwhile, it was from Holland also that the most significant contribution to geographical thought in the period was to appear.

VARENIUS: IN DEFENCE OF GEOGRAPHY

Geographia generalis, a response to challenge

In 1650 with the *Geographia generalis* of Bernhard Varenius (1622–50) there occurred the first determined effort to relate geography to current developments in scientific thought in the time of Descartes, and it is significant that when Humboldt later become concerned with the reform of general geography in writing *Kosmos* he was highly conscious of the standards set by Varenius. The theories of Varenius deserve closer examination from this point of view, and for that purpose it is necessary to return to the original Latin of his major work, which he completed in Amsterdam at the age of 28, in the year of his death.[35] A need is apparent for a new edition of this important text, since those produced in the seventeenth century are now relatively rare and no satisfactory English translation exists. As Baker in a paper on Varenius pointed out in 1955, the last English version dates from 1765 and it, like earlier ones, is unreliable.[36] With the assistance of J. T.

CVII. a.

The terreſtrial Sphere.   *ſphæra terreſtris.*

The earth is round,
& therefore to be repreſented
by two Hemiſpheres, a . . . b
The circuit of it (grees
is three hundred & ſixty de-
(whereof every one maketh
fifteen German miles)
or 5400 miles ;
and yet it is but a prick,
being compared to ý world,
whereof it is the Centre.

They meaſure the longi-
tude of it by Climates,
and the Latitude
by Parallels. 2.   (it about,

The Ocean 3. compaſſeth
and five Seas waſh it,
the Mediterrane Sea, 4.
the Baltick Sea, 5. the Red
Sea, 6. the Perſian Sea, 7.
and the Caſpian Sea, 8.

Terra eſt rotunda,
fingenda igitur
duobus *Hemiſphæriis.* a . . . b
   Ambitus ejus,
eſt *graduum* CCCLX.
(quorum quiſque facit
Milliaria Germanica XV)
ſeu Milliarum VMCCCC :
& tamen eſt Punctum,
collata cum Orbe,
cujus *Centrum* eſt.

   Longitudinem ejus
dimetiuntur *Climatibus*, 1.
Latitudinem,
lineis *Parallelis.* 2.

   Eam ambit *Occanus,* 3.
& perfundunt V *Maria,*
Mediterraneum, 4.
Balticum, 5. *Erythræum,* 6.
Perſicum, 7. *Caſpium.* 8.

Fig. 2 The terrestrial sphere, from Comenius' *Orbis Sensualium Pictus* (1659). In common with most geography texts of the time, Comenius located the earth in the centre of the universe.

Fig. 3 The title-page of Bernhard Varenius' *Geographia generalis* (1650).

Christie, Principal of Jesus College, Oxford, Baker in that paper provided a new translation of some extracts from the *Geographia generalis*; however, these selections are limited in scope, the main concern of his essay being with the attitude of Varenius to special geography. Frequently also the Latin is interpreted here, as in the last English edition, in terms of a theory of science that does not seem to accord with the outlook of Varenius himself. An effort has been made to correct this in the present study with a new translation of the introductory sections from his *General Geography*.[37]

An active response to the ideas of Descartes, Bacon and Galileo is evident in the *Geography* of Varenius, for although he did not refer to those authors by name it is clear he intended to produce à kind of geography that would qualify as a science in their terms. Already, during his youth in Germany, he had pursued a general interest in the natural sciences: after three years in Hamburg at the Gymnasium of Joachim Jungius (1587–1657), a noted reformer of science and philosophy, he had commenced the study of medicine and mathematics at the University of Königsberg,[38] before leaving for

Holland where, like Descartes, he found a refuge for his work during the devastating Thirty Years War, until their death in the same year, 1650. Along with Descartes he stressed the importance of mathematics in the search for general principles that marks a science, and he saw this as indispensable in geography. Following Bacon, moreover, he classed geography as a branch of mixed mathematics and stressed its dependence on experience in providing explanation.

In attempting at the outset a precise definition of geography Varenius, following the model of Ptolemy, emphasized its concern with those aspects or affections of the earth that depend on quantity, and he included among these size, shape and motion – the primary qualities listed by Galileo as those measurable aspects of matter to which science should be restricted:

Geography, called one of the mixed mathematical sciences [scientia Mathematica mixta], teaches those affections of the earth and its parts which depend on quantity, namely shape, location, size, motion, celestial phenomena and other related properties.

On that account Varenius went on to question the kind of regional descriptions produced by many writers:

By certain people it is less strictly taken as merely the description of regions of the earth [regionum Telluris] and their distribution.

Strictly, also, his definition would seem to exclude what he called political descriptions; Varenius, however, was reluctant to take such a step in practice and defended their inclusion for reader interest:

By others on the contrary it is too widely extended, when they add a political description of individual regions. These however are easily excused since they do this to retain and arouse the interest of their readers, who are generally bored with a bare enumeration and description of regions without an explanation of the customs of the people.[39]

This disarming appeal to convenience did not remove the underlying problem of an inability to reconcile the new scientific theories of Galileo and Descartes with an advanced study of geography which included man in its scope. Faced with that difficulty Varenius accepted the solution offered by earlier writers and maintained from the beginning a division between general or universal geography, dealing with the whole earth mainly in terms of mathematical relations but with little reference to man, and special or particular geography, concerned with 'individual regions of the earth' and including a study of the inhabitants [2]. In that way he solved the methodological problem for general geography, according to the Cartesian view at least, but transferred it to special geography.

To a large extent in his proposals Varenius actually went little beyond earlier texts which had carried on Strabo's regional descriptions of the inhabited world, while preceding this with a general section more or less modelled on the geography of Ptolemy. Similarly he used well-established names in adopting the terms general and special for the two divisions of the subject.[40] He considered these more apt than the 'improper' classification into *exterior* and *interior* geography suggested by Golnitzius, author of a small and traditional *Compendium Geographicum*, published seven years before his own by Elzevir at Amsterdam.[41] Instead, like the mathematics professor David Christiani (1610–88) of Marburg, whose *Systema geographiae generalis* appeared in 1645, Varenius followed Keckermann in attempting to formulate a systematic general geography. More significant, however, was the way in which Varenius attempted to relate these traditional approaches in geography to the powerful seventeenth-century theories of scientific method and so establish a basis for the recognition of geography as a modern science. In that effort he faced two major tasks: on the one hand to justify the inclusion of special and, in effect, human geography, in opposition to the rationalist view of science; and on the other hand to overcome the reluctance of geographers themselves to develop the theoretical basis of their study. He made a vigorous response to both challenges.

Special geography: towards a more empirical rationalism

Varenius showed a strong interest in special geography, devoting to it the longest section in his introductory chapter, under the heading *Affections*, a term suggesting a study of earthly conditions, occurrences, and changes, which in the traditional Aristotelian view were contrasted with the changeless regularity of the heavenly spheres. Here he assigned human factors an important place in regional studies:

There are, it appears, three kinds of things, namely *terrestrial, celestial*, and *human* [*Terrestria, Caelestia*, et *Humana*], which in individual regions deserve consideration, and indeed in special geography make explanation of individual regions possible, with profit to students and readers [3].

The *celestial* are 'those which depend on the apparent motion of the sun and stars', and include the effects of latitude on the weather. Among *terrestrial* aspects he included the boundaries, shape, and size of the region, its mountains, waters, 'woods and deserts', its produce, mining and animals. In discussing the third aspect, 'the *human*, which depends on men, that is, the inhabitants of regions', he proposed a consideration of their appearance, food and drink, their occupations and sources of income, including trading. Then, departing to some extent from the traditional topics of civil geography – cities, schools, political and religious organization, and historical events – he suggested in addition a large number of topics relating to the social life of the people:

their virtues, vices, learning, intelligence, . . . their customs concerning childbirth, marriages, funerals . . . the speech or language which the inhabitants use . . .

distinguished men, craftsmen and inventions of the inhabitants of individual regions [3–4].

Varenius, then, was by no means apologetic in his approach to this part of the subject. Indeed, his next statements provide fairly strong evidence of his intention to follow the *Geographia generalis* with a work on special geography, to accompany an earlier venture in that field, his previous publication, the *Descriptio Regni Japoniae et Siam* of 1649, a compilation of descriptions by other authors. After the list of contents for *particular geography* he wrote, 'Apart from these we shall add to the special geography many sections about the practical application of geography [Nos praeter illa adjungemus Speciali Geographiae multa capita de Usu Geographiae].' Later he added, 'In our general geography we have explained in general terms certain affections which in the special geography we shall relate to the explanation of individual regions' [4–5]. Although the omission of articles in Latin leaves the possibility that this may have been a general reference to special geography, it seems more likely that Varenius was referring here, as in other contexts, to a specific future project. Altogether there is strong justification for Baker's argument that Varenius considered special geography to be 'an integral part of geography', not simply an aspect to be included reluctantly, as some writers have suggested, in a concession to popular demand.[42]

The critical passage, on which the latter opinion probably was based, follows the summary given by Varenius of the three aspects to be treated in special geography and appears to represent his response to criticism of any inclusion of the *human* aspect:

These are the three kinds of affections to be explained in special geography, and though those which make up the third class are less correctly referred to geography, something however must be conceded to custom and the convenience of students [4].

Far from indicating reluctance on his part, however, the last part of this statement can be interpreted, in conjunction with others on the same theme, as showing the young geographer ready to challenge the strict rationalist view of science. For Varenius the utility argument was sufficiently strong to justify calling into question and even modifying a scientific theory that seemed out of touch with current needs.

Having previously defined geography as a mathematical science concerned with quantity, Varenius already had acknowledged the difficulty of including man in its scope. Ptolemy, 1500 years before, had endeavoured to make sure his geography dealt only with quantity by restricting it mainly to cartographic problems – that is, to those aspects he felt could be handled mathematically – meanwhile assigning everything else to chorography. Varenius, by proposing to incorporate in geography a consideration of human learning, intelligence, language and so on, raised that problem again and furthermore brought the science as he envisaged it into conflict with the more recent Cartesian dualism of mind and matter.

In addition Varenius faced the problem of meeting the accepted requirements for proof in a science. According to the rationalist school, scientific propositions should be drawn from general principles and be capable of logical or mathematical proof, all reliable knowledge, in their view, depending on such *demonstration*. In this context special or local geography, regarded as being concerned with individual instances rather than universals (the *species* as compared with the *genus*), was considered incapable of demonstration. The rather vague notion that geography itself should deal with general statements rather than particulars, with 'wholes' rather than parts, is found in Strabo and also in Ptolemy, who offered the single and rather limited practical suggestion that geographical descriptions should mention only 'the most significant instances of each type' – the large cities and important rivers for example – leaving the small villages and the branch streams to chorography.[43] Even by the time of Varenius there was not much evidence of progress towards establishing an effective procedure to implement this idea since the time when Plato had declared that real knowledge resides only in universals.

Observing that the problem of abstraction is by no means confined to geography, Varenius pointed out that 'a similar difficulty occurs in other branches of philosophy' and that even Aristotle in his study of animals expressed some doubt as to the order that should be observed in discussing the general characteristics of animals and the individual species from which these are abstracted. A comparable dilemma arises in geography, Varenius claimed, in deciding whether to discuss aspects in general terms or in the context of individual regions, and he had already attempted to solve this with the proposal that generalizations established in the general geography should be applied in special geography to particular localities [5].

Without giving much consideration as to how such generalizations might be derived in the first place, Varenius was more concerned with the way in which the propositions of geography are to be proved. Since, in comparison with the mathematical disciplines and astronomy, geography remained at a disadvantage in providing logical demonstrations, it must therefore, he argued, rely on experience as well as on mathematical proof, and here he turned to the views of Bacon for support:

The principles which geography uses to confirm the truth of its propositions are threefold. First, the propositions of geometry, arithmetic and trigonometry. Second, the precepts and theorems of astronomy . . . Third, experience, for the greatest part of geography, particularly special geography, rests solely on the experience and observation of men who have described individual regions [5].

In his section on Method (*Methodus*), Varenius made a remarkable effort to combine the neo-Platonic view of mathematically demonstrable science with the Baconian form of sense-empiricism in geography:

Concerning method, that is, the manner of proving the truth of geographical teachings [*dogmatum geographicorum*], it should be explained that in general geography very many are confirmed by demonstrations properly so-called, above all those concerning celestial affections. In special geography however, practically everything is explained without demonstration . . . since experience and observation, that is, the evidence of the senses, confirm most things, indeed they cannot be proved in any other way.

His solution to the problem of including special geography in the science was to widen the definition of science itself, accepting confirmation by experience as well as by demonstration:

For science [*scientia*] is taken in three ways. Firstly, for any kind of knowledge even if it derives only from what is probable. Secondly, for certain knowledge [cognitione certa], whether this certitude depends on the strength of demonstrations or on the testimony of the senses. Thirdly, for knowledge solely by demonstration: which use is most strict and is appropriate to geometry, arithmetic, and other mathematical sciences, except chronology, astrology and geography, to which the word science in the second sense is applicable [6].

Although he did not attempt to deny the third and 'strict' use of the term *science*, as certain knowledge derived from demonstration, it is his second category that is of more interest, for here he included also Bacon's definition of science with its insistence that the evidence of the senses must be accepted. Although Varenius himself did not use the term *empirical* for this latter kind of science, he was in effect outlining here the framework of method for the natural sciences of the following century, with an attempt to combine two conflicting traditions in the search for 'certain' as distinct from 'probable' knowledge.

Along with astrology, which he included with considerable reluctance, Varenius assigned to the second kind of science both chronology and geography, and it seems clear from this and the earlier references quoted that he intended to include general and special geography together in this category of certain knowledge, indicating a belief that although demonstration is more applicable in the general part, both aspects are dependent to some extent on experience. Basing a contrary interpretation on the 1736 English translation of this passage as quoted in Baker's article, Joseph May suggested that for Varenius '"general" geography is to be understood as a science in the third sense', that is, as a 'rational science', whereas special geography is empirical; however, he appears to have been hampered by the lack of clarity in the eighteenth-century version, commenting indeed that 'the meaning of this passage is not altogether clear'.[44] Although Baker himself also provided the correct Latin for the passage, his interpretation is rather obscure, and no effort is made to relate the whole discussion to the scientific thought of the period.

In his attempt to define science in terms that would encompass all of geography, Varenius reflected a general transition in scientific thought during the seventeenth century, away from the belief of earlier centuries that knowledge is to be confirmed only by logical argument or by mathematical proof, and towards an incorporation of the view favoured by Bacon, that science is dependent on observation and experiment. Varenius himself did not inquire too searchingly into the nature of experience or the possible inconsistencies of the empirical method then emerging, but he made it clear from the outset that he opposed a narrow view of utility or practicality, and his work indicated the significant position of geography in testing the effectiveness of the new theories of science. In this he was clearly a predecessor of Humboldt, those two indeed being distinctive in their awareness of the problem: in turning their attention to the improvement of geography, both were prepared not merely to adapt it to current standards of science, but if necessary to question or even reject as inadequate a philosophy of science that cannot accommodate such a study.

General geography and a more rational empiricism

Like Humboldt after him, Varenius regarded the general study of the earth as an essential part of human knowledge and in his preface, a dedication of the book to the consuls of Amsterdam, he defended with conviction the value of geography: 'although this is concerned only with the earth, which is like a point if compared with the heavens, nevertheless it forms a no less valuable part of learning than astronomy'. Dismissing the argument, derived from Aristotle, that the earth is of a different nature from the rest of the universe and should be studied on that account, he began by emphasizing the significance of the earth to man, not only as 'the home of the human race' and 'the source of our first origin' but also 'because from our earth we contemplate the movements of the heavenly bodies'.[45]

If it seems surprising today that such persuasions were necessary at all, it must be remembered that Bacon's arguments for the utility of knowledge were still relatively novel, at a time when astronomy was almost the only quantitative science to gain wide acceptance among scholars. The situation had changed little from the time of Strabo, who argued that it would be absurd if a man studied astronomy 'and yet paid no attention to the earth as a whole, of which the *ecumene* is but a part'.[46] Like Strabo too, Varenius was ready to recommend geography for the assistance it can give to the statesman. To the consuls he shrewdly went on to point out that geography is not only valuable as a part of human learning, it is also useful: 'knowledge of the earth is not only most worthy of men, but is also most necessary in the republic of letters and for practical life'. Stressing the assistance given to trade and navigation by geographical maps and local descriptions, he passed over the popular arguments for geography as a support for other studies – for 'theology, natural science [Physicae], politics and other disciplines', or for 'history, whose beacons not unjustly are considered to be chronology and geography' [p. 276 below]. The notion of geography and chronology as the sun and moon or the eyes of history was common in this period: it is found in Keckermann and, as E. G. R. Taylor noted, had been given currency even before 1600 by Chytraeus and Hakluyt, the dependent status of geography increasing, indeed, until by 1650, 'Geography had fallen back into the position of a mere adjunct to Mathematics and History, written, not by professional geographers, but by scholars and divines whose main preoccupations were with other fields of knowledge.'[47]

A mathematician himself, Varenius considered the main obstacle to the development of geography in his day to be, not its concern with mathematical calculations, but rather a neglect of the principles on which these are based. While pointing out the utility of geography for the state, especially in providing reliable maps and other navigation aids for sailors, Varenius emphasized that these had not been produced initially by those people who currently were using them – or even teaching their use – without knowing how such formulae were derived. In a defence of the intellectuals who produce such ideas, Varenius reminded the consuls that a technique such as map-making, being derived from 'theoretical geometry', originated from mathematicians, 'since it demands a greater effort of mind than men are willing or able to support, who, intent only on gain, seek merely the application and practical use of knowledge'. The case is very well illustrated, he told them, in arithmetic, now so widely used in the city, even by the youngest tradesman:

But indeed, before the rules, in accordance with which practice is established, were extracted by mathematicians from the inner depths of arithmetic and proved, and an efficient method of applying them pointed out, there was absolutely nobody who could do these things [p. 277 below].

Continued progress depends on such thinkers, for whenever a 'new rule is to be found for a new practice' it is the mathematician who must provide it. Similarly in many sciences, 'it is very easy to show how many benefits have originated from learning for the practical utility of human life'. A science, he suggested, whether mathematics, mechanics, geodesy, or geography, must not be taught and practised only by those technicians who are minimally instructed in its practical application: without a strong intellectual foundation it cannot develop, and ceases to be a science. His arguments in effect amounted to a call for a more rational empiricism and a wider concept of utility.

A major reason for the decline of geography in his day, Varenius declared, was the neglect of general principles and the virtual reduction of the subject to a description of individual regions:

But those who have so far written on geography have discussed at length special geography alone, practically without exception, and have explained very little relating to general geography, with much that is necessary being neglected and omitted, so that young men learning special geography are for the most part ignorant of the bases of this discipline, and geography itself scarcely preserves the title of a science [ipsa Geographia vix Scientiae titulum tueretur] [pp. 277–8 below].

Like Plato, who claimed that it was the unwarranted disdain for philosophy that brought him to its defence, Varenius set out to provide an intellectual basis for the development of geography as a science: 'When I noticed this, in order to provide a remedy for this evil, I began to turn my thoughts . . . to the writing of a general geography.' This, he saw, must come before the completion of his other researches, including those planned 'on natural observations in various parts of the world, on the food and drink of different peoples . . . on the various medicines of different peoples'. The outstanding contribution of Varenius was his recognition of the need to provide a general theory for the subject as a whole.

A theoretical basis for geography

Varenius commenced his *General Geography*, then, with an introductory section in which he discussed the theory, method and history of the 'discipline as a whole', going further probably than any writer since antiquity in attempting to develop the conceptual basis of the subject. Through such *precognita*, he pointed out, 'the reader's mind conceives some idea of the whole discipline, or at least its arguments' [1]. In a similar way Humboldt later placed a great deal of emphasis on his own *prolegomena* as providing the key to the whole of his *Kosmos*, recognizing a primary task in the development of an empirical science to be the clarification of its general concepts, as a means of giving coherence to individual observations. Applying this principle to other aspects of the study, Varenius commented, for example, on the inadequacies of Greek and Roman geography, that 'With regard to the winds their insight was limited to a few varieties, and a general view was altogether unknown to them' [8].

A thorough explication of these ideas, of course, is not to be found in Varenius and indeed it is in retrospect that their significance becomes evident. The introductory chapters of the *Geographia generalis*, although brilliant for their time, show signs if anything of some haste in their preparation. His section on the *Origin of Geography*, for instance, is an interesting attempt to trace some aspects of the history of the discipline from ancient times, but he dealt in it mainly with various early attempts at exploration and did not discuss the great geographers of ancient Greece, making no reference even to the monumental geographical works of Strabo and Ptolemy which must have been readily available to him in Latin translations. The section concludes with a note, oddly out of sequence, that 'Anaximander, however, who lived in about the 400th year before Christ, is recorded as being the first to have tried to measure the earth.'[48]

Varenius either was unfamiliar with, or chose here to ignore, the other contributions to the mapping of the earth made by the early Greeks, including Eratosthenes with his famous estimate of the earth's circumference, and Hipparchus (*c.* 150 B.C.) who is credited with the introduction of latitudinal zones or *climata* as well as the use of circles of latitude and longitude to fix locations on earth, all topics that Varenius himself dealt with at some length. The work of those early geographers had been discussed by Strabo, who himself gave considerable attention in Books I and II of his *Geography* to the previous history of the subject. Altogether, the relative sophistication of thought evident in Strabo or Ptolemy stands in strong contrast to the paucity of such theory in later European attempts at geography. Varenius showed no evidence of direct familiarity with that vigorous early tradition in his subject, a reflection not only of his youth and limited experience, but of the general intellectual decline of geography in his time.

The general geography itself was organized by Varenius, with an evident concern for an orderly arrangement, into three divisions. Of these the *absolute part* dealt with the earth itself, its surface features and atmosphere. The *respective part* considered 'those affections and occurrences which happen to the earth through celestial causes' – latitudinal zones, length of days, seasons, temperatures, differences in time, etc., and here he mentioned also the ancient Greek division of peoples according to the zones they inhabit, these being indicated by the shadow of the sundial. Finally the *comparative part* was intended to contain 'an explanation of those properties which emerge from comparison of the various places of the earth' [2]. In this third part he discussed longitude and the calculation of locations and distances, concluding, after the manner of Keckermann, with several chapters on the construction, loading and navigating of ships that constituted a virtual treatise on navigation. In contrast to the subsequent approach to comparative geography in the nineteenth century, as the systematic comparison of one region or set of phenomena with another, the term *comparative* was used by Varenius here in a very limited sense.[49]

In Chapters V and VI of the *absolute* part, Varenius introduced the Copernican theory of the earth's motion around the sun. This was a significant action on his part, and the way in which he presented these ideas is interesting for the light it throws on his theory of scientific method. Acknowledging at the beginning that the heliocentric view had received the censure of the Church of Rome, he declared his intention to explain the theory briefly 'because to many people this appears to be highly probable' [45]. Here he was using a similar technique to that in his defence of special geography, by pointing out that it had widespread support and for that reason alone deserved attention. In bringing forward arguments in favour of the earth's rotation, he recalled that while Ptolemy's doctrine of a fixed earth remained the common opinion of many astronomers, it must be remembered that long ago the Pythagoreans maintained that the stars keep their place while the earth moves, even though no further mention of this was found until the time of Copernicus, Kepler and Galileo. Whatever the theory used to explain the phenomena, Varenius reminded his readers, the earth remained the centre of their observation. Furthermore, anticipating a modern view of relativity, he suggested that either hypothesis can be accepted without interrupting the work of geography or even astronomy, for, 'whether you follow the Ptolemaic or Pythagorean way of thinking, the firmness and certitude of general astronomy and geography lose nothing' [54–5]. Subsequently of course both these theories were to be modified, and the problem of detecting motion in the so-called 'fixed stars' of the Copernican scheme continued to occupy astronomers into the nineteenth century. At this point Varenius, however, showed a characteristic response, to withdraw from arguments that under present conditions appear futile, and devote effort instead to the constructive development of other aspects of knowledge.

To some extent, in taking the attitude he did, Varenius probably was concerned to divert attention from a particularly contentious issue. Writing at least a quarter of a century after Cluverius or Carpenter, and from the relative safety of a fairly tolerant Holland, he was in a more secure position to incorporate the Copernican theory in his geography, but his approach remained cautious. He was careful to present these views in the way that Copernicus himself had proposed them, not as a set of assertions about the solar system but rather as an alternative hypothesis. At the same time this was more than a technique on his part for avoiding conflict on a particular issue, it is indicative of his whole approach to knowledge. It was characteristic of both his language and his outlook that he should write about geographical *dogmas* or teachings, not about geographical 'facts', and translators who fail to record this do him an injustice.

This is not to say that Varenius rejected entirely the idea of truth and certainty as goals for inquiry. He relied a great deal on mathematical proof, and when he wrote in the Dedication (see p. 277 below) of 'demonstration, on which certainty [certitudo] depends', he evidently intended by *certitudo* more than merely a consensus of opinion. What is more significant is the extent to which he maintained an awareness of the place of the observer in all observations, and the element of probability in a large part of knowledge. In discussing the use of the model globe and geographical maps, for instance, he was cautious about basing proofs on them when their accuracy must be taken on trust: 'In these', he stated, 'we are rather following descriptions composed by geographical authors: the globe and maps serve the purposes of illustration and easier comprehension.'[50] Experience, or the evidence of the senses, also can provide the basis for certain knowledge, Varenius evidently believed, although he did not examine the claims of sense-empiricism too closely. If anything his leanings towards the rationalism of Descartes seem to have helped him avoid the objectivism of Bacon, but his attempt at a synthesis remained tentative, and his effort to assert a consistent theory for geography in the face of the conflicting views of his time appears, in the encounter with science, as a brief foray that was not sustained by those who followed him.

After Varenius: retreat

Varenius himself, then, did not elaborate a philosophy of science in any detail to guide his followers. Moreover, in the first English translation of his work, by Richard Blome in 1682, the Dedication, along with several passages dealing with the theory of science and with the incorporation of recent discoveries, such as those on the tides or on the magnet, were even omitted, effectively concealing, for those who could not read the Latin text, the interest Varenius had shown in those issues.[51] The *Geographia generalis* remained a standard work for the next century, but in a rather static way and without further conceptual development. Although Isaac Newton as Professor of Mathematics made some corrections in the editions he produced for his students at Cambridge in 1672 and 1681, he left the theoretical introductory sections intact.[52] After him, no geographer attempted to expand this aspect of the general geography, for although many copied the work and referred to it with respect, they did not attempt to develop his ideas on general principles. Edmund Bohun, for example, wrote in his preface to his *Geographical Dictionary* of 1688:

Many have desired the first Principles of Geography should have been shortly stated, by way of Introduction, but this has been done so often . . . that I conceive it needless, especially seeing *Varenius* his General Geography (which is perhaps the best Book that was ever Written as to this) is in English, and may be easily had.[53]

A decline in any interest in general principles apparently was becoming, even in Bohun's day, a feature of the new empiricism as it was handled by the majority of practitioners, and it is perhaps ironical that the very theory of empirical method which Varenius called on to justify the recognition of special geography as he envisaged it should eventually impede the development of geography as a whole. General geography as Varenius presented it fell out of favour because it involved a comprehensive approach rather than a specialist one, and was not amenable to the experimental method. Meanwhile special geography, in dealing with man, was considered by the mechanists to be outside the ambit of natural science, so that while many specialist sciences flourished during the next century, in geography the regional studies remained in an ambiguous and even defensive position in relation to the scientific tradition. As the subject in general lost its theoretical basis, then, geography suffered a decline as a science, virtually until the time of Kant.

CHAPTER 3

GEOGRAPHY IN DECLINE: THE AGE OF NEWTON

Although in many respects conditions might seem to have favoured the development of a vigorous geographical tradition in the years after Varenius, this did not come about. The expansion of commerce, the growing application of statistical analysis to such matters as population, economic factors and weather, in addition to a wider public interest in maps and descriptions of various parts of the earth, were all promoting a more active concern with studying the modern world rather than that of classical times, but no adequate theory emerged to encompass those developments and support the reorganization of geography. Instead, the study of natural history expanded markedly by the end of the century.

At present, historical research on this period is scanty and the following discussion will concentrate chiefly on English sources, a number of which have not previously been examined in any detail. In terms of theory, however, neither France nor Germany at that time appears to have been able to provide any significant leadership; indeed, Oskar Peschel, commenting on the low standard of the German contribution in the century prior to 1750, stated frankly, 'The best that German geographers in the preceding century could offer was reflected light, information from French and British researches.'[1] The decline of German geography is illustrated in the *Cosmographia* of Bartoldi Feind, a small and undistinguished book in two parts – *Astrognosia*, an introduction to astronomy, and *Geographia*, 'a description of the earth-sphere' – which reached a fifth edition by 1694, when the author still hesitated to confirm the Copernican hypothesis since it conflicted with both 'sacred writings' and visual evidence. Written in German rather than Latin, it compared poorly in content with the earlier work of Varenius.[2]

ENGLISH GEOGRAPHY, 1650–1700

In England the second half of the seventeenth century saw the achievements of Newtonian physics providing a widely acclaimed model for Western science; yet as far as general publications in geography were concerned it was a time of conservatism and collation, marked by the translation and reproduction of earlier works rather than the search for a new format. This period saw the appearance of some of the last British cosmographies.

The decline of cosmography

One of the most popular English works at this time, judging by the number of new editions, was the *Cosmography* of Peter Heylyn, a sequel to his *Microcosmus* of 1621, and carrying the full title: *Cosmography in Four Bookes, Containing the Chorographie and Historie of the whole World, and all the Principall Kingdomes, Provinces, Seas, and Isles, thereof.* First published in London, in 1652, it was followed by a second edition in 1657, a third in 1666 and others, with corrections and additions, dated 1670, 1674, 1677 and 1682, continuing into the early eighteenth century with a 1703 edition 'Improv'd with an historical continuation to the present time by E. Bohun'.[3] Heylyn's *Cosmography* evidently had a wide circulation, even to America, a copy of the second edition being listed in the records of the Harvard library in this period.[4] As his title indicated, Heylyn's emphasis was on history and regional description, more in the tradition of Abbot than of Carpenter or Varenius. He showed no concern with astronomy, and comparatively little with geography as a science, the name cosmography in his case suggesting a combination of traditional geography and civil history. In an apparent attempt to cater to popular taste, the later editions included 'an Accurate Index' and a number of corrections, the 1682 edition advertising, twenty years after the death of Heylyn, 'Revised and Corrected by the Author himself immediately before his death'.[5]

The work itself was not new in conception, being intended to replace his own *Microcosmus*. For many years, Heylyn explained in the preface to the *Cosmography*, he planned no enlargement of his earlier work, except for the 'amending of Errors', but so many of these were found that after eight editions, 'I could no longer call it mine'. Heylyn went on to recount vividly how in 1649 he was given a further impetus to return to what he called his younger studies, by a stranger – a big gentleman, according to his story – who gave him a shove, with the scornful reminder that 'Geography is better than Divinity.' Soon afterwards, persecuted during the Civil War for his Royalist and High Church affiliations, Heylyn as an unemployed cleric, afflicted by poverty and poor eyesight, retired to write 'a General survey of all the World, the Government, Affairs, and successes of it', with his first premise clearly stated as 'the Creation of the World by Almighty God'. It was to be one of many works during the next century in which an attempt to subordinate geography to divinity put an evident strain on both. Heylyn stated in his preface: 'In the pursuance of this Work, as I have taken on my self the parts of an *Historian* and *Geographer*, so I have not forgotten that I am an *Englishman* and, which is somewhat more, a *Church-man*.' Therefore, he added, 'though the *History* and *Chorography* of the World, be my principal business', the book is also intended to record the history of England and of Christianity.[6]

In those circumstances, a rigorous or novel approach to geography was scarcely to be expected in this work, and it is easy to agree with Eva Taylor's opinion that 'Dr. Heylyn was a better *raconteur* than Cosmographer': as she pointed out, 'his concept of Geography is the limited one which dominated academic textbooks right down to the present century', his concern being to locate such features as provincial boundaries, main towns, and the course of the chief rivers, with little concern for primary sources in his descriptions. It is more difficult to accept Taylor's next statement that Heylyn in his General Introduction 'adumbrates the basis of a Philosophy of Geography'.[7] This view was based apparently on Heylyn's argument that differences between nations and regions may be regarded as an instance of 'the Power and Wisdom of Almighty God' who in that way 'hath united all the parts of the World in a continual Traffic and Commerce with one another'. An interest in promoting international trade, relatively common at that time, was evident also in Varenius, but while this might be said to provide a useful basis for a commercial geography it is by no means a sufficiently wide concept to support a philosophy of geography as a whole, nor did Heylyn offer any consideration of the place of geography in relation to the rest of science and to human knowledge in general. Repeating in his General Introduction the well-established view on the necessity for close links between geography and history, Heylyn pointed to the usefulness of both 'for understanding the affairs of Ages past, as for commerce . . . with the Nations present'. Beyond that, in his *General Praecognita of Geography* he showed if anything less originality than in his *Microcosmus*, and relied heavily on Ptolemy, from whom he claimed to take his definition of geography as 'a description of the whole Earth, or the whole Earth imitated by writing and delineation, with all other things generally annexed to it'. Similarly he repeated fairly closely Ptolemy's views on chorography:

a *Chorographer* doth describe a Country by the bounds, Rivers, Hills, and most notable Cities, and not of the whole Earth universally, which is the proper work of a *Geographer*. So that *Chorography* differeth from Geography as a part from the whole.[8]

More in the tradition of Strabo, however, Heylyn here, as in the *Microcosmus*, included *chorography* as a part of geography, along with *topography*, which he saw as the description of some particular place or city, and as a third branch of geography he again added *hydrography*. For Heylyn, geography was simply 'an aggregate' of these three, concerned only with the Regions . . . their Sites, and several Commodities'. The study of man, on the other hand, his 'Habitations, Governments, and Manners', was assigned by Heylyn to *Civil History,* while a general view, the 'universal comprehension of *Natural* and *Civil* Story', was the domain of *Cosmography* [23–4]. Unfortunately, as cosmography at that time was already becoming discredited, Heylyn's narrow definition of geography itself seems to have proved little short of disastrous as a guide to its development in the ensuing years.

After the Restoration of 1660, with a professedly Anglican monarchy reinstated, the Church of England came to exercise a considerable dominance over British natural science. Heylyn's work remained in favour, as much probably for his religious and political views as for any geographical merit. It was followed in 1679 by the *Cosmographia, or a View of the Terrestrial and Coelestial Globes*, of John Newton; by then, however, the decline of cosmography was virtually complete. In Newton's *Cosmographia* there was no evidence of new research, his astronomy indeed being pre-Copernican. Denying even the daily rotation of the earth, he presented instead the Aristotelian doctrine of the *Primum Mobile* or first motion: 'that Motion by which the several Spheres are moved round the World . . . in 24 hours from East towards West . . .'.[9]

Interestingly enough, John Newton intended his *Cosmographia* as a contribution to educational reform. In his dedication and preface he expressed opposition to the continued emphasis on the teaching of Latin and Greek grammar while studies such as Arithmetic, Geometry, Music and Astronomy were 'neglected . . . in both Universities, as in all Inferior Schools'. Following the advice, as he said, of Sir Francis Bacon, he called for a more practical kind of education, to produce useful citizens fitted for the trades, and he joined the movement in favour of the vernacular: 'All the Arts should be taught our Children in the English tongue, before they begin to learn the Greek or Latin Grammar'; therefore, after his previous treatises on *Grammar, Arithmetick, Rhetorick*, and *Logic* he now offered in English an introduction 'to *Geometry* and *Astronomy*; which I call by the name of *Cosmographia*'.

Newton divided his *Cosmographia* into four Books, the first dealing with *Geometry* – magnitude or continued quantity – and consisting mostly of pages of tables. The second and third books were devoted to *Astronomy* and the calendars, while the fourth, on *Geography*, was also Ptolemaic in outlook. In the preface he described this addition as

an Introduction unto *Geography*, in which I have given general Directions, for the understanding how the habitable part of the World is divided in respect of Longitude and Latitude in respect of Climes and Parallels with such other Particulars as will be found useful unto such as shall be willing to understand History; in which three things are required; The time when, and this depends upon *Astronomy*; the place where, and this depends upon *Geography*; and the Person by whom any memorable Act was done, and this must be had from the Historical narration thereof.

Geography, as John Newton presented it, was again primarily a help to reading history, and though he assigned a consideration of time to astronomy rather than to history, he expressed a common view of his period in believing geography to be concerned with place, but not with time. His definition of geography, furthermore, appears to have been taken directly from Keckermann: 'Geography is a Science concerning the measure and distinction of the Earthly Globe' [411]. In all, his *Introduction to Geography*, in less than fifty pages, provided only the most mediocre repetition of such traditional topics as Circles, Zones and Climates, with notes on the continents and their various provinces. Clearly, in the hands of textbook writers like Newton, geography did not stand to fare very well in the movement for the reform of education that marked the following century, the dismal evidence of their work being perhaps a reflection on a common tendency among reformers, to call for change while themselves remaining painfully uninformed, not only of the most valuable aspects of the tradition they are rejecting, but also of the extent to which they might be perpetuating its mistakes.

Restoration geography: the loss of theory

The general deterioration of the subject and the loss of any cohesion in geographical thought after Varenius is evident in what might be called the Protestant geographies, reproduced in England during the latter part of the seventeenth century. A popular work at this time was Samuel Clarke's didactic compendium, *A Mirrour or Looking-Glass both for Saints, and Sinners*. Like Heylyn in the same period, Clarke (1599–1683), 'sometime Pastour in Bennet Fink, London', attempted in this text to combine divinity and geography, adding a section on *the Wonders of God in Nature*. First published as a small book without a geographical section in 1646 during the Puritan period of the Commonwealth,[10] Clarke's *Mirrour* included in its later editions *a Geographical Description of all the Countries in the known World*, indicating probably an increased demand for such works. The Fourth Edition appeared in 1671, enlarged, as it stated, 'Especially in the *Geographical* Part, wherein all the counties in *England* and *Wales* are Alphabetically Described: Together with the Cities, and most remarkable things in them: As also the Four Chief *English* Plantations in *America*'.[11]

This was by no means intended as a serious contribution to research in natural science, and the aggressively moralistic tone of the first part was continued in Clarke's *Geographical Description*. Here he quoted as his theme the lines from Psalm 104 – *O Lord, how manifold are thy Works! In Wisdom hast Thou made them all* – lines that appeared frequently in the natural theologies of this period when the study of nature was commonly justified on the grounds that it provided evidence of divine creation and providence. Although Clarke's subtitle stressed the reliability of his information, claiming it to be 'collected out of the most Approved Authors, and from such as were Eyewitnesses of most of the things contained herein', his chief concern seems to have been with offering his readers the traditional list of curiosities, including the 'famousest Cities and Fabricks', both ancient and modern, and 'the rarest Beasts, Fowls, Birds, Fishes, and Serpents'. His

Geographical Description itself contained neither maps nor any general introduction. Commencing, after the manner of Abbot, with a single statement on *The Division of the World*, that 'the Earth is commonly divided into four parts, *Asia, Africa, Europe*, and *America*', he proceeded to describing Asia first – his only innovation – giving lists of cities along with their classical names. For 'the *Brittish* Islands', he added some guide-book-type notes on the English counties, arranged alphabetically, preceding this with a few general statements of limited geographical value. 'England', he noted, for example, 'is called the *Purgatory of Servants*; the *Hell of Horses*; and the *Paradise of Women*.'[12]

Something more serious seemed to be promised in Clarke's account of the four English colonies in America: Virginia, New England, Bermuda, and Barbados. The title-page, dated 1670, proposed a consideration of 'the temperatures of the Air: The nature of the Soil: The Rivers, Mountains, Beasts, Fowls, Birds, Fishes, Trees, Plants, Fruits etc. As also, of the Natives of *Virginia*, and *New-England*, their Religion, Customs, Fishing, Hunting, etc.'. What followed, however, was by no means an extension of the kind of special geography planned by Varenius, for Clarke offered only a composite of assorted information, collected indiscriminately from others and presented largely for its curiosity value. There was no attempt at an organized geography for critical readers, much less any pretensions to a science that deals meaningfully with normal occurrences.

In 1682 Richard Blome attempted to fill the conspicuous gap in geographical theory for English readers by translating the *Geographia generalis* of Varenius. Along with this, in the absence of a special geography by Varenius himself, he included 'a Geographical Description of all the World, Taken from the Notes and Works of the Famous *Monsieur Sanson*, Late Geographer to the French King'. Blome published these together under the title *Cosmography and Geography* – the dual name indicating apparently an intention to extend his readership to the adherents of both traditions, since he himself seems to have made no distinction between the two. In his preface he stated simply, 'Amongst all those Arts and Sciences which Man ought to have a Knowledge of, the Description of the Earth and Heavens, which is termed *Cosmography* and *Geography* (for the utility and dignity thence arising) ought not to have the least estimate.'[13]

The original Dedication to the *Geographia generalis*, in which Varenius argued for greater attention to the theoretical basis of geography, was omitted from this translation, Blome substituting his own preface. Here he presented the usual arguments for the utility of geography – in commerce between countries; for the *Merchant* and *Navigator* (on this he followed Varenius closely); for *History* which is 'of little use' without geography; and for the statesman who in his negotiations requires a knowledge of the 'Nature . . . Customs, Strength, etc. of the Nation or People'. In his list of beneficiaries, Blome included also 'the Moral *Philosopher* . . . for how can he search into . . . the Genius, Natures, Inclinations, or Studies of Men, and what is most proper for every distinct Nation or People (being his adequate subject) without this Chart to stear by?' This statement is particularly interesting for showing the way in which the scope of moral philosophy, as opposed to natural philosophy, had become defined by some writers since the time of Galileo and Descartes; it is evident that geography, in dealing with the *nature* and *customs* of people, overlapped the field of moral philosophy according to that view. Varenius would probably have gone further to include the products of human intellect in his special geography, but Blome did not pursue that aspect. Geography, he added, is useful also to the *Physitian* who needs it for studying not only 'Drugs . . . from Foreign Parts', but also 'the variety of Bodies or Constitutions, which are habituated according to the Climate and Soil of the Country'. The environmental determinism suggested here was given a stronger expression in the concluding section of the preface, where Blome referred to the *Climates* or zones into which geographers had divided the world, each climate being dominated by a planet which exercises some influence on the people of that region. Clearly not sharing the scepticism Varenius had shown towards astrology, Blome affirmed that '*Planets* have their *Operations* or *Influences* on the *Inhabitants* dwelling under each of the said *Climes.*' Differences between peoples derive from 'this diversity . . . in the *Climates*, the scituation of *Provinces*', this extending even to mental differences: 'the *Air* and *Elements*', he continued, 'discriminate the Constitutions of *Men*, and those Constitutions, their Natures; for the manners of the Mind follow the temperament and disposition of the Body'. So, he went on, '*Northern People* being remote from the Sun . . . are Sanguine, Robust, full of Valor and Animosity' – and he proceeded to include in his preface a classification of peoples according to zones, similar to that put forward by Bodin more than a century earlier.

Closely linked with Blome's environmentalism was his view of the earth itself as a product of divine design, in which, according to the orthodox beliefs of his day, the relationships between man and nature, as well as a clear discrimination between them, had been predetermined by God. Incorporated in this, although seemingly at variance with the doctrine of environmental determinism, was the belief that the world had been created to serve the purposes of man – a view to be found in a secular form in Aristotle and Pliny, and recurring strongly as part of Christian orthodoxy in the modern period where it proved an obstacle to the development of the natural sciences. In terms of that view, the main justification for the study of nature was the contemplation of the divine creation and the demonstration of its perfection, with geography, for example, being justified as little more than a kind of pious substitute for pilgrimage. Such an interpretation, though not suggested at all by Varenius, was expressed by Blome very early in his own preface:

seeing the Earth was created by God to be the habitation of Man, if . . . we cannot so well Travel with the Body, yet at least we would visit, behold, and contemplate it in our Minds; for its beauty, admirable elegancy, and the Honour of the Creator.

This was scarcely an outlook to encourage a critical attitude to geographic inquiry, and the *Geographical Description of the World* by the French geographer Sanson d'Abbeville, included in the same volume, did not seem to offend Blome with its evidences of credulity and its mundane approach. Commencing with Spain and France in Europe, Sanson's description concluded with Brazil. With regard to the American continent he made the sound prediction that 'whole Volumes might be made only touching the Nature and Propriety of their *Grains, Herbs, Plants, Fruits, Fowl, Beasts* and *Fish*, which are all different from ours'; at the same time, he claimed in his own account to have 'comprised all that seemed most necessary concerning America'. His final statement, however, as presented by Blome, offered dubious assurance of the basis used for the selection of data. Referring to the European impact on the native Americans, he concluded:

But of all our *Beasts*, nothing so much astonished them as our *Horses*; and it was near a hundred years in Peru, and other parts of *America*, before those *People* would be persuaded to mount them.   *Finis.*[14]

It is understandable that works of this kind did not achieve a high regard among the serious scientific endeavours of the period. Blome made some effort in this book to comply with the growing interest in the statistical organization of data by adding 'about an Hundred Cosmographical, Geographical and Hydrological Tables', summarizing information on land areas, river lengths and so on, and he included maps, but his basic impatience with the philosophy of science as Varenius approached it is evident in his deletion from the *Geographia generalis* not only of the original dedication but also of certain relevant passages which apparently did not accord with his own outlook. A loss of the dignity that Varenius had given the subject was inevitable in such circumstances, as geography was left to be dealt with by men of the calibre of Blome, who appears at his worst almost as a figure of fun; his own *Dedication* to a member of the English nobility scarcely compared with the original in intellectual content: 'My Lord', wrote Blome, 'When I consider You are the Duke of Albemarle, the very Title is so Great, that it puts a damp on my Quill . . .'[15] Perhaps in his defence it should be remembered that political pressures were extremely strong at this time in an England verging again on civil war, when the power of the ruling class and the established church seems to have put a damp on the quill of many writers.

In 1688, the year of England's 'Glorious Revolution', Edmund Bohun published his *Geographical Dictionary*, having, as he stated in his preface, been asked 'about a year since' to write such a work. His aims were twofold: firstly, 'to raise a desire in the English Nobility and Gentry, to have a fuller and larger work of this Nature', and secondly, 'to make this as General, and as Useful as was possible'. Emphasizing his desire to avoid offence – 'my Observations being purely Geographical and Chronological' – Bohun hoped that any accidental 'Moral reflection' would be found inoffensive, and he avoided any reference to the 'Principles of Geography', with the excuse that these were sufficiently treated in the *General Geography* of Varenius. At the same time he admitted that recent geography had been distinguished by its quantity rather than its quality: 'Geography is an Art which from very small beginnings, has in our days swelled into a vast bulk, and yet it is capable of great improvements.'[16] Similarly, in a letter written in 1691 to Laurence Echard of Cambridge, praising Echard's recent brief introduction to geography, Bohun complained: '*Geography* is become in our Times, since the Invention of Printing, a vast and voluminous Study' so that many are discouraged 'by the meer Bulk of the Writers'.[17]

Bohun himself in his *Dictionary* was more concerned with the improvement of information retrieval than with any development of geography as a discipline; however, he did show some concern with standards of scientific accuracy. Reviewing some of the many previous attempts at a geographical lexicon, and acknowledging his own reliance on the recent works of Baudrand and Hofman which included 'Modern or New Geography' as well as Ancient, Bohun in his preface claimed that he had tried to correct earlier errors, having 'very frequently consulted the Maps, and other Geographers, and some Travellers too'. Admitting that 'I have not had time to measure all the distances', he noted that where sources differed he had adopted a policy of using 'Words which express my diffidence or uncertainty, which was better than to be positive without Evidence, or any Authority'. He added a new *Table of Longitudes and Latitudes*, with Paris as the first Meridian, published in 1687 'by Phil. de la Hire, Professor of the Mathematicks at Paris'. At the same time, referring to the need for more accurate observations by travellers, he expressed an opinion shared by writers as far apart as Ptolemy and Alexander von Humboldt:

As to the Longitudes and Latitudes of Places, it were to be wished that Travellers would very exactly observe that, in all the more considerable Places, and at their return give an account of their Observations, it being the first step to the reformation of our Maps.

Faced, like other writers of his time, with the problem of rapidly proliferating material, Bohun saw the tabulation of information and the provision of an index or dictionary as a solution.[18] For Bohun, the main task of geography was in locating places, both ancient and modern; he showed no concern with its conceptual development, viewing the study primarily as an aid to reading history, along with chronology, which in his opinion dealt with *time*, where geography considered *place*:

Without the knowledge of Times and Places, the best Histories are little better than

Romances, at least they leave faint and confused Notions on the Minds of the Readers, whereas these two Circumstances fix and confirm things.

The study of chronology in his day was still at a comparatively early stage, and Bohun claimed some novelty in his own contributions: 'As to the Chronological Accounts I have added; that is a new and late Design attempted by few, and therefore not easy to be done.' It is significant to observe developments in the writing of modern history also taking place at this time. Bohun himself admitted in his preface that his *Dictionary* was neither a geography nor a history: the main aim, he stated, 'of a work of this Nature, is to be a kind of General Index to Geographical Books and Maps, to shew where any Place stands, but then the Description belongs to the Geographer, and the Fate of those Places to the Historian'. According to his view, then, geography was to give a description of places, history an account of events.

Geography and history: the problem of the unique

It has become almost customary among writers in the history of geographical thought to maintain an emphasis on this kind of contrast between geography and history, as a way of demonstrating a distinctive position for geography in relation to the sciences in general. More significant however, though not often considered, are the remarkable similarities between these two disciplines in their encounter with the scientific method that was formulated in the tradition of Bacon and Descartes. Both historians and geographers found themselves on the defensive, unable to meet the requirements for experiment and objectivity set up by the Baconians, and excluded from the natural sciences also, according to the Cartesian view, as studies dealing only with the particularity of events or places and lacking the capacity to formulate laws.[19]

Varenius, it will be recalled, made a strong attempt to reconcile geography with both philosophies. While stressing the need to develop the mathematical and theoretical basis of geography, he was not prepared to follow the teachings of Galileo and Descartes to the extent of limiting the subject to a study of measurable aspects of the material world and excluding the study of man; in assigning the human aspect largely to special or *particular* geography, he justified its acceptance as a science in terms of Bacon's argument that the evidence of experience provides an adequate basis for knowledge.

It was from this background that the distinction subsequently emerged in geography of *nomothetic* versus *idiographic* studies. The terms themselves, coined evidently in nineteenth-century Germany, were derived from the Greek *nomos*, a law or custom, and *idios*, peculiar or separate, and in the twentieth century their use seems to have been confined largely to the geographical literature, where they became key words in a dispute which in

its extreme form perpetuated the seventeenth-century belief that it is possible to distinguish clearly between a method of inquiry that establishes laws
or generalizations, and one that simply describes individual instances. Few
geographers today, however, are familiar with the philosophical and historical context in which that kind of dispute arose, and the debate aroused
much confusion. Frequently in mid-twentieth-century discussions, for
example, regional geography was described as *idiographic* since it deals with
individual or unique regions, while systematic or general geography was
considered to be *nomothetic*, although the debate persisted on whether
geography can produce exact mathematical laws, and in this respect, prior to
the neo-rationalist movement, human geography was commonly held to be
in a methodologically inferior position to physical geography. Again, R. E.
Dickinson regarded the empirical approach itself as being essentially 'descriptive', even 'ideographic' (*sic*), since in his view it deals with individual
phenomena, in contrast to the nomothetic approach which is 'theoretical'
and 'deductive'.[20] Such a distinction is rejected here, for it is based on a
narrow interpretation of the empirical method, one that assumes observation of an object to be possible without relation of it to prior experience
and hence without theory and generalization as a necessary part of the
process.

Richard Hartshorne, acknowledging in a recent public lecture his responsibility for introducing the terms 'nomothetic' and 'idiographic' into
geographical literature in English, added in retrospect: 'though I sometimes
wonder if this was a good thing'. Explaining his use of *nomothetic* as dealing
with the generic or general, *idiographic* with the individual (as in the
German *einmalig*), he stressed that it was never his intention to suggest that
one can be taken without the other. In his view each individual, 'each place
in its location', is unique, at least in some important respects, and indeed
'this is the heartrock of the basic idiographic concern in geography . . . just
as in history the event is a single non-repeating occurrence'. At the same
time he recognized that the study of such individuals must be accompanied
by the use of generalizations – the selection of what is significant for study
being of course a matter of judgment – although he himself admitted to some
doubt on the place of concepts in geographic theory.[21]

Geography then, in those terms, is both nomothetic and idiographic, and
in that respect of course is in the same position as other sciences, for whether
or not the various objects of study are each in themselves unique, they are
perceived always in terms of relationships, and the process of study operates
through general concepts. Each atom of oxygen might be in some respect
unlike any other atom, but this has not precluded the development of
modern chemistry; all scientific analysis or description can be regarded as
the meaningful discussion of such individual instances. It was not the
uniqueness of individual regions, or of their inhabitants, that hampered the
development of geography, but a lack of the conceptual means for dealing

with them. This is particularly evident in considering the century after Varenius, when geography failed to keep pace with the successes being achieved in physics and chemistry through a series of brilliant intellectual syntheses.

SCIENTIFIC EMPIRICISM: THE NEWTONIAN SYNTHESIS

Newton and the Royal Society

In marked contrast to the deterioration in geographic thought during the Restoration period in England, the experimental sciences of mechanics and chemistry flourished at this time under the aegis of a new scientific society. Founded in London in 1660 for the promotion of *Physico-Mathematical Experimental Learning*, the Royal Society took its name two years later with the receipt of a charter from Charles II, and provided a focus for scientific ventures in England at a time when Oxford was depleted by the royalist dismissal of many intellectuals appointed earlier during the Commonwealth. The influence of Baconian thought on the new Society was pronounced during the early years, being evident especially in the statutes of the Society drawn up by Robert Hooke in 1663, which stipulated that the aim of the Society was 'to improve the knowledge of naturall things, and all useful Arts, Manufactures, Mechanick practices, Engynes, and Inventions by Experiment, – (not meddling with Divinity, Metaphysics, Morals, Politicks, Grammar, Rhetorick, or Logicks)'.[22] Initially, applied science and mechanical questions were of most interest to members; however, during the 1670s Bacon's influence seems to have declined as British scientists paid increasing attention to the astronomical work of Galileo and the mathematical, rationalist approach of Descartes.

A similar trend was evident in the Paris Academy of Sciences, set up under the patronage of the crown in 1666 by Colbert, Minister to Louis XIV, to promote advances in science that would further his policy of expanding French trade and commerce. There too the early influence of Bacon, leading to the production of various 'histories' of machines, plants, and so on, was later supplanted as Cartesian views became more popular, and the French shift from the practical to the literary and philosophical aspects of science became pronounced under Fontenelle (1657–1757), secretary to the Paris Academy from 1699 for forty years. Geography in France received if anything less encouragement with this change in views, and under Colbert's successor, Louvois, previous work on longitude, and on the construction of a map of France, was suspended by the end of the century.[23]

In England the men who were to have the most profound impact on later scientific thought belonged to a group including Christopher Wren (1623–1723), Robert Hooke (1635–1703), Isaac Newton (1642–1727) and Edmond Halley (1656–1742), who were concerned to resume Galileo's work of explaining the movements of bodies in space in quantitative and mechanical terms. Isaac Newton, with his Universal Law of Gravitation, which was published in the *Principia Mathematica* of 1687, provided what was acclaimed as a triumphant justification for the mechanistic view of the universe. According to Newton himself, the First Cause of this universe is certainly not mechanical; however, like Robert Boyle and other members of the Royal Society, he asserted that the concern of science is not with causes but with observed effects, its aim being to calculate and give a mathematical expression to mechanical laws. To some extent, then, he agreed with Descartes in using mathematical analysis to proceed from the known to the unknown. Descartes had applied geometry to problems of motion, and Newton went even further in developing the calculus as a method of studying the mathematical relations between variable quantities. At the same time he diverged strongly from the Cartesian view that scientific proof can be deduced purely from ideas provided by the intuition. Following Bacon instead, Newton attempted to base his assumptions on experiments and observations of phenomena; where direct experiment was impossible in astronomical calculations, he favoured the method of analogy, using, for example, the analogy of a projectile such as a stone in calculating the orbits of planets.

In dealing with ideas, Newton made an attempt to keep his own *philosophy of nature* separate from his scientific work or *natural philosophy*, and he endeavoured to limit the introduction of speculation into experimental science. Without going so far as to try to eliminate all theory from scientific method, he attempted to distinguish between those propositions that he considered to be derived from phenomena, and others, not so secured, which he relegated from science as undesirable hypotheses:

The word 'hypothesis' is here used by me to signify only such a proposition as is not a phenomenon nor deduced from any phenomenon, but assumed or supposed – without any experimental proof . . . hypotheses of this sort . . . have no place in experimental philosophy. In this philosophy, propositions are deduced from phenomena, and afterward made general by induction.[24]

Like Bacon before him, Newton did not inquire too closely into the relation between phenomena and ideas, nor the way in which particulars themselves might give rise to knowledge. Nevertheless, to the satisfaction of his many followers, Newton, as one eminent historian of philosophy expressed it, 'was able to weave together the two main strands of seventeenth century scientific thought: the mathematical rationalism of the Continental tradition . . . and the "physico-mathematical experimental learning" which the Royal Society cultivated', so that 'for a century after Newton nearly all philosophizing . . . was an attempt to come to terms with Newton's thought'.[25] Newton's synthesis gave rise to a new form of scientific empiricism and provided a basis for the physical science of the eighteenth century.

He offered a mechanical view of nature acceptable even to religious authorities by stressing it to be the will of God that the material universe should operate as an autonomous and ordered mechanism, and although in the present century Newton's absolutes of space, time, and absolute motion have all been superseded in the theory of relativism, his ideas for a long time remained dominant, almost to the extent of impeding further speculation. By the end of the seventeenth century Newton's theories were already widely accepted among English scientists; his successor in the chair of Mathematics at Cambridge, William Whiston (1667–1752), supplemented the then popular Cartesian philosophy with the teachings of Newton, and in 1688 a French review of the *Principia* by the philosopher John Locke (1632–1704) helped spread Newton's ideas beyond Britain. During the first half of the eighteenth century they gained wide acceptance on the continent, especially through the interpretation of Voltaire (1694–1778).

For geography the outcome was not so promising. Although Newton himself had produced a new edition of Varenius at Cambridge for his students, the kind of research promoted by Newton's concept of scientific method did not favour the development of geography. Moreover, a separation between the mechanical and the moral aspects of the world was maintained strongly through the next century, and a study that was obliged to deal with both aspects was placed in a difficult position. Deprived of a secure theoretical foundation, geography was left outside the mainstream of scientific activity, while the experimental philosophy with its increasing prestige dominated almost all new research. Geography, then, was not drawn into the mathematically-oriented development taking place in other sciences, and the lead suggested by Varenius in that direction failed to be pursued effectively at this time, even though many of the conditions for it were already in evidence before the end of the century.

Quantification: from qualities to quantities

A movement towards the quantification of geographical data was evident during the seventeenth century, although little of this activity was apparently incorporated into the geography of the time. Regular daily weather records were kept by a number of individuals in England: the diaries of Dr Napier from 1598 to 1635, and of Elias Ashmole from 1677 to 1685 survive among the Ashmole manuscripts (423 and 438) in the Bodleian Library, Oxford, although others apparently have been lost. The measurement of rainfall using rain-gauges was carried out by Townley in Lancashire from 1677 to 1703, and by William Derham at Upminster after 1697, while John Locke not only kept occasional weather records between 1666 and 1692, and attempted to construct a simple scale of winds, but also urged that more constant records should be made throughout England 'with the help of some instruments . . . for exactness', so that 'rules and observations concerning the extent of winds and rains, etc., be in time established'.[26] At the request of the Philosophical Society at Oxford, records were kept during 1684 of the weather, the barometer and winds, and the publication of these was accompanied by the suggestion that similar observations should be made in many foreign parts at the same time to assist in explaining and predicting the weather.

In the same period the beginnings of a statistically-based demography appeared. Although in Europe no reliable census data were available at this time, in England the Bills of Mortality had recorded deaths and christenings in some parishes from the sixteenth century. Using these, John Graunt (1620–74), a London haberdasher, published in 1662 his *Natural and Political Observations . . . made upon the Bills of Mortality*, a work which reached a fifth edition by 1676. In the same period Sir William Petty (1623–87) gave the name *Political Arithmetick* to this numerical analysis of social phenomena. Contrary to the view of Descartes that man's free will places him outside the universal order, there was now a growing recognition of an astonishing degree of regularity in the available population data over various periods of time and in different countries. Analysis of social or economic phenomena, and even prediction on the basis of statistical probability, could now be attempted. Petty, in his *Five Essays in Political Arithmetick* (1683–7), produced estimates of population for England and Wales, and predicted a doubling of population within 40 years. Later, Gregory King (1648–1712), in his *Natural and Political Observations and Conclusions upon the State and Condition of England* (1696), attempted an estimate of population based on the number of houses, as well as a classification according to age and family status, and he calculated the annual increase of population. In addition he compiled a table of national income and expenditure according to various social groups for 1688, providing altogether what one historian called 'the most elaborate demographic balance-sheet produced up to that time'.[27]

This minor statistical revolution, however, appears to have had little impact on the geography of the day, and although the current authors occasionally included commercial statistics or tables from Graunt and King in their geography texts, they seem to have taken no part in the actual pioneering work. Instead, one of the leaders was Edmond Halley, the astronomer and friend of Newton, who became interested in calculating the probability of life expectancy for insurance purposes. He also produced a map of Winds of the Globe in 1684, and in 1700 introduced the technique of isogonic lines that Humboldt in 1817 was to apply to the expression of climatic data. In the interval between these two men, although a greater interest was expressed in population statistics, comparatively little was done to extend the use of quantification into further aspects of geography, and it must be admitted that even Humboldt confined his statistical analysis of economic and population data to what he called his *political essays*.[28]

During the early stages in the development of modern science after the renaissance, geography had exercised considerable leadership in giving a more precise mathematical expression to concepts of the world, with its achievements in cartography. Much of this, however, remained dependent on the intellectual contribution of Ptolemy. Carpenter in 1625, for example, was in large part following Ptolemy when he stated that the main concern of geography was with '*Quantities, measures, distances* . . . assigning to each region its true longitude, latitude, clime, parallel, and Meridian', while it was for chorography or special geography to deal with 'the accidental *qualities* of each place, particularly . . . which places are barren, fruitfull, sandy, stony, moist, dry, hot, cold, or mountainous'.[29] This list of 'qualities' – that is, those aspects not able to be treated mathematically – was to be gradually reduced over the next two centuries, and Humboldt himself made significant advances, applying quantification to the study of economic production, geology, climate, and the high-altitude regions of the earth. But the effort to quantify and measure, which stimulated other sciences during the seventeenth century, was not articulated in any systematic way at that time as a task for geography.

The decline in geography was pronounced in this period, and indeed around 1700 the subject seems to have reached a low ebb. After Varenius, whose call for a sounder mathematical and intellectual basis for the subject found little response, there is a noticeable gap in the chronological record of general geography texts as far as serious contributions are concerned. For nearly a century after 1650, while travellers' descriptions multiplied and useful cartographic work continued, the field of general geography was left almost entirely to the dictionary, the compendium, the reprint of Cluverius, Heylyn, or Varenius. In this lacuna the place of geography was virtually taken over by the expanding study of natural history, at that time receiving its major impetus in England.

THE RISE OF NATURAL HISTORY

During the last decades of the seventeenth century the most ambitious attempts at providing some comprehensive view of the world within the scientific tradition appeared not in the geography texts of the day but in the physico-theology of a group of English writers. Their natural history represented a conservative response, an attempt to pursue scientific inquiry in the empirical tradition of Bacon and at the same time to search for a 'system of the world' compatible with Church teachings. The work they produced was thus often a mixture of religious dogma and useful pioneer contributions to the natural sciences: as Humboldt later commented, 'Faith and knowledge blended with one another to produce the so-called systems of Ray, Woodward, Burnet and Whiston in England.'[30]

Their works nevertheless are of considerable significance to the present investigation, both for their attempt to reconcile conflicting elements in the thought of their times, and as a source from which later geographical studies were to develop. Although largely neglected by historians of geography, except with regard to those theories of the earth recognized as forerunners of modern geology, these works do in effect constitute a significant phase in the history of geographical inquiry. The theoretical contribution made by those early writers of natural history was an extremely significant one particularly in their efforts to provide a unified concept of nature in the face of the dualist approach advanced by the mechanists. Looked at in historical perspective the British natural history movement forms an important link between the empirical theory of Francis Bacon and the scientific geography of Alexander von Humboldt. Occurring at a time when geography along with cosmography appeared to be virtually losing its identity as an intellectual discipline, the rise of natural history marked a critical stage in the encounter between geography and scientific empiricism.

Burnet's *Theory of the Earth*

One of the earliest and most controversial of these works was *The Theory of the Earth* by the English clergyman Thomas Burnet (*c.* 1635–1715). In 1681 Burnet published in Latin his *Telluris theoria sacra*, translating and expanding this in 1684 as *The Theory of the Earth*. This theory, he stated at the outset, 'is not the common Physiology of the Earth', but concerns the origin of the earth and the great revolutions in its history. In defence of his work he took care to explain that 'Philosophy, which is the Contemplation of the works of Nature', is an immense task requiring breadth of vision: 'short sighted minds are unfit to make Philosophers, whose proper business is to discover and describe in comprehensive theories the *Phaenomena* of the World, and the Causes of them'. His own theory, he insisted, was therefore not 'a Romance' or artificial fiction, but a search for 'solid truth', an attempt to replace ignorance with knowledge, and he continued to defend the free use of reason in such speculations: ' 'Tis a dangerous thing to ingage the authority of Scripture in disputes about the Natural World, in opposition to Reason'; he warned particularly, in much the same terms as Galileo and Descartes fifty years earlier, of condemning the heliocentric theory on that basis, 'if within a few Years . . . it should prove as certain and demonstrable that the *Earth is mov'd*, as it is now, that there are Antipodes'. At the same time he declared his own inquiries to be aimed at supporting the scriptures – 'this theory being writ with a sincere intention to justify the Doctrines of the *Universal Deluge*, and of *Paradise*'.[31]

According to Burnet, the present form of the earth is to be explained by the catastrophe of the biblical flood, which destroyed the initial perfection of the earth: 'We have still the broken Materials of that first World, and walk upon its Ruines . . . And this unshapen Earth we now inhabit, is the Form it was found in when the Waters had retir'd.'[32] He denied any continuous change in this 'second face of Nature', stating that 'since the Deluge all things have continued in the same state, or without any remarkable change'. In the future, he concluded, the great *Conflagration* would produce a new face of nature and an ultimate consummation [325–6].

Although the extremes of this theory seemed to justify the Royal Society's determination to avoid such speculation, Burnet himself was by no means entirely opposed to the empirical science of his day or its mechanical hypothesis. Declaring that 'The Course of Nature is truly the Will of God', he added that 'God made all things in *Number, Weight* and *Measure,* which are Geometrical and Mechanical Principles', and he linked these three with the mathematical sciences of *Arithmetick, Staticks,* and *Geometry.* We must use these methods, he stated,

for when we contemplate or treat of Bodies, and the Material World, we must proceed by the modes of Bodies, and their real properties, such as can be represented, either to Sense or Imagination, for these faculties are made for Corporeal things; but Logical Notions, when applied to particular Bodies, are meer shadows of them . . . No man can raise a theory upon such grounds [316].

Rejecting the logic of Aristotle and what he called the 'dry Philosophy' of the scholastic age, Burnet clearly was engaged in a struggle to reconcile the scientific theories of Bacon and Descartes with his own work. While accepting a division between the material and intellectual aspects of the world, he made an interesting attempt to incorporate both in a general concept of nature: 'Universal Nature', in his view, comprises 'all the orders of Being in the Intellectual World and all the Regions and Systems of Matter in the Corporeal'. His concept included the Platonic idea of universal order: 'it is necessary to suppose, that there is an *Idea* . . . according to which this great Frame moves, and all the parts of it, in beauty and harmony'. However, like the experimental philosophers of the day, he himself avoided discussing the intellectual world, which in Christian theology included the angels as intelligent beings, and instead limited his concern to the material universe, which he described as 'one system, made up of several subordinate systems' – including the fixed stars and the planetary system [319–20].

The earth, according to Burnet, has its own system of providence – 'an Order establisht by the Author of Nature for all its *Phaenomena* (Natural or Moral)' – which governs mankind and all changes in the course of nature: 'Both in the Intellectual and Corporeal World there are certain Periods . . . 'Tis Providence that makes a due harmony or Synchronism betwixt these two.' At the same time he suggested that there are two aspects of providence to correspond with the division between rational and material phenomena:

The Providence of Earth, as of all other Systems, consists of two parts, Natural and Sacred or Theological. I call that Sacred or Theological that respects Religion, . . . the government of the Rational World, or of Mankind . . . When we call the other part of Providence *Natural*, we use that word . . . as respecting only the Material World [323–4].

Despite his concept of 'universal nature', then, Burnet identified nature on earth with the material aspect and excluded the intellectual. The place of man in this theory clearly presented a problem. According to Burnet, man as a living being is part of nature: 'Man and other living Creatures . . . make the Superior and Animate part of Nature.' All aspects, however, of man's social life – 'Laws, Government, natural Religion, Military and Judicial affairs . . . which make an higher order of things in the Civil and Moral World' – are not strictly to be included in nature, although they are dependent on it:

As the Animate World depends upon the Inanimate, so the Civil World depends upon them both, and takes its measures from them: Nature is the foundation still, and the affairs of Mankind are a superstructure [244–8].

While Burnet regarded mankind as the superior part of nature, he did not take the doctrine of *Natural Providence* to mean that the universe was made for man: 'We must not by any means', he declared, 'imagine, that all Nature, and this great Universe was made only for the sake of Man, the meanest of all Intelligent Creatures that we know of; nor that this little Planet . . . is the only habitable part of the Universe.' In a rebuke to the Baconians with their confident assertion that knowledge is power, Burnet stressed man's ignorance and his vanity: 'Vain Man', he stated, 'hath no power over external Nature, little over himself . . . His Birth and Education generally determine his fate here, and neither of those are in his own power' [310–11].

The *Theory of the Earth* is interesting in this regard. Just as Burnet took a stand against the widely-held view of man as master of nature, so also he refused to support any assumption of European superiority over other races. He offered a rare plea for tolerance of those groups often ridiculed as savages: 'An Indian hath more reason to wonder at the European modes, than we have to wonder at their plain manner of living. 'Tis we that have left the tract of Nature.' The concern with the state of nature that was to become so prominent in eighteenth-century thought is thus already evident in his writings. Similarly, although Burnet avoided dealing with the civil world in this work, his comments on the influence of custom suggest those of Hume a half-century later:

'Tis a strange power that custom hath upon weak and little spirits; whose thoughts reach no further than their Senses; and what they have seen and been us'd to, they make the standard and measure of Nature, of Reason, and of all *Decorum* [249].

Despite his attempt to advance scientific inquiry while supporting Church teachings, Burnet's theory soon aroused strong criticism both from churchmen and naturalists, for as new discoveries in natural science multiplied during the ensuing years, particularly with the dramatic evidence of fossil remains in rock strata, the debate over the scriptural record concerning the origin of the world took on increased urgency. In 1690 Burnet produced a short *Review of The Theory of the Earth*, as well as the promised sequel on the biblical prophesies of the end of the world.[33] Both these were included in the second edition of his *Theory of the Earth* (1691), along with *An Answer to the late Exceptions made by Mr Erasmus Warren against the Theory of the Earth*, in which Burnet argued that 'it is no fault to recede from the literal sence of Scripture' when that is 'inconsistent with Science, or experience . . . for Scripture never undertook nor was ever designed to teach us Philosophy, or the Arts and Sciences'.[34] Two years later his theory of the flood was attacked on scientific grounds in John Beaumont's *Considerations on a Book called 'The Theory of the Earth'* (1693).

John Ray: *The Wisdom of God*

A more substantial contribution to natural science, and one in which a number of major issues affecting geography through the next century found an early expression, was the *Wisdom of God*, published in 1691 by John Ray (1627–1705). As one of the leading naturalists of his day, Ray made pioneer contributions to the modern classification of plants and animals. He was elected to the Royal Society in 1667, five years after resigning his fellowship at Cambridge because of his refusal to subscribe to the Act of Uniformity of 1662, and in 1686–7 his two-volume *History of Plants* in Latin was published under the direction of the Society. Over the next years, feeling obliged as he said to write something in Divinity, he contributed several works to a tradition of natural theology that already included Hakewill's *Apologie, or Declaration of the Power and Providence of God in the Government of the World*, 1627, along with later treatises by John Wilkins (1614–72), later Bishop of Chester, Robert Boyle, Henry More, Ralph Cudworth, and numerous clerics, among them Tillotson, Archbishop of Canterbury, and Bishop Stillingfleet, who all defended the biblical cosmology, rejected an atheistic materialism, and generally opposed any extension of the mechanistic hypothesis to the animate world. On this theme Ray published in 1692 his *Miscellaneous Discourses concerning the Dissolution and Changes of the World*, producing a second edition the following year under the title *Three Physico-Theological Discourses*. His main work in that field, however, was *The Wisdom of God Manifested in the Works of the Creation*, which after its first appearance in 1691 was brought out in several new editions during his lifetime and continued to a twelfth edition in 1759.[35]

In this work Ray attempted to collect many observations on nature under a unitary view by showing the entire world as the expression of divine power and providence: 'my Text warrants me to run over all the visible Works of God in particular, and to trace the Footsteps of his Wisdom in the Composition, Order, Harmony, and uses of every one of them'. He admitted that such a task probably defied completion, being beyond 'the joint Skill and Endeavours of all Men now living, or that shall live after a thousand Ages, should the World last so long', and in this he seemed to be moderating Bacon's confidence in what the right method along with community of effort would achieve in science. Ray, however, showed himself a follower of Bacon in emphasizing the collection of 'Particulars', along with reliance on fact and experiment, as the special feature of his own work.[36]

The study of natural history was recommended by Ray, not only as an obligation on the devout to praise the works of God, but also in terms reminiscent of Bacon, as a necessary counterpart to verbal learning. It is not enough, said Ray, to 'content ourselves with the Knowledge of the Tongues, or a little Skill in Philology, or History perhaps, and Antiquity', and he objected strongly to those who disparage the study of natural history, 'in Comparison whereto that of Words and Phrases seem to me insipid and jejune. That Learning . . . which consists only in the Form and Pedagogy of Arts, or the critical Notions upon Words and Phrases . . . is only so far to be esteem'd, as it conduceth to the Knowledge of Things.' Words are but images; we must also study nature itself:

Let it not suffice us to be Book-learn'd, to read what others have written, and to take upon Trust more Falshood than Truth; but let us ourselves examine Things as we have Opportunity, and converse with Nature as well as Books. Let us endeavour to promote and increase this Knowledge, and make new Discoveries . . . Let us not think that the Bounds of Science are fix'd [169–72].

In the face of that task, Ray acknowledged the limitations of the human mind in studying the whole universe:

The Mind of Man being not capable at once to advert to more than one thing, a particular View and Examination of such an innumerable Number of vast Bodies, and the great Multitude of *Species*, both of animate and inanimate Beings, which each of them contains, will afford Matter enough to exercise and employ our Minds, I do not say to all Eternity, but to many Ages . . . [171–2].

He even suggested that acquiring such knowledge might be a suitable task for the Eternal Life. Meanwhile he encouraged the pursuit of this new science: 'I know that a new Study at first seems very vast, intricate, and difficult; but after a little Resolution and Progress, after a Man becomes a little acquainted . . . with it, his Understanding is wonderfully clear'd up and enlarg'd.' Expressing regret for the neglect of the natural sciences at Cambridge, he recommended these as private studies for those with both leisure and ability:

I am sorry to see so little Account made of experimental Philosophy in this University, and that those ingenious Sciences of the Mathematicks are so much neglected by us; and therefore do earnestly exhort those that are young, especially Gentlemen, to set upon these Studies: They may possibly invent something of eminent Use and Advantage to the World; and one such Discovery would abundantly compensate the Expence and Travel of one Man's whole life [173–4].

The model of the gentleman-scientist, already embodied in the membership of the Royal Society, found a number of adherents during the next century, one of the most prominent being Alexander von Humboldt, as a leader not only in research but also in the movement to effect a transition from the kind of Baconian natural history advocated by Ray, towards a new pattern of sciences.

With regard to religion, Ray defended natural science as a suitable propaedeutic even for divinity: 'I do not see but the Study of true Physiology may be justly accounted a proper, or Προπαιδεία Preparative to Divinity', and he vigorously opposed those scientific theories which challenged Christian belief in a created universe governed by providence [175]. A rejection of the mechanistic view was fundamental to his whole work, so that from the outset he condemned 'those Systems which undertake to give an Account of the Formation of the Universe by Mechanical *Hypotheses* of Matter, mov'd either uncertainly, or according to some Catholick Laws, without the Intervention and assistance of any superior immaterial Agent' [29–30]. He ridiculed the atheistic belief of Epicurus and Democritus that the world was produced by a fortuitous concourse of material atoms in infinite space, and for this purpose used the arguments of Lucretius and Cicero: the order and beauty of the world could not be the product of chance. The ancient atomists, Ray acknowledged, believed at least that 'this regular *System* of Things' had emerged from initial confusion and would return to disorder, as one among innumerable worlds. He was more contemptuous of the *mechanick Theists* led by Descartes who declared the universe and all its creatures to be the product simply of matter created and set in motion by God according to laws, the present order being achieved automatically without further divine intervention [41–2]. In discussing these questions, Ray took issue with Descartes for attempting to exclude 'all Consideration of final Causes from Natural Philosophy, upon Pretence, that they are . . . undiscoverable by us; and that it is Rashness and Arrogance in us to think we can find out God's End' [38]. It was clear to Ray that acceptance of such a restriction at that time implied also an acceptance of the Cartesian conceptual framework of mechanism.

Arguing that the mechanistic hypothesis is inadequate to explain the whole of nature, Ray supported the vitalist view:

many *Phenomena* in Nature, . . . being partly above the Force of these *mechanick Powers*, and partly contrary to the same, can therefore never be solved by them, nor without final Causes, and some *vital Principles* [43].

As examples of those phenomena which in his opinion are 'not *mechanical* but *vital*', Ray instanced gravity, and also the tilting of the earth's axis to produce the variety of the seasons. Above all, however, he pointed to the failure of the mechanists to account for the complexity of animal life:

But the greatest of all the particular *Phaenomena*, is the Formation and Organization of the Bodies of Animals, consisting of such Variety and Curiosity, that these mechanick Philosophers being no way able to give an Account thereof from the necessary Motion of Matter, *unguided by Mind for Ends*, prudentially therefore break off the System there [44].

Taking the empiricist position, Ray was highly critical of the rationalists for attempting explanation of natural phenomena without sufficient reliance on experience, criticizing Descartes in particular for his failure to recognize the regular muscular constrictions of the heart:

Natural Philosophers, when they endeavour to give an Account of any of the Works of Nature, by preconceived Principles of their own, are for the most part grossly mistaken, and confuted by Experience; as *Des Cartes* in a Matter that lay before him, obvious to Sense, and infinitely more easy to find out the Cause of, than to give an Account of the Formation of the World; that is, the Pulse of the Heart [45].

Ray argued that some *vital Principle* must be called on to explain such regular motions as that of the heart; matter alone could not sustain these actions without the guidance of some intelligent agent, and he derided the idea of matter giving rise to intelligence: 'Let Matter be divided into the subtilest Parts imaginable, and these be mov'd as swiftly as you will, it is but a senseless and stupid Being still, and makes no nearer Approach to Sense, Perception, or vital Energy.' Ray was suspicious also of the theory that matter, unguided, nonetheless obeys external laws, agreeing with Boyle on this point, that laws are not physical but moral causes, regulating only the actions of intelligent individuals. He argued for some *efficient* cause:

And, as for any external Laws, or establish'd Rules of Motion, the stupid Matter is not capable of observing, or taking any Notice of them, . . . neither can those Laws execute themselves: Therefore there must, besides Matter and Law, be some Efficient, and that either a Quality, or Power, inherent in the Matter itself, which is hard to conceive, or some external intelligent Agent, either God himself immediately, or some *Plastick Nature* [47–50].

To avoid occasionalism – the doctrine of constant divine intervention into all actions of nature – Ray suggested that God must use the subordinate agency of *plastic nature*, a theory, as he pointed out, already proposed by Dr Cudworth in his *Intellectual System of the Universe*. Ralph Cudworth (1617–88), one of the Cambridge Platonists, in that work of 1678 had traced back to Empedocles, Plato, and Aristotle, the idea of nature as an executive instrument which acts with regularity but remains always subordinate to the perfect intelligence.[37] Ray himself used the theory of plastic nature to explain the behaviour of animals, and rebuked Descartes for his view of beasts as machines: 'I should rather think, Animals to be endu'd with a lower Degree of Reason, than that they are mere Machines'; even in the case of plants, Ray argued that the complexity and regularity evident in their growth and reproduction could not be the product of mechanical principles alone – the *Plastick Principle,* he claimed, 'must preside over the whole Oeconomy of the Plant, and be one single Agent, which takes care of the Bulk and Figure of the Whole, and the Situation, Figure, Texture of all the Parts' [52–5].

Ray's concept of nature: constancy in variety

The idea of the whole is important in Ray's thought. Just as he considered the plant from the point of view of its whole *economy*, so he emphasized the existence of balance and harmony in nature. For him this was proof of intelligent design and the superintendence of a wise Architect, for although he was prepared to accept the atomic hypothesis to account for inanimate bodies, he vigorously rejected the materialism of his day that attempted to explain the entire world as the product of chance. Pointing to the regularity and uniformity evident in the motions of the earth and other bodies in space, he claimed 'we see nothing in the Heavens which argues Chance, Vanity, or Error; but, on the contrary, Rule, Order and Constancy; the Effects and Arguments of Wisdom' [67]. Constancy is observable throughout nature, Ray argued; indeed, he adopted an orthodox view, upheld by many churchmen of his day, that there is no sign of error or imperfection in the world, even in man's body [238]. 'Nothing so contrary as Constancy and Chance' was his dictum – one much in contrast to the growing preoccupation, even in his own lifetime, with stochastic processes and the calculation of probabilities.[38]

Simplicity was seen by Ray as a further characteristic of nature, indicative of intelligent design, and he considered it to be a task of science to recognize this inherent order and simplicity in the universe: 'For we see, by how much the *Hypotheses* of Astronomers are more simple and conformable to Reason, by so much do they give a better Account of the Heavenly Motions.' The Ptolemaic theory, or as he called it, the 'old *Hypothesis*', involved 'many *Eccentricks, Epicycles*' and suchlike to account for celestial phenomena, whereas 'in the New *Hypothesis* of the Modern Astronomers, we see most of those Absurdities and Irregularities rectified and removed; and I doubt not but they would all vanish could we certainly discover the true Method and Process of Nature' [63–4]. In answering objections to the heliocentric theory that 'it is contrary to Sense, and the common Opinion and Belief of Mankind', and even to 'some expressions in *Scripture*', Ray adopted a Platonic view: 'our Senses are sometimes mistaken, and what appears to them, is not always, in Reality, so as it appears' [195].

The immense variety evident in nature, according to Ray, could not be the product of necessity alone, while the presence of such constancy in multiplicity was for him proof of divine wisdom. How incredible is it, he asked, 'that Constancy in such a Variety, such a Multiplicity of Parts, should be the Result of Chance? Neither can these works be the Effects of Necessity or

Fate' [239]. Although these arguments were directed towards supporting the doctrine of providence, it must be noted that in terms of method, Ray was arguing that in the new natural science, the diversity of nature must be viewed in relation to a concept of the whole, and it is interesting that the quotation he selected [245] from Pliny to reinforce his interpretation – *Naturae vero rerum vis atque majestas in omnibus momentis fide caret* – was later inscribed on the title-page of *Kosmos*. The passage itself, used by Pliny in presenting his own cosmography, expressed a conception of nature that guided the work of both Ray and Humboldt: 'Truly, the power and majesty of nature in all its instances seems without meaning if one considers only the parts and does not comprehend the character of the whole.'[39]

One of Ray's major contributions in studying the diversity of 'this material and visible World', was his attempt at classification. Commencing with the classical division into animate and inanimate, he refused to accept also the Greek theory of the four elements (fire, water, earth, and air), saying that he inclined 'rather to the Atomick hypothesis' [59]. He classified living beings according first to the '*Genus* or Order of Beings', then to 'the *Species* contained in it', listing 'four great *Genera*' of animals: *Beasts* 'including also Serpents', *Birds*, *Fishes*, and *Insects*. He attempted to form an estimate of the number of species likely to be discovered in each – noting, on the definition of species, that he considered 'all *Dogs* to be one Species . . . The Breed of such Mixtures being prolifick' [21–2]. In addition, following Aristotle without question, Ray divided animate bodies into those with 'a *Vegetative* Soul – as Plants', a '*Sensitive* Soul, as the Bodies of Animals', or a '*Rational* Soul, as the Body of *Man*', while adding to the last group, with reference to Christian theology, 'and the Vehicles of *Angels*, if any such there be' [59].

The question of intelligent behaviour in animals was discussed at some length by Ray, and he added his own observations on such phenomena as migration, nest-building, and care of the young. While denying that animals are mere machines, he regarded them as essentially passive in these situations, acting according to instinct and directed entirely by 'the Providence of Nature, or more truly the God of Nature'. Even the many instances he observed among animals of adaptation to function, or fitness for purpose, were for him evidence, not of evolution, but of wise design: 'So that we see every Part in Animals is fitted to its Use, and the Knowledge of this Use put into them.' Quoting directly from Aristotle the assertion that animals act without skill, inquiry or deliberation, he denied that any knowledge comparable with human reason can be attributed to 'brute animals', for, 'as Dr. *Cudworth* saith well, they are not Masters of that Wisdom according to which they act, but only passive to the Instincts and Impresses thereof upon them' [126–8].

Similarly, in dealing with man as a rational being, Ray stressed that man is 'a Dependent Creature' [184], one who for all his 'Wit and Industry' [215] is not the architect of his own progress. Even the art of writing was in Ray's opinion 'divinely invented' [280], while in manipulating such materials as timber and metals, man was, in his view, simply finding the uses already designed for them, being 'endued with Skill and Ability to use them, and . . . by their Help . . . to rule over and subdue all inferior Creatures' [160–1]. On the last point, Ray advised humility, arguing that the world was made for man, but not only to serve him: man has been placed by the Creator in 'a spacious and well-furnish'd World' and provided with hand and reason to produce a 'comely Order', with cities, gardens, meadows, 'and whatever else differenceth a civil and well-cultivated Region, from a barren and desolate Wilderness' [161–5]. Unlike Burnet, Ray had no doubt that a civilized country, 'improved to the Height by all Manner of Culture', was to be preferred to 'a rude and unpolished *America*, peopled with slothful and naked *Indians*' [165].

Ray: geography and a static world view

In his concern with the cultivated landscape, as well as with the various features of civilization, Ray showed an interest in many aspects that were incorporated into human geography. Man, he pointed out, has been made 'a sociable Creature', constructing cities and public works through 'mutual Help', and improving his own understanding through 'Conference, and Communication of Observations and Experiments'. Consideration of all such social matters, however, was in Ray's view entirely the concern of *politics*, while natural history or *physiology* comprised for him the kind of study of plants, animals, and rocks that later provided the foundation for biogeography and geology. Geography on the other hand implied for Ray simply capes and bays, the location of places. A natural desire in man to study strange countries, he wrote, leads to the advancement of knowledge in all three sciences:

in *Geography*, by observing the Bays, and Creeks, and Havens, and Promontories, the Out-lets of Rivers, the Situation of the Maritime Towns and Cities, the Longitude and Latitude, etc. of those Places: in *Politicks*, by noting their Government, their Manners, Laws, and Customs, their Diet and Medicines, their Trades and Manufactures, their Houses and Buildings, their Exercises and Sports, etc. In *Physiology*, or Natural History, by searching out their Natural Rarities, the Productions both of Land and Water; what *Species* of Animals, Plants, and Minerals . . . are to be found here, what Commodities for Bartering . . . [163–4].

This pattern for the natural sciences was still strongly established at the time of Humboldt who, as already noted, called his own classic regional studies of Mexico and Cuba *political essays*, rather than geography.

In his own work, Ray did a great deal to widen the scope of natural theology in order to justify the serious study of natural phenomena and at the same time provide a coherent view of the world. However, like other churchmen of his time he was committed to a static world view, and he made a constant effort to contain his researches within that framework. Thus he was reluctant to give up his belief in the fixity of species, even when in private correspondence he acknowledged that the unidentified fossils recently discovered must be remains of once-living plants and animals, so that identification of extinct forms seemed likely 'to overthrow the opinion generally received, and not without good reason, among Divines and Philosophers, that since the first Creation there have been no species of Animals or Vegetables lost, no new ones produced'.[40]

Ray vigorously opposed the hypotheses of other naturalists who questioned this concept of a perfectly designed, stable world in which everything had its planned use. In the *Wisdom of God* he rejected any view of the earth as 'a Heap of Rubbish and Ruins', and added a defence of mountains as both ornamental and useful, incorporating here Halley's theory of ridges as the source of streams, as well as his own close observations of alpine dairying in the Juras [215–20]. Privately, in his letters, Ray dismissed Burnet's theory as 'no more or better than a meer chimaera or Romance', and pronounced it 'fundamentally overthrown' by Beaumont.[41] At the same time he was sceptical of Beaumont's own assertion that no mountain on the earth remains in its original form, and condemned his support for Aristotle's theory of a continuous alternation between land and sea.[42] He was equally critical of the theory put forward by John Woodward (1665–1728), professor of medicine in London's Gresham College, in his *Essay towards a Natural History of the Earth*, 1695. Woodward's argument – that present rock strata, along with their fossil deposits, were formed by sedimentation after the catastrophe of the biblical flood – was dismissed by Ray as merely 'a plausible conjecture'.[43] Another attempt, by William Whiston (1667–1752) in *A New Theory of the Earth*, 1696, to explain the flood and the formation of the earth's crust as the effects of comets approaching the earth, was criticized by Ray as 'pretty odde and extravagant and . . . borrowed of Mr. Newton in great part'.[44] Whiston himself was removed from his Cambridge chair for unorthodox beliefs in 1701, his book however continued to a fifth edition during his lifetime,[45] with a sixth appearing posthumously in 1755.

Derham's *Physico-Theology*: a survey of nature

Following Ray's death, a further effort to advance natural science without offending Christian orthodoxy was made by William Derham (1657–1735), Rector of Upminster in Essex, in his *Physico-Theology* of 1713. The text was based on sermons Derham preached in a London church in 1711–12 when he delivered the Boyle lectures, instituted in Boyle's will of 1691 for the proof of the Christian religion. Derham's more venturesome statements, along with a record of his own attempts at a scientific procedure, were contained in the notes, added later for publication, which he described as 'Anatomical

Notes', many of them on dissections and experiments, with observations 'of my own and others', and including also the inevitable counterpoint of classical scholarship in numerous '*Citations* of the *Ancients*', which he retained since they 'may be acceptable to Young Gentlemen at the Universities, for whose Service these Lectures are greatly intended by me'.[46]

Derham set out to provide a 'Survey of the Works of *Creation,* or (as often called) of *Nature*'. Leaving the *Heavens* for a later work, he began here with 'our *Terraqueous Globe*' [3], dealing first with the atmosphere, light and gravity, next with the globe in general – its shape, size, motion in space, and the distribution of land and water – then with soils, rock strata, landforms and their origins. The majority of the book was concerned with what he called the *Inhabitants* of the earth – chiefly with the various kinds of animals, including man, and more briefly, with plants. Derham's natural history, like that of Ray, included not only pioneer observations on many topics that over the next century were to provide a basis for the special sciences of zoology, botany, geology, and climatology, but also a good deal that was to be incorporated in modern geography. In the course of his survey, Derham in particular covered much of the traditional field of geography, apart from cartography and chorography, and indeed on several points he referred to Varenius [41, 50, 53], as well as to Bohun's *Geographical Dictionary* [52]. At the same time, by extending his discussion to include Newton's theory of gravity, along with current research on the atmosphere, the surface structure of the earth, and the adaptation of organic life to the environment, he offered a model for the kind of physical geography produced in the nineteenth century. In this, Derham can be seen as a forerunner of Humboldt. Dealing with the atmosphere, for instance, he began with the concept of an 'aequipoise or balance of the Atmosphere', referring to statistics from his own records obtained with the barometer, thermometer and rain-gauge, and calling for further regular observations of the weather to encourage a search for natural causes of such phenomena as clouds and precipitation [14–23]. Furthermore, Derham's enthusiasm for the empirical method did not lead him to exclude man from his survey of nature: he devoted a complete book to considering, not only the human body, 'this admirable Machine' [285], but also the human mind and its inventions.

The long section in the *Physico-Theology* on particular classes of animals commenced with a chapter on *The Soul of Man* [264–82]. In Derham's discussion the soul, while still based on the Aristotelian notion, was restricted to the cognitive faculties – given by him as understanding, thought, invention, wisdom and memory – although from the outset he declared, 'I shall not dwell on this, tho' the superior part of Man, because it is the least known' [265]. His intention instead was to deal especially with the extent of man's invention, this implying for him 'a View of all the Arts and Sciences, the Trades, yea the very Tools they perform their Labours and Contrivances with' [267]. He included therefore several pages, chiefly in his notes, on the history of learning; following a pattern well-established by his day, when the comparison of the Ancients and Moderns was a popular theme, he contrasted previous 'Ages of *Learning* and *Ignorance*' with the more recent 'Improvement made in all Arts and Sciences' during 'the last Century, and the few Years of this'. Rejecting a mechanical explanation of such movements, Derham suggested an organic analogy: 'There is (it seems) in Wits and Arts, as in all things beside, a kind of circular Progress' – through birth, growth, flourishing, and fading, to 'Resurrection and Reflourishing'. For the numerous inventions of mankind, he referred the reader to the Scriptures, and to various classical authors, including Pliny, as well as to recent writers on natural theology such as Hakewill and Grew, but his work pointed also to the style of a modern history of science as he reported on the latest discoveries on the magnet, referring to recent issues of the British *Philosophical Transactions* and the '*Memoirs de Physique et de Mathematique*, published by the French *Academie des Sciences*' [275–7].

In his notes, Derham approached an internally consistent explanation of the historical sequence of human inventions: referring to the introduction of printing he declared it to be 'manifest, how great an Influence (as it was natural) this Invention had in the promoting of Learning soon afterwards'. However, in the text he maintained that such improvements, including 'the Progress of Christianity' itself, must be considered as evidence for 'the Superintendence of the great Creator and Ruler of the World; who oftentimes doth manifest himself in some of the most considerable of those Works of Men' [280–1]. This theme, of nature and history as the universal spirit made manifest, was developed throughout the next century by the German idealists, being vigorously promoted by Hegel during Humboldt's lifetime, although it was carefully avoided in his own history of scientific thought in *Kosmos*.

Derham himself preached the 'Divine Nature and Operations' of man's soul or intellect, as 'The Copy of the Divine Image in us' [264–5]. Man as a rational being has been designed to fulfil the 'Creator's Vice-regency' on earth – 'to employ the several Creatures; to make use of the various Materials; to manage the grand Businesses; and to survey the Glories of all the visible Works of God!' [274]. While this view, by delegating to man the powers of vice-regent, seemed to allow more independence of action for man than Ray had done, Derham continued to emphasize that all human attainments are divine gifts:

Men are ready to imagine their Wit, Learning, Genius, Riches, Authority, and such like, to be Works of Nature, things of Course, or owing to their own Diligence, Subtilty, or Secondary Causes; that they are Masters of them, and at Liberty to use them as they please . . . But it is evident, that these things are the Gifts of God, they are so many Talents entrusted with us by the infinite Lord of the World, a Stewardship . . . [282–3].

To this stewardship, said Derham, belong all our various occupations and our inclinations for them: whether for the Church, or 'the more secular Business of the Gentleman, Tradesman, Mechanic, or only Servant', or for contributing to 'History, Mathematicks, Botany, Natural Philosophy, Mechanicks etc'; therefore, he declared, all these should be carried out with diligence [283]. Applied in this way, the design theory supported an attempt to harmonize religion and science, while it offered at the same time a clever combination of the two powerful goals of piety and utility, using Christianity, in the manner of St Paul, to maintain social stability.

Under divine management: the designed earth

Above all Derham, like Ray, opposed the idea that the condition of the world could be due to chance rather than wise design, and argued that evidence of planned utility would be found eventually in all things. Mountains are noble and useful, not disordered, he said, quoting Ray, and he expressed approval of Woodward's suggestion that '*Vulcano's* and Ignivomous Mountains' might be useful in preventing earthquakes [69–74]. With regard to the human body, noting that 'it would be endless to proceed upon Particulars', he proposed instead the general view that the whole structure is orderly and adapted to use, with 'no Botch, no Blunder . . . no signs of Chance' [296]. Man's erect posture, he claimed, is convenient in freeing the hands for activity, but is first of all appropriate, as earlier theologians agreed, 'for a Rational Creature, for him that hath Dominion over the other Creatures' [285]. He went on to the interesting comment that man's size has been determined by 'his Relation to the rest of the Universe', suggesting with Grew that giants would be restricted by the availability of food [295–6]. However, he was emphatic that every aspect of the body, including 'the Harmony between the Parts' [307], must be seen as the work of the great Creator:

Should we so abuse our Reason, yea our very Senses; . . . as to attribute one of the best contrived Pieces of Workmanship to blind Chance, or unguided Matter and Motion, or any other such sottish, wretched, atheistical Stuff . . .? [314–15].

The transition to a secular interpretation of nature, after the earlier efforts of Keckermann and Varenius, was to be a long and difficult one, accompanied by indignant opposition from clerics until even after the publication of Darwin's theory of natural selection in 1859. As John C. Greene pointed out in his excellent study of that process, men like Newton, Boyle and Ray 'attempted to weld into a single philosophy of nature two not entirely compatible conceptions' – the biblical idea of the designed universe, and the mechanical view, 'the idea of nature as a law-bound system of matter and motion'; in the following years, however, the gradual acceptance of change in nature, along with 'a vastly extended time perspective and a sense of the relativity of human conceptions', inevitably challenged the old doctrines of creation and revelation, and the adaptations of living organisms were recognized as a product of dire necessity rather than wise design.[47]

An attractive feature of the natural theology of Ray and Derham is their affirmation of the goodness and beauty of the world, its unity and purpose. All these early works of natural history gained a certain strength and coherence from their attempt to relate particular observations to a concept of order, harmony and uniformity in nature. Clarence Glacken indeed saw the design argument, with its recognition of wider interrelationships in nature, as the origin of modern ecological theory; while acknowledging that 'the idea of a unity in nature is very old', he considered the contribution of Ray and Derham in their ideas of the world to be extremely significant: 'In this grand design of nature . . . God, living nature and the earth, and human knowledge were indissolubly joined.'[48]

At the same time it must be remembered that the world view of natural theology was essentially a static one, whereas modern ecology is associated closely with the more dynamic view of a functioning world-system, a concept suggested by Carpenter and touched on by later writers, but emerging clearly in the geographical works of Humboldt. In addition, some of the ideas perpetuated in the natural histories can be seen as detrimental to the development of ecological theory with respect to man's place in the system. The ancient belief that the world was made for man – although moderated to some extent by writers such as Ray and Burnet – contributed to the view of man as master of nature that in a secular form remained dominant through the nineteenth century and was condemned as destructive by Marsh and the early conservationists. That view, combined with the idea of an infinitely bountiful nature in which a perfect balance of all species is maintained by providence, evidently encouraged a high degree of irresponsibility towards the effects of human population growth and use of resources. Derham in his *Physico-Theology*, for example, questioned the assumption that *all things were made for Man*, and even dismissed this as a 'narrow Opinion', disproved by new discoveries of the infinitely greater extent of the universe, although previously favoured by 'most of the Antients . . . as *Aristotle*, *Seneca*, *Cicero* and *Pliny*' [55]. At the same time, in discussing the great variety of things with which the world is stocked, he repeated without hesitation the biblical doctrine of the inexhaustible fullness of creation, a doctrine which Lovejoy called the 'principal of plenitude' and traced to Plato's *Timaeus*.[49] To Derham the idea of a world food shortage was inconceivable:

the Munificence of the Creator is such, that there is abundantly enough to supply the Wants, the Conveniences, yea almost the Extravagances of all the Creatures, in all Places, all Ages, and upon all Occasions [55].

From his own familiarity with statistics, Derham was aware of evidence for a significant rate of increase in population. He even included in his notes a table, collected mainly from Graunt and King, but also from 'my Register of Upminster', showing a higher ratio of births to burials for a number of English and European centres, with an estimate for England as a whole of 1.12 to 1; however, he considered this surplus, like that of males over females, to be simply 'an admirable Provision' to compensate for losses from diseases or from 'War and the Seas', and also to populate colonies [175–7]. Throughout the earth, he declared, 'divine management' assures constant order in both the distribution of the various creatures and in their numbers [167]. That is achieved mainly, he suggested, by regulating the length of life: 'by this means the peopled World is kept at a convenient stay, neither too full, nor too empty' [170–3]. For Derham 'this Harmony in the Generations of Men' and the balance in 'every species of Animals' is proof of design, for how otherwise, on the theory of increases by doubling, 'the World should not be over stocked can never be made out' [178–9]. It is the task of the infinite Creator to '*balance* the several Species of Animals, and conserve the *Numbers* of the Individuals of every Species so even, as not to over or under-people the Terraqueous Globe!' [261].

This faith in divine management still lay behind the Malthusian theories on population increase at the end of the eighteenth century, although, like the notion of design itself, it had by then been considerably undermined as the static world picture of natural theology gave way to the more dynamic outlook of evolutionary thought. The transition is evident not only in the idea of organic evolution, but also in the theory of cosmic evolution proposed in Kant's *Natural History and Theory of the Heavens*, 1755, and the notion of gradual evolution in landforms, expressed definitively in Hutton's *Theory of the Earth*, 1795. Moreover, the idea of social evolution emerged also, as enlightenment views in the same period came to emphasize social and moral progress as both a goal and a responsibility of man himself.

During the eighteenth century as natural history became unwieldy in its scope it was gradually superseded by more specialized studies, and each of the new natural sciences participated in the movement to find internally consistent explanations for the particular phenomena and active processes being examined. No science, however, emerged to take over the more general function of natural history and provide a new concept to replace the design argument, by considering the pattern of interrelationships for the entire world, including man, in terms other than those of divine providence. Geography lagged behind in that respect until the time of Humboldt. The difficulties were considerable: social thought in the eighteenth century tended to diverge strongly from the mechanistic determinism of Newtonian science; meanwhile, the rise of scientific materialism and positivism reinforced an approach to science that did little to encourage the development of geographical research and theory.

CHAPTER 4

EIGHTEENTH-CENTURY EMPIRICISM: LOCKE, BERKELEY AND HUME

Empiricism developed in the seventeenth century largely as a British movement, receiving a strong impetus from Francis Bacon and culminating in the successes of Boyle and Newton. Incorporated in Newton's experimental philosophy, of course, were important elements of Cartesian thought – a mechanistic view of the world, an emphasis on mathematics in the search for universal laws, a dualism of mind and matter. The eighteenth century saw the vigorous extension of the new form of scientific empiricism, both in terms of geographical extent, into Europe and America, although with a focus in France, and also intellectually, into areas of inquiry excluded from Newton's physical science: the study of living organisms, and especially of man himself, the operations of his mind, and problems of social relations.

For a long time, geography remained almost entirely outside the new scientific movement. Although a growing interest was shown in civil geography and demography, the expansion of social science was inhibited at this time, while even those geographical texts promising complete *New Systems* did little more than reorganize traditional material. During the second half of the century, however, a marked increase of activity in the field, and the interest shown in the theory of the subject by Kant as a leading philosopher, indicated a significant change, and by the end of the century serious efforts were made to include geography unequivocally in the scientific process.

THE AGE OF ENLIGHTENMENT

To the historian of ideas the eighteenth century appears a complex and interesting one, especially rich in the fields of social literature and the life sciences. Any detailed record of the scientific or literary achievements of this time must be filled with an impressive number of names and publications that belong to what has been widely called the age of enlightenment. In his competent study of this period L. G. Crocker argued that in the movement known as the enlightenment, which he himself dated from 1680 to 1790, major social and political changes were involved; in his view, although the enlightenment reached a focus between 1746 and 1770 with Voltaire and the French philosophers of the 1740s, it can be seen taking shape during previous generations in the work of many writers who 'sought to enlighten men . . . to free minds from prejudices and unexamined authority . . . to explore the ills of society and to devise remedies'. The enlightenment in his opinion must be seen as part of 'the total matrix of the eighteenth century . . . class interests, political and ecclesiastical authority, religious convictions, the progress of science, the challenge to values, changing views of history, new economic theories, competing views of man, political dissatisfaction', these together generating tensions on crucial issues: 'the role of nature and culture, the claims of individuals and society, freedom and order, traditionalism and revolt'.[1]

In the kinds of issues involved here, and in the intensity of response to them, a remarkable parallel can be drawn with the late twentieth century, and with regard to geography it is interesting that both periods have been associated with an eventual revitalizing of the study. Whether conditions of social and intellectual ferment actually favour geography, however, is a question that cannot be pursued here; within the limits of the present inquiry the intention is to trace, through the complexity of the enlightenment period, the continuing formulation of empiricism in philosophy and science, and its encounter with geography, these being themes that have scarcely been considered previously in conjunction with each other. The first task, then, is to follow the development of Baconian views in the philosophy of later British empiricists, for their work provided the main basis for the scientific empiricism of the eighteenth century, which was marked by the rise of positivism in the context of sensationist and materialist views. Those ideas not only helped form the climate of thought in which Humboldt at the end of the century commenced his work, but, in the kind of science they encouraged, also contributed to many of the problems his own age was trying to resolve. Central among such problems was the evident inability of the various newer sciences to produce any coherent view of the world to replace the mechanist–design theories, and the inadequacy of traditional geography in the face of that situation.

The second task therefore is a consideration of eighteenth-century geographical thought from some of the surviving publications, English and German sources being used here. On the majority of these works surprisingly little information is available, and apparently not even a detailed bibliography has been prepared comparable with that produced by Taylor for English geography to 1650. The present discussion does not in any way pretend to completeness, but is intended as a contribution to a neglected field. Both Alan Downes, in his short paper on Georgian geography, and Harry Robinson, in an earlier brief article on the Dissenting Academies, emphasized the lack of research and almost complete absence of data on the state of geography in this period.[2] Throughout most of the eighteenth century, while other studies such as chemistry, biology or geology underwent revolutionary changes, geography remained almost entirely in the hands of textbook writers who produced compilations of regional descriptions and showed little concern with the theory of the subject or its relation to significant issues in the scientific thought of the time. When examined, then, the general record as far as geography is concerned proves scarcely inspiring, and it is understandable that in recent histories of science it has become customary for almost all reference to geography to be omitted without comment when scientific achievements in the modern period are reviewed. Evidently, no historian in the twentieth century has been prepared to take the approach suggested by Humboldt and consider this loss of intellectual vigour in geography as an indication of a serious deficiency in the scientific tradition as a whole. The decline of geography in the century after Varenius coincided with the rise of Newtonian science and it can be seen in part as an outcome of the scientific methodology of the day. It was only in the latter part of the eighteenth century, as increasing challenges came to the dominance of the Newtonian form of scientific empiricism, that geography again became the focus of active intellectual concern.

The third task, then, in the present investigation, is to examine the positivist model of science that emerged in the course of the eighteenth century, and to consider also some of the major alternatives to that model and the whole mechanist–materialistic view which were being advanced in the same period, especially those arising from the beliefs of German idealism and French romanticism. Increasingly, the mechanist outlook came into conflict with organic theories of life, and with enlightenment views of society and history, while the positivist trend in scientific empiricism was challenged by new theories of knowledge, culminating in the objective idealism of Kant. Coincident with this were signs of a revival in geography, led chiefly by German thinkers, and engaging the commitment of Humboldt in the last years of the century. For Humboldt the issue to be faced was more than the reorganization of an individual discipline – whether geography or the wider study of cosmography: it was a question of maintaining coherence in the entire Western intellectual tradition. The division between natural and moral philosophy, instituted by Galileo and his followers as an effective means of securing relative freedom of inquiry from Church control for the natural sciences, was now seen from Humboldt's standpoint as an impediment to the advancement of knowledge, while the continued fragmentation of the natural sciences themselves was an additional cause for concern. In that situation the revival of geography took on renewed significance; however, the kind of reform that Humboldt envisaged was to encounter a serious obstacle in a firmly entrenched notion of scientific procedure, based on the empiricist doctrine of sense-experience. The formulation of that theory during the eighteenth century received considerable support from the philosophy of the British empiricist John Locke and his followers, George Berkeley and David Hume. A consideration of their work and concurrent developments in scientific thought is the subject of the present chapter.

EMPIRICAL PHILOSOPHY: LOCKE AND BERKELEY

Locke: empirical psychology

One of the most influential thinkers in this period was the English physician John Locke (1632–1704), whose liberal political views and rather narrow empirical theories were adopted with equal enthusiasm on two continents. His philosophy, in the opinion of Isaiah Berlin, 'exercised undisputed sway over the ideas of the entire eighteenth century'; in particular, Locke's *Essay concerning Human Understanding*, published in 1690 and translated by Pierre Coste (1668–1747) into French in 1700, brought him wide recognition, giving him, in Crocker's view, a place along with Newton as one of the two idols of the enlightenment.[3]

Locke himself was closely involved with the scientific and political activities of his day. A student and fellow in Oxford until 1683 when he became a political exile for his Whig associations, he had been made a Fellow of the Royal Society in 1668 and his friends included Boyle, Hooke and Newton. His political writings, the *Letters on Toleration* and *Two Treatises of Government*, published after his return with William of Orange in 1688, were intended to justify the English revolution against absolute monarchy; in the following century they were to be used to justify revolutions in France and America. Locke's arguments for freedom of opinion, individual rights, and the limitation of governmental authority, were based on the idea of the social contract – the notion that society is formed by individuals who agree to surrender their liberty in the state of nature for their own mutual benefit. Thomas Hobbes (1588–1679) in his *Leviathan*, 1651, had argued that the social contract granting sovereignty to the state is a necessity, for otherwise anarchy and self-interest would lead to a condition of war, making man's life in the natural state merely 'solitary, poor, nasty, brutish and short'.[4]

Locke's view of the natural state was more optimistic and in his *Essays on the law of Nature*, written in Latin by 1664 although not published in his lifetime, he attempted to find some absolute basis for morality by postulating 'a universal law of nature binding on all men'.[5] This law, which he believed to be self-evident, was to provide 'a rule of morals', a pattern for man's life imparted by God: man alone cannot be exempt from the laws that govern all other operations in the world. He argued that the providence of a divine being must be taken for granted, 'for it is by His order that the heaven revolves in unbroken rotation, the earth stands fast and the stars shine' [109]. By the end of the century Locke had given up his defence of the geocentric theory; similarly, he seems to have given up the attempt to conceptualize nature as the source of absolute moral principles. However, the method of inquiry proposed in these early essays for the discovery of natural law was retained in large part in his later works, and it clearly represented a development of Baconian empiricism. Opposing the Platonic or Cartesian theory of innate ideas, he declared that knowledge of such a law is not implanted in man, but is discoverable by the 'light of nature', that is, by the proper use of man's own faculties of 'understanding, reason, and sense-perception'. In particular, following Bacon closely, he stressed that sense-perception is 'the basis of our knowledge of the law of nature'. To begin with, such knowledge 'is derived from those things which we perceive through the senses'. After this, in his view, 'reason makes use of these elements of knowledge, to amplify and refine them, but it does not in the least establish them' [123–5]. The reason was assigned no role in perception.

In considering 'the way in which primary notions and the elements of knowledge enter the mind', Locke was concerned chiefly with direct sense-experience, for while admitting also the importance of instruction from others, like Bacon he referred to this as 'tradition', and showed some suspicion of its operations. He denied in particular that conflicting local traditions could give rise to knowledge of a universal law: most persons 'have no other rule of what is right and good than the customs of their society and the common opinion of the people among whom they live'; therefore, on the basis of tradition,

it would be difficult to decide completely what is true and what is false, what is law and what is opinion, what is commanded by nature and what by utility, what advice reason gives and what instructions are given by society [125–33].

The growing recognition of society as the source of morals, along with the assertion that societies themselves are created by individuals and can be changed by them, was to have far-reaching effects during the eighteenth-century. The idea that man can achieve social progress through political change, and through conscious programmes of education, was to gain strength with the decline of belief in divine authority as the source of all personal enlightenment and of all organization in the world. A tendency to replace divine determinism with social or economic determinism was something, however, that Locke himself evidently had no wish to encourage. He preferred to minimize the social aspects of knowledge and emphasize that the surest way to truth is for the individual to pay proper attention to the evidence of his own senses.

In his *Essay concerning Human Understanding*, for which he commenced drafts as early as 1671 and continued to make additions until 1700, Locke set out to inquire into the origin, certainty, and extent of human knowledge. Interested primarily in the way in which the mind is used, he refused to become involved with speculations regarding the physical nature or essence of the mind, or the dependence of ideas on matter, claiming such questions to be ruled out by his own 'historical plain method'.[6] His intention clearly was to apply the scientific method of his day to a study of mental operations, producing in effect what Bacon had called a natural history of the mind. Bacon himself had not explored at all thoroughly the relationship of ideas to experience, and it was Locke who provided an extension of Baconian empiricism by developing what was virtually an empirical psychology.

Locke began by rejecting again the doctrine of innate knowledge: denying that there are 'certain innate principles, some primary notions, κοιναι εvvoιαι; characters, as it were, stamped upon the mind of man', he set out to show 'how men, barely by the use of their natural faculties, may attain to all the knowledge they have, . . . and may arrive at certainty, without any such original notions' [I, 2, 1]. He followed Aristotle and Bacon in favouring the *tabula rasa* theory of the mind:

Let us then suppose the mind to be, as we say, white paper, void of all characters, without any ideas; how comes it to be furnished? . . . To this I answer, in one word, from experience.

Experience, for Locke, depends primarily on the senses, although he recognized another source of knowledge in the functioning of the mind itself:

Our observation, employed either about external sensible objects, or about the internal operations of our minds, perceived and reflected on by ourselves, is that which supplies the understanding with all the materials of thinking.

These two aspects of experience he called *sensation* and *reflection*:

External material things, as the objects of sensation; and the operations of our own minds within, as the objects of reflection, are to me the only originals from whence all our ideas take their beginnings [II, 1, 1–4].

The strong tendency to objectivism already noted in Bacon's thought is evident again in Locke's epistemology: objects, impinging directly on the senses of the individual, give rise automatically to perception. This view in effect emphasized the subjectivity of knowledge: like Bacon before him Locke considered the basis of scientific activity to lie in the individual's own immediate contact with the particulars of nature, and he was reluctant to admit that ideas gained from others form an essential element in experience. Furthermore, accepting in addition the distinction made by Descartes between mind and matter, Locke proceeded to make an even more simplistic division between an external, material world, and an internal one, the mind of the subject. External objects furnish the mind with ideas, in his view, and in that process the mind is merely passive.

Sensation, according to Locke, produces in the mind various simple ideas, an 'idea' being 'whatsoever is the object of the understanding when a man thinks' [I, 1, 8]. Regarding the idea, then, as the element of thought, he went on to apply the atomic theory of physical science to the mind itself:

Though the qualities that affect our senses are, in the things themselves, so united and blended that there is no separation, no distance between them; yet it is plain the ideas they produce in the mind enter by the senses simple and unmixed.

These simple ideas cannot be divided further: 'The mind can neither make nor destroy them.' Man the microcosm reflects in his own mind the structure of the external world:

The dominion of man, in his little world of his own understanding, being much-what the same as it is in the great world of visible things; wherein his power, however managed by art and skill, reaches no farther than to compound and divide the materials that are made to his hand; but can do nothing towards the making the least particle of new matter, or destroying one atom of what is already in being [II, 2, 1–2].

Though Locke considered the mind to be passive in receiving simple ideas, he allowed it an active role in combining these to form complex ideas, such as those indicated 'by the names obligation, drunkenness, a lie, etc'. He drew attention to the importance of language in ascribing constant names to complex ideas, pointing out that these are closely related to the customs of the societies concerned, so that languages themselves change as new terms are annexed to new combinations of ideas. Locke was thus an early proponent of linguistic analysis, recognizing the function of words in enabling ideas to be 'represented to the mind'. Indeed, he stated, 'now that languages are made, and abound with words standing for such combinations, an usual way of getting these complex ideas is by the explication of those terms that stand for them' [II, 22, 1–3].

Like Bacon, however, Locke was evidently not confident in dealing with complex ideas and the whole problem of vicarious experience. He noted that such ideas 'are called notions, as if they had their original and constant existence more in the thoughts of men than in the reality of things'; comparatively few complex ideas, in his view, are obtained 'by experience and observation of things themselves'. In the case of inventions, he pointed out, the ideas are framed first, and here, conscious of his own ideas for political reform, he included ideas that lead to social change:

for it is evident, that in the beginning of languages, and societies of men, several of those complex ideas, which were consequent to the constitutions established amongst them, must needs have been in the minds of men, before they existed anywhere else.

Most complex ideas, he acknowledged, are communicated by others through language to 'our imaginations', prior to direct experience, 'by explaining the names of actions we never saw, or notions we cannot see'. At this point, Locke seems to have found difficulty in relating these notions to his own dualism of material world and subjective mind. His only response was to assert again that all 'complex modes' can be resolved into simple ideas, received initially from sensation or reflection, although his own attempt to analyse as an example 'that complex idea we call a lie' was a rather dismal failure, and he ended by leaving the reader to do it for himself [II, 22, 2–9].

Locke's difficulty with complex ideas has an important bearing on his attitude to scientific knowledge. In line with his time, he accepted the view that science must lead to certain knowledge, and in indicating how this was to be achieved he attempted to combine the prescriptions of both Bacon and Descartes, retaining the Cartesian preference for proof by mathematical demonstration, while arguing as an empiricist that certainty or truth can be attained also through experience, by the reception of simple ideas that correspond directly to elements of external reality. On this account complex ideas presented something of an obstacle to his theory.

In Book IV of the *Essay concerning Human Understanding*, Locke considered the place of complex ideas in reasoning, noting again 'the imperfection and uncertainty of our ideas of that kind': words, the signs for these, being clearer and more distinct, our tendency is to use words in reasoning without considering that the 'real essence' for which they stand may be merely some confused or obscure notion [IV, 5, 4]. Although not well defined, Locke's notion of essence, like that of substance, seems to represent a combination of medieval views on essential natures and substrates, along with seventeenth-century theories of atoms and matter. Adopting the materialist view, Locke considered all study of the external world to involve forming ideas of substances, and here he encountered a serious problem, since in terms of his own theory of knowledge, such ideas must themselves be simple ones if certainty were to be achieved, yet to obtain simple ideas would require a knowledge of the essences of material objects, a task he accepted as being itself beyond the capacity of the human mind.

Where complex ideas refer to substances, then, Locke saw a major difficulty standing in the way of 'true and clear knowledge', since these are not amenable to mathematical demonstration: 'Experience here must teach me what reason cannot', yet the limitations of man's faculties make it impossible 'to penetrate into the internal fabric and real essences of bodies'. While agreeing therefore that 'experiments and historical observations' may lead to useful discoveries and improvements in human life, he was sceptical about a science of nature:

This way of getting and improving our knowledge in substances only by experience and history . . . makes me suspect that natural philosophy is not capable of being made a science [IV, 12, 6–10].

At the same time, Locke stressed that he did not want to 'be thought to disesteem or dissuade the study of nature', and he even agreed with the current writers of natural theology that 'the contemplation of his works gives us occasion to . . . glorify their Author'. But he warned against reliance on 'doubtful systems', 'unintelligible notions', or 'precarious principles'. In studying the 'phenomena of nature', man should limit himself to what is given by the senses, examining particulars and applying specific experiments:

In the knowledge of bodies, we must be content to glean what we can from particular experiments; since we cannot from a discovery of their real essences, grasp at a time whole sheaves, and in bundles comprehend the nature and properties of whole species together [IV, 12, 12–13].

Progress in the study of the external world, then, according to Locke, depends primarily on attention to the evidence of the senses, which can provide direct contact with particulars, and his view that the mind is passive in perception was to remain for more than two centuries a basic tenet of scientific methodology; as Bronowski and Mazlish pointed out, 'It lasted, although attacked by Kant and Hegel on philosophical grounds, until the emergence of Heisenberg's uncertainty principle and Einstein's relativity physics led to a new scientific view: the view that the observer plays an essential part in the discovery of nature.'[7] Locke's theory of knowledge itself, as indicated earlier, can be traced largely to the work of Francis Bacon, whose phrases, although rarely acknowledged by name, recur frequently in Locke's writings. Many writers, including even Isaiah Berlin (1956), have failed to draw attention to this; however, one of the most capable editors of Locke's works, James L. Axtell, noted the 'belligerently Baconian notions of the scientific process' in Locke's writing after 1687, when the publication of the *Principia* led him to appreciate Newton's mathematical interpretation of the mechanistic concept of the world; indeed, although the *Principia* itself in later centuries became celebrated for its mathemetical–deductive method, at this time 'even Newton viewed his own work as an example of the famous inductive method, so powerful was the contemporary myth of Bacon and the experimental–inductive method'.[8]

The problem of induction

The problem of induction, according to Locke's theory, is not insuperable. Following Bacon, he declared in the *Essay* that his intention was to go beyond the deductive reasoning of the scholastics with their dependence on the Aristotelian syllogism, in order to help find 'new and undiscovered ways to the advancement of knowledge', free from mere imitation or obedience 'to the rules and dictates of others'. The mind, he claimed, 'has a native faculty to perceive the coherence or incoherence of its ideas', and so is capable of forming a 'chain of ideas' [IV, 17, 4]. In Locke's view, such chains of ideas can lead to knowledge, where these are composed of simple ideas derived initially from sense-experience and representing some element that has an existence in nature.[9] Induction, then, depends not only on the efficiency of this native faculty of the mind, which with some inconsistency he introduced into his empirical theory, but also on the possibility of identifying ideas with elements in the external world. In spite of his admitted inability to deal with complex notions Locke continued to argue, as Bacon had done, that basically *we reason about particulars*, and he extended this to include particular ideas:

the immediate object of all our reasoning and knowledge is nothing but particulars. Every man's reasoning and knowledge is only about the ideas existing in his own mind, which are truly, every one of them, particular existences . . . So that the perception of the agreement or disagreement of our particular ideas is the whole and utmost of all our knowledge [IV, 17, 8].

Locke's aim, in applying the atomist doctrine to ideas, was to provide a theory of knowledge compatible both with commonsense and with scientific inquiry. Yet by restricting knowledge in this way to the ideas in the mind of the individual, which itself was considered to be set apart from the external world that such ideas are supposed to represent, Locke produced a representationalist theory of knowledge involving a dualism that ultimately tended to undermine his whole empiricist position. Like Bacon he started by affirming that all knowledge is derived initially from the senses, and he linked this with a materialist view, attempting even to incorporate the theory of primary and secondary qualities favoured by Galileo and Newton, so that he gave precedence to those sensations produced by the bulk, texture or motion of material objects. He ended, however, with a position that has been called subjective idealism, since it asserted that all general propositions refer only to ideas or their relations in the mind of the knowing subject, and can therefore have no proven reference to an outside world.

Berkeley: subjective idealism

That position was carried to an extreme by George Berkeley (1685–1753), an Irish bishop, who in his own treatise on *Human Knowledge*, 1710, denied Locke's theory of a material substance that exists independently of experience. According to Berkeley's thoroughgoing empiricism, experience is the only reality. Knowledge is only of ideas; ideas can exist only in a mind: therefore, he concluded, the 'supposed *originals* or external things, of which our ideas are the pictures or representations', are ideas also, and have their existence only in being perceived.[10] Their essence or being is only in perception: 'Their *esse* is *percipi*, nor is it possible they should have any existence out of the minds or thinking things which perceive them.' At this point, Berkeley began to diverge sharply from the commonsense view of reality:

It is indeed an opinion strangely prevailing amongst men, that houses, mountains, rivers, and in a word all sensible objects, have an existence, natural or real, distinct from their being perceived by the understanding.

But, he asked, 'what are the forementioned objects but the things we perceive by sense? and what do we perceive besides our own ideas or sensations? and is it not plainly repugnant that any one of these, or any combination of them, should exist unperceived?' [I, 3–4].

It is evident that Berkeley's opinion would be repugnant to the ordinary person, and particularly to the scientist concerned with studying some aspect of the external world; scientists invariably find themselves obliged to adopt some form of realism, in that they assume the independent reality of the world. Berkeley himself allowed for some objective reality by suggesting that things not perceived by a human mind may continue to exist in the mind of God: he considered it to be self-evident, that

all those bodies which compose the mighty frame of the world, have not any subsistence without a mind; that their *being* is to be perceived or known; that consequently so long as they are not actually perceived by me, or do not exist in my mind or that of any other created spirit, they must either have no existence at all, or else subsist in the mind of some Eternal Spirit [I, 6].

The introduction of a divine mind to support his theory is one of the many inconsistencies that have been criticized in Berkeley's thought, but his work remains of considerable importance. He is not only highly regarded by philosophers for his contribution to logic, but in his rejection of Newton's postulates of absolute space, time, and motion has been seen by Karl Popper as a forerunner of Mach and Einstein with the theory of relativity.[11] Berkeley's assertion that all meaningful statements concerning external objects are reducible to statements about ideas derived from sensation persisted in modern empiricist theories of perception, including the positivist. His empiricism is acknowledged furthermore as a source for what is now called phenomenalism, in suggesting the view that so-called physical objects are mental constructs, combinations of sensible qualities such as hardness, colour or sound.[12]

Later in his *Treatise* Berkeley tried to answer the objections of those who might feel that according to his principles 'all that is real and substantial in nature is banished out of the world, and instead thereof a chimerical scheme of *ideas* takes place'. It is only 'the familiar use of language', he wrote, which makes it seem ridiculous 'to say we eat and drink ideas, and are clothed with ideas', his proposition being simply that 'we are fed and clothed with those things which we perceive immediately by our senses': such objects of sense, he agreed, may be called *things*, according to custom, instead of ideas [I, 34–8].

Berkeley's declared aim was to eliminate the difficulties put in the way of useful knowledge and plain common sense by previous philosophers with their doctrine of abstract general ideas, and the sub-title of his *Treatise* promised an inquiry into 'the chief Causes of Error and Difficulty in the *Sciences* with the grounds of *Scepticism, Atheism,* and *Irreligion*'. However, his attempt to provide a basis for natural philosophy that would counteract both atheism and scepticism was far from successful, nor was his theory attractive to practising scientists. From the time of Berkeley it is necessary to draw a distinction between empiricism in its development as a philosophic doctrine, and the kind of scientific empiricism that came to be adopted in the natural sciences. Throughout the eighteenth century there was a return, on the part of the growing number of men engaged in scientific research, to those doctrines of Locke and his predecessors that appeared to provide a more satisfactory support for their studies. The problems of subjective idealism seem to have become an issue mainly for philosophers and not for the majority of those involved in studying aspects of the external world. As contemporary philosophy raised doubts about man's knowledge of the outside world, scientists tended to reject questions of epistemology altogether, to deny all forms of idealism and retreat to the kind of realism supported in Baconian theory.

HUME: PHILOSOPHY AND SCIENCE

Leading philosophers, meanwhile, continued to grapple with the problems raised by the theories of Descartes, Bacon, and their followers. David Hume (1711–76), the Scottish philosopher, was aware that a logical extension of the empiricist position must lead to a denial of all certainty in man's knowledge of external events; however, his *Treatise of Human Nature*, in which he first published his theory in 1739, met with a disappointing response, and he seems to have achieved greater fame during his lifetime for his *History of England*, 1761. In the introduction to his *Treatise*, Hume acknowledged the growing denigration of philosophy by both scientists and scholars:

Principles taken upon trust, consequences lamely deduced from them, want of coherence in the parts, and of evidence in the whole, these are everywhere to be met with in the systems of the most eminent philosophers, and seem to have drawn disgrace upon philosophy itself.

At the same time he pointed to 'the present imperfect condition of the sciences' where disputes proliferated without any discipline; from this, in his opinion, 'arises that common prejudice against metaphysical reasonings of all kinds, even amongst those, who profess themselves scholars'.[13]

Hume's solution was to make philosophy into a science – 'the science of man' – its aim being to explain 'the extent and force of human understanding'. This was an aim already expressed by Locke, but Hume went further in declaring that moral philosophy should be based, like any other science, on experience and the comparison of experiments, even if these must take the form of 'a cautious observation of human life . . . in the common course of the world'. Claiming that it was not surprising that 'the application of experimental philosophy to moral subjects' should come more than a century after that to natural, he compared the distance between 'my Lord Bacon' and such recent English philosophers as Locke, with the interval between Thales and Socrates in 'the origins of these sciences' [xix–xxiii]. In their new form, he declared, his philosophical researches would be indispensable to the sciences, all of which are dependent on human knowledge:

all the sciences have a relation, greater or less, to human nature . . . Even *Mathematics*, *Natural Philosophy*, and *Natural Religion*, are in some measure dependent on the science of Man; since they lie under the cognisance of men, and are judged of by their powers and faculties.

The 'connexion with human nature', he continued, is even 'more close and intimate' in the other sciences of *Logic*, *Morals*, *Criticism*, and *Politics*; therefore, in proposing 'to explain the principles of human nature, we in effect propose a complete system of the sciences, built on a foundation almost entirely new'. For Hume, since 'the science of man is the only solid foundation for the other sciences', it must itself be based on experience, so that philosophers should point out the limitations of human reason, and admitting their ignorance of the essence of the mind, avoid 'imposing their conjectures and hypotheses on the world for the most certain principles'. However, he insisted, philosophy is not under any special disability here:

if this impossibility of explaining ultimate principles should be esteemed a defect in the science of man . . . 'tis a defect common to it with all the sciences, and all the arts, . . . whether they be such as are cultivated in the schools of the philosophers, or practised in the shops of the meanest artizans. None of them can go beyond experience, or establish any principles which are not founded on that authority [xix–xxii].

Hume, then, denied that the sciences can claim access to any special way to truth: all human knowledge depends on experience, and is subject to its limitations.

In considering experience, Hume, like the earlier empiricists, was preoccupied with tracing ideas to their origin, and like them, he located this in the sense-impressions of the individual. In the *Treatise*, where he followed Locke to a large extent, he made a departure in attempting to distinguish ideas from the perceptions themselves which he called impressions. Accepting Locke's division of simple and complex ideas, he extended this also to impressions, asserting furthermore that 'ideas and impressions appear always to correspond with each other'. In the case of complex ideas, he admitted, 'the rule is not universally true', since it seems that 'many of our complex ideas never had impressions, that corresponded to them, and that many of our complex impressions never are exactly copied in ideas'. He offered the example of a city: 'I have seen *Paris*, but shall I affirm I can form such an idea of that city, as will perfectly represent all its streets and houses in their real and just proportions?' [2–3].

Hume, however, failed to pursue this matter, and his famous scepticism did not arise from this problem; indeed, he stated that as 'all simple ideas and impressions resemble each other; and as the complex are formed from them', the relation between ideas and impressions required no further examination. Using the genetic approach, he declared that the resemblance between the two is not to be attributed to the dependence of impressions on ideas, since according to 'the order of their *first appearance*', the simple impressions always take precedence [4–5]. Even more than Bacon, he was convinced that 'all impressions are clear and precise' [72]. Hume was emphatic that 'we have no idea, that is not derived from an impression' and that 'reason alone can never give rise to any original idea' [155–7]. He agreed with Berkeley that there can be no abstract general ideas which are unrelated to particular impressions and he extended this to include even the symbols of mathematics [17–23].

It was the problem of generalizing from experience that chiefly occupied Hume and gave rise to his scepticism. Logically, he pointed out, 'there can be no *demonstrative* arguments to prove, *that those instances, of which we have had no experience, resemble those, of which we have had experience*'. In his view, 'the only connexion or relation of objects, which can lead us beyond the immediate impressions of our memory and senses, is that of cause and effect'; yet, he argued, what we call cause and effect is simply an inference from the *constant conjunction* of events observed in the past, even the principle 'that the course of nature continues always uniformly the same' being itself open to question. Strictly too, he showed, probability must be rejected as grounds for demonstration, since already 'probability is founded on the presumption of a resemblance betwixt those objects, of which we have had experience, and those, of which we have had none' [89–90].

The questions raised here by Hume were recognized by subsequent philosophers as profoundly significant. As Isaiah Berlin observed, although Locke was vaguely aware of the problem of induction, 'its importance in modern philosophy dates from Hume', and indeed 'Kant's stupendous effort to deal with its consequences inaugurated modern philosophy.' Berlin went on to criticize the pessimism of Hume and later philosophers, arguing that their scepticism arose from an attempt to turn induction into deduction: 'Hume himself is largely to blame: in common with his contemporaries, he regarded deduction as the only authentic form of true reasoning; and therefore attributed our inferences from cause to effect to the "imagination", the source of irrational processes.'[14] In endorsing these comments on Hume's analysis of causation and the inductive method, two points should be noted: firstly that his acceptance of the faculty theory of the mind requires further examination, and secondly that Hume himself did not use the term induction in his discussion.[15]

The role of imagination

With regard to the imagination, Hume seems to have encountered the same difficulties already noted in the theories of Aristotle and Bacon, who assigned the imagination to the sensitive rather than the rational faculty, viewing it as both the source of fanciful ideas and as the means by which sense-perceptions are formed into images in the mind. Hume, however, went further, and made the imagination also the source of our notion of the continued existence of objects, arguing as Bacon had done earlier that such an attribute cannot come from the senses, since they provide only individual impressions, while on the other hand, following Berkeley, he noted that the understanding shows us by 'philosophical principles' that a belief in the existence of body apart from perceptions is unreasonable, even though it may be shared by 'children, peasants, and the greater part of mankind'. It is the imagination, acting on 'the coherence and constancy of certain impressions', that ascribes independent existence to external objects [193–5].

In assigning this important role to imagination, Hume admitted the difficulty of constructing an adequate theory of the external world. Like Berkeley, he rejected the materialist doctrine of substance, and was aware that a consistent phenomenalism cannot accept a theory of 'the double existence of perceptions and objects' [211]. He had already argued that the ideas of space and time are derived from our impressions, time being identified with 'the succession of our perceptions' [34–5]. Now he extended this approach to question even the notion of personal identity, the idea of the self, declaring that 'what we call a *mind*, is nothing but a heap or collection of different perceptions, united together by certain relations' [207]. At the same time he acknowledged that although it cannot be logically proved, the existence of an external world must be taken for granted in order to avoid a '*total* scepticism'. Faced with this impasse, Hume responded with a kind of existentialist view: reason, he suggested, and 'all the rules of logic require a continual diminution, and at last a total extinction of belief and evidence', yet in the course of daily living man cannot tolerate this rational interpretation. Hume therefore proposed the hypothesis '*that all our reasonings concerning causes and effects are deriv'd from nothing but custom; and that belief is more properly an act of the sensitive, than of the cogitative part of our natures*'; where conflicting reflections or sensations might 'reduce the mind to a state of total uncertainty', the imagination enables us to '*retain a degree of belief, which is sufficient for our purpose, either in philosophy or common life*' [183–5]. In this way, 'nature breaks the force of all sceptical arguments in time' [187].

The imagination, then, apparently represented for Hume a kind of natural impulse, overriding human rationality in order to sustain action. The association of imagination with action has already been noted in Bacon's theory; in addition Hume seemed to imply that it is the imagination, bringing impressions into relation as it runs on in its 'train of thinking', which gives coherence to 'that connected mass of perceptions, which constitute a thinking being' [207–13]. This led H. H. Price, Professor of Logic at Oxford, in a penetrating analysis to comment that for Hume the imagination 'seems uncommonly like the permanent self which he has rejected'; in his opinion, 'the imagination is even more fundamental in Hume's theory of knowledge than he himself admits'. Hume's exposition was by no means thorough or consistent, and Price noted that the role assigned to the imagination arose in part from Hume's concern to defend the common attitude to knowledge against the representationist theories of scientists and philosophers in his own time – 'the Cartesians and the followers of Locke'. At that point Price however added, rather oddly, that this backing of the vulgar against the philosophers, which is evident also in Berkeley and Reid, may be seen as 'the first faint beginnings of the excesses of the Romantic Movement', and he went on to note, with evident regret, 'the distressing affinity between Hume's philosophy and Rousseau's'.[16] Price did not elaborate on his reasons for condemning any indication of a link between Hume and Rousseau or the romantics. This is significant, for as noted earlier, the eighteenth-century romantic movement itself can be considered as a reaction to the growing prejudice at that time, both in scientific theory and in philosophy, against the imagination and feelings as having any part in gaining reliable knowledge. The followers of Descartes favoured the detached, rational understanding and the logic of mathematical deduction; the followers of Baconian objectivism put their faith in the senses. Hume showed the inadequacies of both groups, and although he did not attack the faculty model itself his assertion of a new role for the imagination indicated his search for a more adequate theory. His purpose, it must be remembered, was to rescue philosophy from the inferior position it was being accorded in relation to the sciences, and he did this by denying the possibility of certain knowledge to scientists, as well as to philosophers. Neither group has any special access to truth that sets them apart from ordinary people: 'all knowledge resolves itself into probability, and becomes at last of the same nature with that evidence, which we employ in common life' [181].

The limits of rationality

Hume defended this statement, even with regard to the certainty claimed for the mathematical sciences. Earlier he had accepted 'algebra and arithmetic as the only sciences, in which we can carry on a chain of reasoning to any degree of intricacy, and yet preserve a perfect exactness and certainty'. Geometry cannot attain this perfect precision, he argued, since all its abstractions such as triangles are derived from particular cases: 'its first principles are still drawn from the general appearance of the objects; and that appearance can never afford us any security, when we examine the prodigious minuteness of which nature is susceptible'[71]. Now, Hume extended his scepticism to any application of mathematics: 'In all demonstrative sciences the rules are certain and infallible; but when we apply them, our fallible and uncertain faculties are very apt to depart from them, and fall into error' [180]. The view of mathematics as a system of deduction, which achieves its internal consistency simply by conforming to its own initial postulates, has become more widely accepted since Hume's day, and is well expressed in Einstein's dictum: 'In so far as the statements of geometry speak about reality, they are not certain, and in so far as they are certain, they do not speak about reality.'[17]

Having pointed out that 'demonstration is subject to the control of probability', Hume went on in a manner reminiscent of the sixteenth-century French sceptic, Montaigne, to emphasize what he called the natural fallibility of the judgment. Here he questioned the Cartesian idea of detached human rationality, even rejecting also the assumption, scarcely queried since Aristotle, that man's reason is absolutely unlike any of the sensitive qualities he may share with the animals:

The common defect of those systems, which philosophers have employ'd to account for the actions of the mind is, that they suppose such a subtility and refinement of thought, as not only exceeds the capacity of mere animals, but even of children and the common people in our own species; who are notwithstanding susceptible of the same emotions and affections as persons of the most accomplish'd genius and understanding.

In contrast, Hume declared it to be obvious 'that beasts are endow'd with thought and reason as well as men'. For those who expressed astonishment at 'the *instinct* of animals', he had a disturbing rejoinder: 'reason is nothing but a wonderful and unintelligible instinct in our souls, which carries us along a certain train of ideas'. This instinct, he stated, arises from past observation and experience – from habit – but then 'habit is nothing but one of the principles of nature' [176–9].

Human reason, like the imagination, in Hume's view is a natural impulse: 'Nature, by an absolute and uncontroulable necessity has determin'd us to judge as well as to breathe and feel' [183]. By replacing a divine with a natural determinism, Hume did not solve many problems, especially as he admitted subsequently that when it comes to definition there is no word 'more ambiguous and equivocal' than nature. *Nature*, as opposed to miracles, he pointed out, can include 'every event, which has ever happen'd in the world'; it is also commonly opposed to the 'rare and unusual', and sometimes to *civil* or *moral*; frequently, '*nature* may also be opposed to artifice', but here Hume took a stand against the separation of man and his works from nature:

We readily forget, that the designs, and projects, and views of men are principles as necessary in their operation as heat and cold, moist and dry: But taking them to be free and entirely our own, 'tis usual for us to set them in opposition to the other principles of nature [474–5].

Hume's appeal to nature had some of the characteristics that were to become marked in the subsequent work of Rousseau and those later writers who came to be associated with romanticism. Like them, he showed a particular concern with the feelings and the imagination, and also with what had been called the moral faculty. Books II and III of the *Treatise* dealt with the neglected problems of human *Passions* and *Morals* and, like Bacon, he linked both these as aspects of sense. Sensation, according to Hume, gives rise to all bodily pains and pleasures, while the emotions are derived from these through *reflexion* [275]. Moral distinctions, he went on to argue, are derived from the same source: the moral sense, through the feelings of pleasure and pain, distinguishes between virtue and vice [470]. The reason, he claimed, is impotent in this regard since it can influence neither passions nor actions [457]. His discussion of these topics indicates the strong hold which the Aristotelian model of the mind still retained in his time. The division of the mind into a rational and sensitive faculty was too strongly entrenched for him to reject entirely, but at least, like Pascal in the previous century, Hume was able to recognize some of the problems arising from that view and question the paramount position assigned to reason:

Nothing is more usual in philosophy, and even in common life, than to talk of the combat of passion and reason, to give the preference to reason, and to assert that men are only so far virtuous as they conform themselves to its dictates.

He set out to show the fallacy of those systems of philosophy, both ancient and modern, that emphasized 'this suppos'd pre-eminence of reason above passion' in asserting 'the eternity, invariableness, and divine origin of the former' as against 'the blindness, unconstancy and deceitfulness of the latter' [413]. Hume's assertion that the reason is natural in its origin, and by no means paramount in relation to the imagination or the moral sense, must be seen in this context. Acknowledging the enormity of the stand he was taking in his philosophy, he wrote mournfully: 'I have expos'd myself to the enmity of all metaphysicians, logicians, mathematicians, and even theologians' [264].

Hume might as well have added the natural scientists to his list, for although his reliance on the senses was thoroughly in accordance with the empiricism of his day, neither his arguments on the fallibility of human reason nor his insistence on the function of imagination and the emotions in knowledge and morality was likely to be welcomed by scientists. The current theory of scientific method asserted that incontrovertible data can be obtained through the senses and subsequently ordered by the reason, provided the emotions and imagination are not allowed to intervene. Indeed, according to the model of the mind then adopted, scientific research was held to involve only the senses and the understanding, all the rest of the sensitive faculty as then conceived, including the moral determinants, being therefore excluded, so that the affective aspects of human life, along with all questions of aesthetics or morality, were considered to be of no concern to science. Hume's scepticism, his assertion that demonstration, even in mathematics, is subject to probability, and his determination to deny to the sciences any more certainty than to philosophy, all constituted a threat to the growing autonomy of science.

Hume's moral philosophy: mental geography

In 1748 Hume published a further work, *An Enquiry concerning Human Understanding*, intending it as a definitive restatement of his views. Here he resumed his defence of the study of 'moral philosophy, or the science of human nature', referring to it as 'this mental geography' – Scotland apparently being one of the few places in Europe at the time where a respect for geography was sufficiently high for moral philosophy to benefit by comparison. In a passage reminiscent of Bacon's *Novum Organum* [I, 84], Hume asked:

shall we esteem it worthy the labour of a philosopher to give us the true system of the planets . . . while we affect to overlook those who, with so much success, delineate the parts of the mind, in which we are so intimately concerned?

There cannot, he claimed, 'remain any suspicion that this science is uncertain and chimerical; unless we should entertain such a scepticism as is entirely subversive of all speculation, and even action'.[18]

All the objects of human inquiry, he continued, may be divided into two kinds: '*relations of ideas* and *matters of fact*'. Every affirmation to which certainty can be ascribed belongs to the first kind, and here he included the mathematical sciences. On the other hand, 'all reasonings concerning matter of fact seem to be founded on the relation of *cause and effect*', but since 'every effect is a distinct event from its cause', this relation can never be deduced by reason or stated with certainty: 'all the laws of nature, and all the operations by bodies without exception, are known only by experience'. On that basis, Hume warned against attempting in the sciences to find ultimate causes in nature:

The utmost effort of human reason is to reduce the principles, productive of natural phenomena, to a greater simplicity, and to resolve the many particular effects into a few general causes, by means of reasonings from analogy, experience, and observation. But as to the causes of these general causes, we should in vain attempt their discovery; . . . we may esteem ourselves sufficiently happy if, by accurate enquiry and reasoning, we can trace up the particular phenomena to, or near to, these general principles [24–30].

In this he was giving a further justification, of a logical kind, to the approach already advocated by Newton.

An echo of Hume's view is to be found, over a century later, in Humboldt's conclusion that in the study of natural phenomena, 'the investigation of a partial *causal relationship* and the gradual increase of *generalizations* in our physical knowledge are for the present the highest objects'.[19] Humboldt himself did not acknowledge Hume as a source; indeed, it is interesting that while frequent references to Francis Bacon occur in *Kosmos*, the later British empiricist philosophers are not mentioned at all by name, although there is evidence that their inquiries contributed to the formulation of Humboldt's thought. Hume's concept of nature, for example, and especially his denial of the long tradition that set nature in opposition to human reason and art, found support in *Kosmos*, where the implications for science are considered.[20] In his own philosophy, Humboldt of course went on to develop what was more a Kantian or even Hegelian approach to knowledge by considering the function of ideas in social and historical terms. Hume on the other hand, like Berkeley and Locke before him in their epistemology, adopted an approach that was both genetic and subjective, and in contradiction to the maxim of the new empirical science – that function rather than cause should be the object of study – it was the origin of ideas rather than their function that he sought in the mind of the individual.

At the same time Hume made a salutary effort to point out that the natural sciences are not independent of the human mind and so cannot ignore questions of epistemology. In this he was continuing the tradition of the previous empirical philosophers, all of whom – Bacon, Hobbes, Locke, even Berkeley – were deeply influenced by the science of the day and were concerned primarily to provide a theory to sustain this: if anything, they tended to identify philosophy almost exclusively with the theory of knowledge. Hume, however, set out to reduce philosophy itself to a kind of scientific psychology, limiting it, in effect, to the prevailing conception of scientific method. Therefore he was in danger of making a straitjacket for thought by introducing the current scientific outlook, with its suspicion of all unproven 'conjectures and hypotheses', into philosophy as well. Kant, subsequently, was to break from this by directing philosophy to study the general concepts through which human thought proceeds, but by then the separation between philosophers and scientists was becoming not only more pronounced, but even institutionalized, and it seemed that no single philosopher would be able to change this trend.

Many philosophers still actively defend a separate sphere for philosophy. In his commentary on the philosophers of the enlightenment Isaiah Berlin, for example, condemned the attempt to identify philosophy with empirical science as the major fallacy of eighteenth-century thought, yet he showed little interest himself in the question of scientific method; as a philosopher his chief concern was to define a separate sphere for philosophy, one involving only those 'unresolved (and largely unanalyzed) questions, whose generality, obscurity, and, above all, apparent (or real) insolubility by empirical or formal methods, gives them a status of their own which we tend to call philosophical'. Unfortunately Berlin neglected to explain what he meant by the empirical method, except to distinguish it from 'deductive reasoning as it is used in the formal disciplines' of mathematics, logic, or grammar, while his suggestion that empirical sciences deal only with questions that are themselves, as he put it, 'empirical (and inductive)' failed to clarify this at all, leaving the impression if anything that the sciences, both empirical and formal in his terms, can be progressively cut off from philosophy.[21]

A major difficulty in dealing with empiricism is the absence of critical discussion on many central issues in the present period, when philosophers have tended to avoid any serious concern with the scientific movement or even with the historical development of ideas. Meanwhile almost all recent philosophy of science, in which most of the leaders have been physicists or mathematicians, has reflected their own special interests, with a consequent neglect not only of studies like geography which so far this century have failed to produce any outstanding philosophers, but also of the question of science as a whole and its relation to society in general. Thomas Kuhn, who himself made a transition from theoretical physics to the philosophy of science, was one of the first to recommend a more serious concern with history in this context. The historian's function in his view is not merely to reinforce the textbook tradition in science and chronicle the accumulation of past discoveries, but to promote an awareness of the scientific process and contribute in forming a new image of science; with the increasing adoption of this role for history, he believed it is already possible to discern the first stages in a 'historiographic revolution in the study of science'.[22]

One of the characteristics of the new movement in the philosophy of science is the recognition that historical research, far from being considered to have only an antiquarian interest, has an active role to fulfil in the continuing development of ideas. In the light of recent conflicts in geography between the proponents of positivism and phenomenology, of scientific impartiality and social action, an examination of the emergence of modern geography in the context of those very disputes in the eighteenth century can contribute to an understanding of the problems involved.

CHAPTER 5

ON THE MARGINS OF SCIENCE: EIGHTEENTH-CENTURY GEOGRAPHY TEXTS

CRAMBE RECOCTA: BRITISH GEOGRAPHY TO 1750

For geography the period from Locke to Hume was an undistinguished one, especially with regard to the theory of the subject in the textbooks of the time, and although the present discussion is confined to an examination of British sources, the situation in this respect appears to have been little different elsewhere in Europe.[1] In England most of the new texts were considered little more than *crambe recocta* – a rehash of previous works – while the actual reproduction of earlier books, even those from the mid seventeenth century, continued for some time. The *Geographia generalis* of Varenius, reprinted in a Latin edition by Jacob Jurin at Cambridge in 1712, formed the basis for a rather poor English translation by Dugdale in 1733, a work which nevertheless reached a fourth edition in 1765, the second and third editions 'with large Additions' having appeared in 1734 and 1736.[2] In America the Harvard College curriculum in 1723 included 'a system of Geography' using Newton's 1672 edition of Varenius as a text, while as late as 1756 at the University of Pennsylvania the study of Varenius' *Geography* was being recommended as part of an ideal liberal yet utilitarian education.[3] Other works of even less distinction from the period around 1650 also seem to have retained their popularity in England, where an edition of Heylyn's *Cosmography* appeared in 1703, and Samuel Clarke's book, under the title *A New Description of the World*, was published in 1712, almost thirty years after the author's death. Two smaller texts, from the last decade of the seventeenth century, by Laurence Echard of Cambridge, and the Scottish author, Pat Gordon, also continued to dominate the textbook market for some time.

Echard's *Compendium*: the universal scholar

An interesting contribution in many respects was the *Compendium of Geography, General and Special* (1691) of Laurence Echard (*c.* 1670–1730), which continued to an eighth edition in 1713. Pocket-sized and organized in note form this book was traditional in its brief general part, and even more compressed in its notes on individual countries, yet altogether quite competent, and although composed some ten years before the beginning of the eighteenth century (the second edition, 'Improv'd by Laurence Echard', was dated 1691) it nonetheless tended to look ahead to problems that were to become a dominant concern of geographers during the enlightenment period. Echard dedicated 'this small *Trifle*' to Dr John Covel, Master of Christ's College, Cambridge, 'that this *Science* may be more esteemed in our *University*' and he was clearly aware of obstacles to be overcome before such esteem might be achieved.⁴ Echard himself had no revolutionary approach to recommend. Defining geography simply as 'a *Science* which teacheth the Description of the *Earth*', he accepted the seventeenth-century division into general and special parts. In an Appendix, however, he raised some questions of theory, expressing concern in particular about what he called the 'Universality' of geography which required a geographer to be almost 'an *Universal Scholar*', one with

Considerable skill in all Arts and Sciences . . . an *Etymologist*, an *Astronomer*, a *Geometrician*, a *Natural Philosopher*, a *Husbandman*, an *Herbalist*, a *Mechanick*, a *Physician*, a *Merchant*, an *Architect*, a *Linguist*, a *Divine*, a *Politician*, one that understands the *Laws*, and *Military Affairs*, an *Herald*, an *Historian* and what not? For this is a *Science* so general, as it is defin'd to be a Description of the Earth, so it may be said to be a Description of all things in the Earth; so that there can be no Art or Science . . . or anything that deserves the Observation of the curious, but may be well comprehended under the name of *Geography*, except *Astronomy* alone [222].

Some element of exaggeration is to be expected here, and in part the exasperation of a young man may be indicated too – Echard explained in the preface to his 'small Manual' that 'a compleat *Book* . . . needed a person of riper years' – but his comments indicate, nonetheless, the acuteness of the problem facing geographers in an age of increasing specialization. Already the growing volume of information had created pressures, so that by the eighteenth century the *Compilation* can be seen giving way to the *Compendium* with its tables and index; Echard himself, while accepting the location of places as his basic aim, stated that he found it advisable for the third impression of his *Compendium* to publish his European tables separately in a Gazetteer.⁵ In addition, with the Baconian model for natural history still being given implicit acceptance, the mounting collections of particulars threatened to become even more unmanageable, and geographers became aware that they could not expect to find refuge, as other scientists did, in the progressive limitation of their subject matter. On the contrary, the attempt to keep pace with new developments in European exploration and settlement, along with a continuing expansion of interest in the field of human geography, was widening the scope of their concern. Echard, for example, in his *Special* part referred to the regions of *Terra Australis Incognita* as well as the New World and claimed his *Compendium* to be 'Collected according to the latest Discoveries'. Meanwhile, his appended 'Rules to make a larger Compleat *Geography*' – all of which prove to be rules for describing a *region* or country – covered a range of topics comparable with modern studies in extent. Commencing in these Rules with such traditional topics as the climes and visible stars, almost in the manner of Varenius,[6] Echard proceeded through a number on landforms, the weather, productivity and manufacturing, local animals, the *'Wonders of Art and Nature'* to a last and much longer section on the inhabitants, where the human geography that he envisaged was to consider social, economic, and educational aspects, including urban geography, and even recreation, as well as transport, languages, religions, government and history [213–20].

Clearly, without a strong conceptual framework to guide research and organize the selection of data such a programme was likely to encounter great difficulties, yet it was a long time before any sustained effort was to be made in the direction of providing an effective modern theory. In general, geographers seem to have been isolated in terms of the scientific thought of the times, and their increased involvement with civil geography, itself one of the distinctive trends in the subject during the century, placed them farther outside the orbit of the physical sciences. To change this would have involved a direct confrontation with the increasingly materialist and fact-oriented scientific empiricism of the day, and the task of modifying that view of science has required a long and concerted effort into the late twentieth century.

The geographers of the enlightenment were in no way prepared for this kind of intellectual encounter; indeed, it is tempting to suggest that the prospect of adding epistemology and the skills of the philosopher to his list for Echard's universal scholar might well have precipitated him into the Cam. At the same time the parallel with the mid twentieth century is remarkably close, for, then also, geographers were too overwhelmed with the volume of research and the task of quantification to participate actively in re-examining the philosophy of science.

In default, then, geography seems to have proceeded after the seventeenth century without close contact with the more highly respected specialist sciences of the time, and in effect without the discipline of an adequate scientific method. Local descriptions increased and the production of elementary texts for schools or popular interest continued, for as Echard himself concluded, the wide appeal of geography out of all the sciences gives it a special status: *'This being the only one that comes under the Capacity of all Mankind'* [236]. To cope with the increasingly vast bulk of information two alternatives seem to have been followed. The first was to systematize and compress, imposing a kind of external order, and this gave rise to the characteristic school texts of the period: the abstract, the introduction, and the geographical grammar. The second alternative, for those writers who aspired at all to advance the field, was expansion, the result of this being those multi-volume productions, often with a minimal intellectual contribution in the introduction and even less in the way of a conclusion, aptly named by John K. Wright the 'bibliographic dinosaurs' of the century.[7]

Pat Gordon's *Geographical Grammar*

The *Geography Anatomiz'd or, the Geographical Grammar* of Patrick Gordon, first published in 1693, probably as a competitor to Echard's book, remained one of the most popular texts in Britain during the first half of the century, continuing after its second edition in 1699 through to a twentieth in 1754, and setting a pattern for similar works in the following years. Criticizing earlier works as either so *voluminous* as to 'fright the young student', or else '*confused* (being writ without any due Order or Method)', Gordon aimed to provide 'a just Mean betwixt . . . the large *Volume* and a *narrow Compend*', in order 'to present the younger sort of our Nobility and Gentry with a Compendious, Pleasant and Methodical Tract of *Modern Geography*'. In spite of an attempt in the title to associate this work with 'exact analysis' and the scientific method of anatomy, it proved to be completely traditional with its division into general and particular parts. The first or general part, 'Being a Compendious System of the true Fundamentals of Geography', referred the reader to Varenius for an elaboration on its numerous definitions and theorems, while the second part promised 'a *Particular View* of the Terraqueous Globe . . . a clear and exact Prospect of all remarkable Countries, and their Inhabitants', although it too proved to be relatively unchanged even from Münster's approach and was distinguished chiefly by its *Analytical Tables*. These, 'the main Business of this Book', were intended, as Gordon explained in the preface, 'To present to the Eye at one View, a compleat *Prospect of* a Country', although the prospect offered was actually an extremely restricted and statistical one, for in his opinion, 'the Business of a *Geographical Tract* is not so much to heap up a vast multitude of Names, as to shew the *Divisions* and *Subdivisions* of every Country, with the principle *Town* in each of them, and how all such are most readily found'. To an objection that his work was merely a *rehash –* not 'a *new Discovery* in the Science of *Geography*, but only a bare *Crambe recocta* of those who have gone before us' – he replied that whereas previous tables, '*English, French*, or *Dutch*' had given no more than 'a bare Catalogue of Names confusedly set down without any due Order or Method', his own included directions to relate places quickly to maps.[8]

Gordon's *Grammar* was frankly an abstract from other authors, including Echard, Heylyn, and Cluverius, and he denied plagiarism in using information from their works, unaltered 'when I found it succinctly worded by a credible Pen'. He had no programme for research in the subject; in his view '*Geography*, which, with some small Taste of *Chronology*, may be deservedly termed *The Eyes and Feet of History*', was chiefly 'a needful preliminary Study' for Modern History, although he suggested it also as a cure for idleness and vice. Commenting finally in his preface to the 1741 edition on the need for revising his geography since its 'ready Admittance into many of our publick Schools', Gordon promised this to be his last revision, since he planned instead 'a compendious Body of *Ancient Geography* . . . a work extremely wanted' for the schools.

An indication of one rather significant trend is to be found in Gordon's book, this being a growing tendency to associate geography with both European imperialism and Christian evangelism. In his Dedication, to Thomas, Archbishop of Canterbury, Gordon noted that in considering religion in his geographical treatise he had found, in the known world, 'scarce Five of Twenty-five Parts thereof are *Christians*' and this, he suggested, with appropriate loyalty, should provide a task of conversion for the '*best Church* upon Earth'. His Appendix included not only an account, after the manner of Clarke (1671), of European plantations abroad, but also a set of '*Proposals* for the Propagation of the *Blessed Gospel* in all *Pagan Countries*'. In the course of the eighteenth century the geography text was to become more frequently a vehicle for teachings related not only to Christian piety but also to the expansion of Christian domains.

In the shadow of Locke: mid-century texts

Meanwhile another factor evidently inhibiting the development of geographical theory was the influence of John Locke, whose dominance in enlightenment thought extended in some measure to geography in the early part of the century, although no previous research seems to have been made into this question. In the first years of the eighteenth century few new works seem to have been produced, one of these being *A Treatise of Antient and Present Geography* (1701) by Edward Wells, Lecturer in Rhetoric in Oxford, a text that continued to a fifth edition in 1738.[9] More widely known was the *System of Geographie* (1701) of Herman Moll (d. 1732), a Dutch geographer, who came to London in 1698, publishing there numerous maps and geographical works. A copy of Moll's *Geographie* was in the possession of Locke, who referred to that work in a 1703 composition, *Some Thoughts concerning Reading and Study for a Gentleman*. There, in offering advice on the reading of geography, Locke began by repeating the familiar dictum, 'To the reading of history, chronology and geography are absolutely necessary'; then, indicating an opinion of geography that was by no means high, he continued:

In geography we have two general ones in English, Heylin and Moll; which is the best of them I know not, having not been much conversant in either of them. But the last I should think to be of most use because of the new discoveries that are made every day tending to the perfection of that science . . . These two books contain geography in general; but whether an English gentleman would think it worth his time to bestow much pains upon that . . .[10]

History, Locke had declared earlier in *Some Thoughts concerning Education* (1693), 'ought to be the proper Study of a Gentleman', geography being merely preparatory to that, and therefore a more suitable study for children. While recommending with some enthusiasm that the child's education should begin with *Geography*, Locke quite clearly regarded this as scarcely more than the location of places on maps or the globe, a task involving at the outset in his view only the visual sense and the memory, but not, according to the faculty psychology of the day, the reason or judgment which supposedly developed later in children:

> For the learning of the Figure of the *Globe*, the situation and Boundaries of the Four Parts of the World, and that of particular Kingdoms and Countries, being only an exercise of the Eyes and Memory, a child with pleasure will learn and retain them.[11]

Subsequently Locke produced a sequel to this instruction, dealing with the earth, its place in the solar system, and its atmosphere, waters, rocks, plants and animals, while going on to conclude with chapters on the Five Senses and his theory of 'the Understanding of Man'. Written around 1698, probably with the assistance of Newton, whose views were clearly expressed in the first four chapters, it was titled not geography but *Elements of Natural Philosophy* since it discussed all these subjects as elements rather than in terms of their distribution on the earth.[12]

A similar work of this kind, produced by Locke's friend Jean Le Clerc of Amsterdam, was added by Moll to later editions of his own geography, under the title, 'a natural History of the Earth, translated from the *Physica, sive de Rebus Corporeis* of Monsieur *le Clerc*'.[13] The custom of including in descriptive geographies an introductory general or astronomical part written by another author became quite common throughout the century. During this time, Locke's pronouncements on geography became a kind of authority for the writers of school texts, reinforcing if anything its decline as an academic discipline; in universities and academies it appears from the limited evidence available that the teaching of geography tended to become even more perfunctory and more frequently subordinated to astronomy, while public interest was satisfied with books of travel. In England, for example, Moll's *System of Geography* seems to have been one of the few general works written for adults in the early part of the century, and it did little more than continue the tradition of descriptive chorography. Even in the 1711 edition of this elaborate work, the introductory section, or as Moll called it, 'the Introduction which relates to the Scientifical Part', produced by 'Dr. Gregory, late Astronomy Professor at Oxon', was taken in large part – with open acknowledgment – from Varenius.[14] Moll's more compact version, *The Compleat Geographer*, remained a more substantial work than the school texts of the time and included a number of maps by Moll himself, although in attempting to provide an account of the chief cities and provinces of every state in the world he admitted to relying entirely on the writings of others: 'The whole Book is nothing else but the Words of the most credible Travellers and Historians, and most judicious Geographers, disposed in a regular local Method.' Moll included a detailed list of these authors, but without giving any indication of the basis used for determining credibility except to state his belief that 'true Knowledge of the Earth' is to be obtained only 'by consulting the Travellers that have been on the Spot'. For the 1723 edition Moll added descriptions of Asia, Africa, and America from the accounts of travellers, especially, he stressed in his 'Advertisement', those from the previous thirty or forty years, noting that 'an intended New Edition of Dr. *Heylin's Cosmography*', which was to have included such recent observations, had since proved incomplete.

In the absence of local initiatives during the early part of the century, translations of French works were used to meet the need for school texts, one of these being *A Short and Easy Method to understand Geography*, translated, around 1715, from a work of Nicolas de Fer (1646–1720) by an anonymous 'Gentleman of Cambridge' who defended this 'borrowing from the French' – currently popular in Europe – by stating that 'French Writings, and above all, French Geography are the safest, as well as the best Commodities they can send us'. An unimpressive book, still solemnly teaching the daily rotation of the heavens, its regional descriptions were confined mainly to Europe with a concluding reference to the discovery of *Terra Australis* in 1697.[15] Equally mediocre was *The Geography of Children*, a translation of the 1736 *Géographie des enfans* of Nicolas Lenglet Dufresnoy (1674–1755), which was published at London in 1737, the anonymous translator noting that 'a Geography adapted to the use of Children has long been wanted'.[16] The popularity of this book was remarkable: after the tenth edition in 1776 had incorporated a new section, 'a method of learning geography without a master, for . . . grown persons', it went through more than 24 London editions by 1825, the 'twentieth English and first Kentucky edition' having been reprinted at Lexington in 1806, while other editions under the main title *Geography for Youth* appeared in Philadelphia as late as 1798, and Dublin, where it reached a sixteenth edition by 1806.

Among the British efforts in the early part of the century was Gawin Drummond's *Short Treatise of Geography* (1708) which continued to a third edition at Edinburgh in 1740 – a small book without maps, little more than a gazetteer with an emphasis on administrative divisions and the geography of the Roman empire. Drummond explained that his decision to produce a short 'Compendium of ancient and modern Geography' had resulted from 'observing that so necessary a Study is much neglected in the Schools of our Country, which proves a great Impediment to young Scholars in understanding the Roman authors and Modern Histories', and he added that most of the texts available 'exceed the Bulk and Rate of ordinary School Books'.[17] The standard of his commentary is indicated by the vagueness of his opening statement on the solar system: 'The Terraqueous Globe is situated according to *Ptolemy* and *Tycho*, in the centre of the World, but according to *Copernicus*, between the *Orbs* of *Mars* and *Venus*' [1]. Of no more distinction, the *Manual of Modern Geography*, written by J. Gregory in 1739, avoided any reference to the Copernican system and in lieu of geographical theory added a respectful reference in the preface to Locke's *Thoughts Concerning Education*; Gregory's fourth edition in 1760 incorporated a new section on *Hydrography*.[18]

In contrast to the chorographical approach of such works, many student texts in that period gave renewed emphasis to instruction in astronomy and the globes. This trend has been noted by Robinson in the dissenting academies of Britain where, along with other 'modern' subjects, geography had been added to the curriculum in an effort to provide a more utilitarian education than the verbal, classical one – although even here it seems that geography, like history, was neglected in comparison with the sciences, and in many cases evidently survived as an adjunct to astronomy. Similarly, Warntz commented on a closer link between geography and astronomy or navigation in American colleges in the course of the eighteenth century.[19]

A text evidently designed to cater for that trend in Britain was *The First Principles of Astronomy and Geography*, published at London in 1726 by Isaac Watts, a Doctor of Divinity who in his preface explained, like Heylyn a century before, that geography and astronomy were not his 'special Province', this being rather 'the Knowledge of God, the Advancement of Religion'; however, believing a knowledge of 'these Mathematical Sciences' to be necessary for appreciation of the Scriptures and the works of creation, he had revised a youthful work written more than twenty years earlier. In spite of its mediocrity his book reached an eighth edition by 1772. Cautious in his repetition of traditional topics, Watts did indicate a break with Ptolemy by commenting that terms such as Amphiscii 'are only *Greek* Names invented to tell how the Sun casts the Shadows of the several Inhabitants of the World, and are not worth our present Notice'; stating also that 'the latest and best Astronomers' confirmed the heliocentric theory, he added: 'yet to make these things more easy and intelligible . . . we shall here suppose the *Sun* to move round the *Earth* . . . as it appears to our Senses'.[20]

Similar in many respects was George Gordon's *An Introduction to Geography, Astronomy, and Dialling* (1726). It was, the title-page promised, 'Adapted to the Meanest Capacity', being intended, as the author went on to explain, 'for the Use of some young Persons of Distinction, for whose tender years, Systems of *Geography* and of the *Celestial Physicks*, would have been too hard a Study'. Small and old-fashioned, with few diagrams, this book, like the work of Watts, provides a rather depressing commentary on the teaching of geography in British schools at the time. Under the inside title of 'A Compendium of Cosmography: Containing the Principles of Geograhhy [*sic*] and Astronomy' Gordon offered little more than the traditional Greek definitions of circles, zones, climates and so on, with notes on continents and islands reminiscent of sixteenth-century texts, followed by lists of names of countries. With an evident attempt to give his work some authority Gordon justified his addition of a *Compendium of Chronology* by referring to the words of 'the learned Mr. Locke, . . . in his extraordinary Thoughts concerning *Education*'.[21] The quotation that followed proved to be nothing more than Locke's comments on the hackneyed theme of linking geography and chronology as a kind of preparative or propaedeutic to history: 'without *Geography* and *Chronology*; I say, History will be very ill-retained, and very little useful, but be only a jumble of Matters of Fact, confusedly heaped together . . .'[22] Within the context of that view of geography as a source of information about places, it is scarcely surprising to find the subject being presented in texts such as Gordon's as itself little more than 'a jumble of Matters of Fact', invigorated by no fresh insights or challenging ideas.

Locke himself had offered a more provocative line of thought in his *Essay concerning Human Understanding* (1690) with his efforts to link the concepts of place, space and time in discussing the measurement of duration and expansion: time, he argued, can be measured only by a succession of experiences, for instance 'by the motion of the great and visible bodies of the world', while 'to measure motion, space is as necessary to be considered as time'.[23] Yet such a move away from the Newtonian absolutes of space and time, towards a space–time concept more consonant with relativity physics, was to find no response in the geographies of the day and Locke himself apparently made no application of these ideas to geography.

Even in the mid twentieth century when the notion of geography as a *science of space* became widely accepted, strong resistance remained to the introduction of ideas of time and process. The old distinction of history–time, geography–space was rarely questioned, so that H. C. Darby, for example, in 1953 expressed a minority viewpoint for his day in arguing that we cannot draw a line between geography and history: 'All geography is historical geography.'[24] Subsequently David Harvey went on to discuss the methodological problems associated with what was regarded as the subjectivity of time according to the earlier positivist view of rational scientific explanation, and the beginnings of his own break with a rigid positivism can be discerned in the surprise with which he pointed out how relatively recently the discovery of time in science since the eighteenth century has brought into question the static renaissance view of man and nature, the timeless truths of Descartes and the unchanging mechanism of Newton.[25] The process that Arthur Lovejoy called the 'temporalizing of the chain of being' and its conversion into 'the program of an endless Becoming'[26] was a pronounced feature of eighteenth-century thought; thus the absence of any discussion of such issues in the texts of the day draws attention to the extent to which the geography of that period appears to have remained insulated from the surrounding ferment of ideas.

Salmon's *New Grammar*

An attempt to introduce into a geography text at least some new developments in research is found in Thomas Salmon's *A New Geographical and
Historical Grammar* (1749). Intended evidently to replace Pat Gordon's
Geographical Grammar, Salmon's book claimed to provide a geographical
part that was 'truly *Modern*' and indeed the introductory notes on the figure
and motion of the earth included a reference to the report recently published
by Maupertuis on the confirmation of Newton's idea of the earth as an oblate
spheroid by the 1736 French expedition to measure a degree of the Meridian
at the Polar Circle. Moreover, Salmon expressed support for the Copernican
theory, despite 'Ecclesiastical Censure', and pointed out also that a new set of
corrected maps, 'engraved by Mr. Jefferys, Geographer to His Majesty' was
included in his book: 'For, since the Days of my Friend *Moll* the Geographer,
we have had nothing but Copies of Foreign Maps, by Engravers unskilled
in Geography, who have copied them with all their errors.'[27]
   The division into general and special geography, still used in Pat Gordon's
Grammar, was dropped by Salmon, who placed the traditional content of
the general part into three short introductory sections on the shape and
motion of the earth, on geographical definitions – commencing with a rather
cumbersome one on *geography* as 'a Description of the Surface of the
natural Terraqueous Globe, consisting of Earth and Water, which is represented by the Artificial Globe' – and a third section on earth and water,
winds and tides [15–37]. The majority of the book describes the continents
by their political subdivisions, including, along with lists of rivers, capes and
bays, a good deal on trade as well as political and social conditions, pointing
to a revived interest in commercial geography [37–615]. The book itself
proved successful, passing through twelve editions in London and one at
Edinburgh by 1772, although it is perhaps an interesting indication of the
growing emphasis on astronomy in England that the last London edition in
1785 should appear under the changed title: *Salmon's Geographical and
Astronomical Grammar*.
   Altogether, the lack of vigour in British geography at this time was to
show little improvement in the second half of the eighteenth century when
the lead in exploring new directions was to be taken by German geographers.
Already at least one German text was finding a ready market throughout
Europe, despite its shortcomings.

A German model: Hübner's *Short Questions*

Referring to the dependence of German geography on French and British
research in the first half of the eighteenth century, Oskar Peschel instanced
the *Elementa Geographiae Generalis*, 1712, of Johann Georg Liebknecht,
and more particularly the undistinguished text that achieved considerable if

undeserved popularity, Johann Hübner's *Kurze Fragen aus der alten und neuen Geographie (Short Questions from Ancient and Modern Geography)*, published in 1696 and updated in a new edition 'for the present times' at Leipzig in 1721. As Peschel noted regretfully, 'In inexhaustible succession at that time the editions of Hübner's geographical questions were repeated, even being translated into several languages, although they contained almost nothing that one could not find in maps.'[28]

An English translation of Hübner's book, produced in London by James Cowley, 'Geographer to his Majesty', reached a second edition in 1742, and by that time, as Cowley stated in his preface, in addition to a French translation, over thirty German editions had already appeared. Writing evidently before the 1737 English version of Lenglet Dufresnoy's work using questions and answers, Cowley added that although popular in France this method had not yet been used in an English *Introduction to Geography*, although he considered it appropriate in a text intended for 'the Use of Schools' and 'adapted to the Capacity of all Ages and Conditions of both Sexes'. In his opinion, '*Geography* is a Science not only useful but very agreeable', and therefore it is 'surprising to see how shamefully it is neglected amongst us', by both children and grown persons. For Cowley, however, geography remained an introductory study: it has, he stated, repeating the rather quaint phrase used by Pat Gordon, 'been justly term'd *the Eyes and Feet of History*', and he continued with the standard acknowledgment to the pedagogical authority of the day: 'the late celebrated Mr. *Locke*, in his excellent Treatise on Education, recommends the Study of it . . . and is of Opinion, that Children ought to begin with it, as being introductory to all other Studies'. Along with a set of maps and an index of places mentioned, the book included a dictionary of names from ancient geography, and it was apparently popular enough to justify a further edition in 1746. The text of this work, with its series of questions on the rivers, provinces, commodities, and so on, of each country, was uninspiring and tradition-bound. Geography, defined here as 'a Description of the Surface of the Earth', was divided once again into *universal* geography, considering 'the whole Earth in general', and *particular*, describing 'each distinct Country by itself'.[29] In contrast, a conscious effort to upgrade the subject in line with modern developments, a move evident in Salmon's *New Grammar* of 1749, was to become a feature of the second half of the century, particularly in Germany.

NEW SYSTEMS OF GEOGRAPHY 1750–1800

The German contribution: Büsching's *Erdbeschreibung*

A major attempt to provide a new system of geography was made by Anton Friedrich Büsching (1724–93), professor of philosophy at Göttingen from

1754 to 1761, whose lectures on geography became the focus of his major work, his lengthy *Neue Erdbeschreibung*, which appeared in eleven volumes from 1754 to 1792 as well as in numerous other editions during his lifetime, although it had not been brought to a conclusion at the time of his death. The replacement of the Greek term *Geographia* with the German equivalent *Erdbeschreibung* was a feature of his work, which itself was associated with a period of growing national consciousness in which Germans were to claim a significant position in geography. One of Büsching's outstanding contributions was his incorporation of statistics into geography. In this he was following the demographic work of Johann Peter Süssmilch, whose book *Die gottliche Ordnung in denen Veränderungen des menschlichen Geschlechtes (The Divine Order in the Variations of the Human Race)*, Berlin, 1742, had earned him recognition as the founder of population statistics in Germany, where according to Peschel, the idea of the scientific application of statistics was first published by Gottfried Achenwall (1719–72) in a Göttingen dissertation of 1748, which was followed with a work on European political science in 1749; Büsching's own innovation was the application of data on area and population numbers to regional descriptions, and the recognition of population density as a geographical problem.[30]

In other respects Büsching was largely conservative in his approach. Like Süssmilch he was an advocate of natural theology, while his system of geography remained, in the tradition of Strabo, basically a description of the various parts of the earth according to political units, with greatest weight being given to the Empire of Germany, after the manner of Münster. Büsching presented this as an account of the civil or political state of the earth; avoiding the popular seventeenth-century distinction between general and special geography, he nevertheless retained a division between natural and civil aspects, as his definition indicated: 'By geography we understand a well-grounded account of the natural and civil state of the known earth-surface [*Erdbodens*]'. His use here of the term *Erdboden* rather than *Erde* indicated his intention, like Hübner before him, to deal with surface features rather than the earth as a whole, and he further limited his description by excluding the unknown lands near the poles. He stressed also the scope of geography as narrower than that of cosmography, a study currently undergoing something of a revival in Germany with the formation of a Cosmographical Society; again, Büsching showed a preference for a Germanic term, *Weltbeschreibung*, rather than the classical one: 'As our earth is only a part of the universe, so geography [*Erdbeschreibung*] is only a part of cosmography [*Weltbeschreibung*] (*Cosmographia*), with which it is closely related.'[31]

The scope of Büsching's geography remained vast, however, since he accepted the view that it should provide 'a well-grounded account of all things', and he attempted to set up some guide-lines for method. According to the particular aim of the author, he suggested, such accounts may be short or lengthy, 'but at all times the useless and the trivial must be eliminated, in order that books of this kind should not swell to an inordinate and unwieldy size'. A simple style, a concise manner, and an ordered approach are essential, he claimed, since a geography text should contain more than maps or lists of names. Above all, he stressed the need for accuracy in the sources used: 'Its sources must not be other geographies, but good descriptions of particular lands and places, and specific investigations' [10–11].

This emphasis on the need for a critical selection of sources, combined where possible with personal research, was one of Büsching's significant contributions to theory in his day. Its importance was featured, for instance, in the English translation of his work in 1762 which, while reducing much of the original preface, retained Büsching's defence of this approach. Like Moll he preferred 'accounts written on the spot', and he added his own enquiries from his travels: 'This I look upon as the only means to bring Geography to a greater degree of perfection than it has hitherto acquired.' His criticism of earlier treatises applied equally well to English geographies of the time:

My predecessors in this Science, indeed, generally copy from each other, . . . they either had not, or could not have recourse to the best sources; or, . . . did not use them with a proper degree of care and impartiality. Hence a person who has the least skill in geography, or knowledge of the Terraqueous Globe, has reason to complain, that the Systems of Geography hitherto published are of very little service.[32]

Büsching asked for his own mistakes to be corrected, and he was himself of course by no means immune from error. His account, for example, of the achievements of learning in Germany included the assertion, 'It was a *German*, namely *Martin Behaim* of *Nurenberg*, that first discovered the fourth part of the world, afterwards called *America*; and it is the *Germans* that have publish'd the best books on *Geography*.' To which his English editor added the courteous footnote, 'I am afraid our author's amiable partiality for his countrymen has carried him a little too far here. *Columbus*, the *Genoese*, at least, is reputed the discoverer of America' [IV, 23].

Nevertheless, Büsching's *Neue Erdbeschreibung* marked in some respects a turning-point for geography in his century. An indication, for instance, of an approach to the subject later made famous by Humboldt was to be found in Büsching's concern to achieve a single consistent view: 'I purpose to bring together in one view the best chorographical and topographical descriptions extant.' Books and manuscripts, as well as private correspondence with men of learning in different parts of the world, provided his sources, and he introduced population statistics, with estimates of total numbers, as well as births and deaths from bills of mortality. At the same time he expressed regret that he was working alone in this endeavour: 'To write a system of *Geography*, or in other words, to give a Description of the Earth, is a very difficult, laborious and important task, and requires the united efforts of whole societies' [I, v–vi]. In Büsching's case no philosophical reference to the sociology of knowledge is to be imputed to this last statement: the effective arrangement of accurate descriptions was his aim, and he evidently assumed that the single view he sought was to be achieved by the mere contiguity of information. As a Lutheran minister himself, his own conceptual framework was the doctrine of the designed earth, and in this he brought to geography the viewpoint presented earlier in the natural history of such British writers as Burnet or Ray: he saw divine providence as the source of all order on the earth, this being reflected in the diversity of regions, the boundaries and political constitutions of nations, as well as all commerce between them, including even colonization, for which Büsching claimed virtues as well as vices. Man in his view is the instrument of God, subduing and perfecting nature with gardens and cities.

Of considerable significance, especially as a model for later German efforts in this field, was Büsching's further attempt to provide some kind of organizational structure for geography, within his primary division of the subject matter into natural and civil aspects. The main bulk of his own work was concerned with civil or political (*bürgerlichen*) features, namely a description of individual countries, according to the traditional categories of size, strength, government, institutions, inhabitants, churches, cities, and so on. With regard to what he called natural (*natürlichen*) features, Büsching suggested little more than an outline for their future treatment, under the two headings of mathematical and physical geography; together these were to cover those aspects discussed in the general geography of Varenius, while extending beyond this in one important respect, to include in physical geography a more extensive consideration of man. Mathematical geography in Büsching's plan was to consider the earth as a body in the universe, and its place in the solar system according to the theories of Ptolemy, Tycho Brahe and Copernicus. Traditional topics from classical geography were also included: latitude and longitude, zones and climates, even the Greek terms for inhabitants of different zones – askioi, antipodes, antioeci – while some detail was to be given on the use of the globes, an important aspect of geography at this time. More significant for future developments was his plan for physical geography.

Büsching's physical geography: a search for new theory

Physical geography (*physikalische Erdbeschreibung*), concerned with 'the natural state of the earth' (Greek *physis*, nature), in Büsching's opinion 'as yet is very imperfect'. It should deal first, according to his plan, with the earth's atmosphere, its density, pressure, temperature and winds, in different locations and in relation to the convenience of inhabitants. Selecting as an example a region that was to feature in the researches of Humboldt at the turn of the century, he noted that 'the inhabitants of *Quito* in *America*, who dwell on the highest part of our Earth hitherto known, breathe the purest Air' [I, 37]. Büsching outlined also a section on water or hydrography, dealing with fresh water in springs and rivers, and with what he called mineral water, under which he discussed the sea, its area, waves and currents. The most interesting part of his physical geography for the present discussion, however, was his section on the earth in general. Commencing with mountains, valleys, and deserts or uninhabited places, he went on to consider the three *natural kingdoms* into which, as he noted, natural philosophers customarily divide the earth: the mineral, plant, and animal kingdoms. His geology was extremely primitive even for his time, and he showed little concern with plants, except as curiosities: 'It is not consistent with my plan to enlarge on the Vegetable Kingdom' [I, 46]. Again, with regard to animals, he was interested only in describing the rarest animals in each country; in the animal kingdom his chief concern was with man, for according to his view of divine design, man was seen as part of nature:

I shall take a general view of the Human Species only, as they are the noblest and most important Creatures on the Earth, and are appointed by God to acquire the knowledge of, and dominion over it [I, 46–7].

Büsching's outline for a discussion of man pointed the way to the human geography of the following century, while his theories on population predated those of Malthus. Beginning with a consideration of population growth and estimates of numbers based on available figures, he suggested that the constant increase in the human race indicated by the excess of births over burials is subject to certain checks, among which he included plague, war, famine and celibacy. He went on to consider differences among people in colour and in civilization, indicating here an enlightened humanitarian outlook unusual in the geography of his time. Differences in colour are to be ascribed in his opinion to variations in climate, diet, or way of life, the fairest complexions belonging to those in temperate zones: 'But whether these are the most beautiful among the species, or whether a well proportioned *Moor* or Black may not be reckoned as beautiful, I leave to the impartial determination of others' [I, 48]. Similarly with regard to culture, Büsching rejected the assumption that some races may be intrinsically inferior in mental ability:

But as to the difference in their intellectual faculties, we are not to look for that in their nature, or climate, but in the greater or less opportunity they have of improving and exercising their mental powers. An inhabitant of *Greenland* or *Lapland*, a *Moor* or a *Hottentot* is in his way as intelligent as one among the more civilized nations, and if the former had the same opportunities of improving his understanding and regulating his passions as the latter enjoys, he would not be at all inferior to him.

Like Burnet in the previous century and Kant in his own day, Büsching was aware that notions of what is admirable depend on custom, and he argued that other groups should not be ridiculed on that account:

Many nations and individuals, who pretend to rank themselves among the civilized part of their species, have so many odd and absurd customs as might justly expose them to the ridicule of others, who are called Barbarians and uncivilized [I, 48].

Although Büsching himself did not develop this outline for physical geography, his programme can be seen taking shape to some extent in the efforts of isolated geographers after him, in a tradition that provided the background for Humboldt's work. Moreover, the long survival of Büsching's model is indicated in the appearance of a plan remarkably similar to his, a full century later, under the heading 'The Elements of Physical Geography', in the *Handbook of Geography* published by K. F. R. Schneider, another German supporter of the divine design theory.[33] Büsching's physical geography provided a challenge for his followers; however, among those who preferred to adopt a secular approach to science, it seems that the current conflict over the place of man in a science of nature tended to retard the development of any unified study. During Büsching's lifetime considerable advances were being made in such fields as ethnography, anthropology and the science of languages, but these on the whole were not incorporated effectively into geography, except in the context of regional descriptions that remained largely outside the range of recognized scientific activity.

Immanuel Kant (1724–1804), whose lectures on geography at Königsberg appear to have owed a considerable amount to Büsching's example, especially at their commencement in 1756, later moved after 1772 to separate *anthropology,* as the study of human consciousness, from his *physical geography.* Kant's lectures were continued until 1796 and attracted considerable interest, one of his more famous students being Johann Gottfried Herder, although the publication of Kant's *Physical Geography* did not occur until shortly before the author's death; this work will be discussed in connection with Kant's philosophy, in the following chapter.

The approach of both Büsching and Kant to physical geography marked a considerable departure from that of the Frenchman, Philippe Buache (1700–73), in the same period. In his 'Essai de géographie physique' of 1752, Buache outlined three main areas of geography: natural or *physical, historical*, and *mathematical*. He divided physical geography into an *exterior* part, concerned with terrestrial features – landforms, mountains, rivers, lakes, etc. – along with such marine features as seas, straits, and islands, in contrast to the *interior* part which should deal with the relationships of land and sea, the origin of springs, the nature of minerals and rock strata, the interior of the sea, the movement of ocean currents, and other topics important for navigation or otherwise 'useful to human society'. Buache's *Géographie physique* did not include a study of man: this was to be dealt with in historical geography, which in his view should consider the first tribes and the migrations of different peoples or *nations* across the earth, the extent and location of empires, kingdoms and republics, and the knowledge of geography in different historical periods. In addition to such political geography, he suggested that *la Géographie historique* should include ecclesiastical, military, and commercial geography. Finally, mathematical or theoretical geography, as the third main aspect, is concerned with the science of projections and cartographic methods, as well as astronomical observations which show the relationship of the terrestrial globe to the heavens and in his view form the basis of a geographic science. His own *Essay*, dealing only with physical geography, concentrated on a study of world mountain chains – an aspect, he noted, previously neglected by geographers. Illustrating his argument with a polar projection of the northern hemisphere, he outlined a series of natural regions formed by mountains and drainage basins, proposing these as a basis for geographical description. In this significant attempt to outline what he called a *natural system*, he emphasized the continuity of relief on the globe, extending to submarine mountain chains and valleys; however, his general view did not encompass the place of mankind in such a natural system.[34]

Other attempts to widen the traditional range of geographical inquiry emerged during the following decades, as indicated by the work of Torbern Bergmann (1735–84) in Sweden, who incorporated recent research on such matters as earth temperatures and tides in his *Physikalische Geographie* of 1773, and that of Johann Christoph Gatterer (1727–99) in Germany. As professor of history at Göttingen, Gatterer lectured on geography and, like Büsching, contributed to the strong Göttingen school of geography led by Johann Michael Franz (b. 1700), founder of the German *Cosmographical Society*, who became the first professor of geography at Göttingen in 1755 and promoted the view of geography as a support to history. Gatterer himself explored new developments: the application of statistics to world economics was continued in his *Ideal einer allgemeinen Weltstatistik* (Göttingen, 1773), while his *Abriss der Geographie* (Göttingen, 1775) considered three divisions of geography, into *Länderkunde*, the description of countries, grouped according to natural divisions, *Staatenkunde*, or political geography, and *Völkerkunde*, the study of peoples. The idea of geography as providing 'a picture of the earth surface (*ein Gemälde des Erdbodens*)', expressed already by Gatterer, was developed by Johann Georg Müller in his own attempt to provide a new model for geography, the *Versuch über das Ideal einer Erdbeschreibung*, 1789. Müller's concern with this higher point of view, a consideration of the heavens, earth and mankind, and their influence on one another, was seen by Schneider, in his 1857 *Handbuch der Erdbeschreibung* [I,1], as a landmark in the development of geography towards achieving the status of an independent science instead of remaining an aggregate of unrelated information, the handmaid of history.

By the end of the eighteenth century in Germany, the place of geography in the sciences was attracting considerable attention, not only from the writers of geography texts, but also from intellectuals of the calibre of Kant, Herder and Humboldt, opening the way for a vigorous development of theory. In Britain, meanwhile, the textbook tradition showed no such stimulus.

British texts: bourgeois geography

The period when Büsching in Germany was introducing his *Neue Erd-beschreibung* was a remarkably empty one for geographical thought in Britain, where the chief development seems to have been the emergence of what might be called a bourgeois geography, appealing to the general curiosity and current imperialist interests of the well-to-do classes and their children. Apart from further editions of earlier texts, few new works appear to have been published and of these, two were geographical dictionaries. In 1759 Andrew Brice produced a two-volume work, *A Universal Geographical Dictionary; or Grand Gazetteer; of General, Special, Antient and Modern Geography*, compiled, he claimed, 'with the utmost accuracy from the most approved Travellers, Geographers, Historians, Philologers etc.', and offering 'a comprehensive View' of the countries of Europe, Asia, Africa and America, their 'Soil, Extent and Situation', their products, their 'Governments, Arts, Manufacturers, Traffic, Genius, Manners and Religion', as well as 'Curiosities (natural and artificial)', and a good deal on their history. Maps were included, with longitude and latitude given for most places in the text, but no general introduction at all was offered, and Brice's emphasis on a description of 'British Dominions and Settlements throughout the World' indicated the association of this work with imperialism as much as with geographical inquiry.[35]

A similar work from this period was John Barrow's *New Geographical Dictionary*. Well illustrated and incorporating a virtual atlas, it offered a full account of 'all the Empires, Kingdoms, States, Provinces, etc.' in the four known continents, with an emphasis on commercial and historical aspects, but it made no pretence at a scientific approach. According to Barrow, 'Geography is the Art by which we describe the figure, magnitude and position of the several parts of the surface of the earth', and although he promised 'An Introductory Dissertation, explaining the Figure and Motion of the Earth, the Use of the Globes, and Doctrine of the Sphere', this proved to occupy scarcely a dozen pages, and included the classical teachings on zones and climates, along with more recent information on winds and the Newtonian explanation of tides.[36] Barrow, a teacher of mathematics and the author of a dictionary of medicine (1749) and another of the arts and sciences (1751), as well as a collection of voyages (1762), was typical of those writers who catered to the popular market for geography in Britain for over a century.

An interesting development at this time was the appearance of *The Young*

Ladies Geography: or, Compendium of modern Geography, published by
Demarville in both French and English at London in 1757, and later modified as *The Young Gentleman and Lady's Geography* in a London edition of
1765, reprinted in Dublin the following year with the popular title 'Lady's
Geography' on the spine.[37] In the dedication to Queen Charlotte the editor
offered this work as an 'improvement in female literature' and as an

endeavour to entice from the hands of the Fair, obscene and ridiculous novels,
(which serve only to vitiate their morals, inflame their passions, and eradicate the
very seeds of virtue) by persuading them to the study of a science both useful and
amusing, and without some knowledge of which they cannot read even a public paper
of intelligence with pleasure or advantage.

The *Introduction to Geography* which followed was clearly not intended to
inflame the passions. Commencing with a section, taken sometimes ver-
batim from Varenius, on the origins of geography and definitions of Greek
terms such as perioeci and amphiscii, the author added a useful account of
developments in cartography or, as he called it, the 'progress of geography'
since 1500, and he went on to some well-informed statements on the recent
confirmation of Newton's theory on the earth's axis being shorter than the
equinoctial diameter, and the general acceptance, despite earlier 'ecclesiastical censure', of the Copernican hypothesis which, he noted,

is most agreeable to the tenor of nature in all her actions, for by the two motions of
the earth, all the phenomena of the heavens are resolved, which, by other hypothesis, are inexplicable without a great number of other motions, contrary to philosophical reasoning [xvi].

In the main section of the book the regional description – beginning with
England, its counties, rivers and wonders – was largely conventional,
although it included a sympathetic account of the Irish, as a population of
two million, three-quarters of them papists, who 'have made noble struggles
for the support of their liberties and the preservation of their country' [21].

Some improvement in the standard of textbook production is evident in
several other works from this period, one of these being *A brief Survey of the
Terraqueous Globe*, published at Edinburgh in 1762 by the schoolmaster
John Mair (*c.* 1702–69), author also of popular texts on Book-Keeping and
Arithmetic. Commencing with an affirmation of the heliocentric theory, and
illustrated with coloured maps by T. Kitchen, Mair's *Survey* was republished
in 1775 with additions 'by an able hand'.[38] Coloured maps were a feature also
of a brief yet remarkably efficient book, *A View of the Earth*, published at
London in 1762 by the Reverend Dr Richard Turner, 'Late of Magdalen-
Hall, Oxford, now Rector of Comberten . . . and Teacher of Mathematics
and Philosophy at Worcester'. Addressing his book to 'young gentlemen
and young ladies (However titled, or otherwise distinguish'd) Thro' all Parts
of the British Empire', Turner began with a confident statement on the
elliptical orbit of the earth around the sun, and proceeded to give a clear and

concise commentary on the effective regional maps, one marked however by the kind of racist views frequently associated with imperialism: 'The Europeans in general', he noted, 'are well made, and tolerably fair', but in north America the natives are 'generally of a *brown* Complexion . . . the *Mexicans* are civil and docile, the rest savage and cruel', while the inhabitants in 'the *Unknown Parts* of the World . . . near the Pole . . . are very few, and these savage, low in Stature, and of ugly Mien'.[39] In presenting technical information Turner showed considerable resourcefulness, providing a number of diagrams and 'a new and curious *Geographical Clock*' with a moveable central circle, showing time in different parts of the world. He concluded with a number of geographical problems and paradoxes, accompanied by solutions, suggesting finally that after this study of the elements of geography the student should turn 'for a more *particular* Description of the *Manners, Customs* and *Constitutions* of the different Nations and Countries . . . to Mr. *Salmon's* Geographical Grammar' [36]. Three years later Turner produced another short text on modern astronomy, *A View of the Heavens*, while his *View of the Earth* went on to a fourth edition by 1787.

Meanwhile, in a return to the tradition of natural theology, Benjamin Martin, 'optician', published in 1769 his *Physico-Geology: or, A New System of Philosophical Geography*, a reprinting, with a new introduction, of a section from the second volume of *The General Magazine of Arts and Sciences*, printed at London in 1764. After a stereotyped introduction on the globe and its zones, with the inevitable lists of geographical definitions and problems, followed by a more modern account of tides and currents, Martin went on to assert a close connection between *Natural History*, which he defined as the study of the earth's productions, and *Cosmography*, seen as a combination of geography and hydrology:

The Knowledge of the Earth, in regard to its Figure and Dimensions, and the Division and Distribution of the several Parts of the Surface, constitute that Science, which is properly called *Geography*; or rather, *Cosmography*; for the Science of Geography, in a proper Sense of the Word, signifies nothing more than a Description of such Parts of the Surface of the Earth as are really, Land; the other Part, which describes the Waters being called *Hydrology*.[40]

He promised a fresh approach in describing the natural history of each country: 'Nor shall we here pursue the old beaten Track of other Geographers, but shall attempt to strike out one entirely new.' His lengthy regional descriptions, however, were mediocre, except for the competent maps of Emanuel Bowen, and his interpretation was conditioned by the doctrines of divine design: at the outset, in an apparent reference to Derham, he cited 'the Reflections of a late pious and elegant writer' on the earth as a 'Specimen of the Divine Skill and Munificence', which displays always 'an exact Oeconomy and boundless Profusion . . . and though it has accommodated a Multitude of Generations, it still continues inexhaustible'

[34–42]. Martin was no innovator here: as Europe entered its period of great demographic expansion the dangers of excessive population growth, which had already been raised by a number of leading writers including even the pious Robert Wallace in *The Various Prospects of Mankind, Nature and Providence* of 1761, were to become the focus of the influential *Essay on Population* of Thomas Malthus in 1798.[41]

Guthrie's *Grammar*

One of the most famous texts produced in Britain during the last part of the century was the *New Geographical, Historical and Commercial Grammar* of William Guthrie (1708–70), first published in 1770, and reproduced with additions more than 25 times until 1843. Continuing the tradition of Pat Gordon's *Geographical Grammar,* Guthrie replaced Salmon's text, which had also included *Historical* in its title, adding to his own in turn the word *Commercial* to indicate his concern with 'commerce, the prime mover of the oeconomy of modern states'.[42] The introductory section on astronomy was written by James Ferguson, the Scottish shepherd who had commenced his own studies with Gordon's *Grammar.* Alan Downes has drawn attention to the position of Guthrie as one of a number of Scots, working in London, who contributed geographical texts during the Georgian period, and he saw in Guthrie's work a reflection of the ideas of the French author Montesquieu whose famous *Spirit of Laws,* with its relation of social institutions to the total living conditions of people, appeared in 1748.[43]

Guthrie's work also owed a good deal to Büsching's *Geography,* and the authors of a competing text wrote some rather caustic comments on this in 1796:

That *System of Geography* which has been long in the most general use in this country, is well known to have been originally an Abridgment of the ponderous *Geographical* work of the laborious *Busching*, executed by the late Mr. John *Knox*, with some assistance from the works of James *Ferguson*, the celebrated self-taught astronomer, and from the pen of William *Guthrie*.[44]

Büsching's division of natural from civil or political geography was accepted by Guthrie, who noted in terms reminiscent of Büsching that:

The science of natural geography, for want of proper encouragement from those who are alone capable of giving it, still remains in a very imperfect state; and the exact divisions and extent of countries, for want of geometrical surveys, are far from well ascertained.

In these circumstances, 'a general idea of the subject', added Guthrie, echoing a phrase from the English translation of Büsching, 'is all indeed we can attain, until the geographical science arrives at greater perfection'. Meanwhile, he declared his intention of avoiding 'those fabulous accounts . . . which . . . swell the work of geographers'.[45]

Like Büsching also, Guthrie was concerned with the level of education as a significant factor in what he called 'the character of nations'; however, he did not move to incorporate this aspect as part of natural or physical geography, but rather defended its inclusion in what he called *moral or political geography*, where he added a section on 'the history and present state of learning' in each of the countries described. His comments on this topic are interesting: 'There is a nearer connection between the learning, the commerce, the government, etc. of a state, than most people seem to apprehend'; all these in his opinion are 'absolutely necessary for enabling us to form an adequate and comprehensive notion of the subject in general', yet such aspects 'till of late, seldom found a place in geographical performances'. Writers have been hampered, he added, since few travellers have taken note of such matters, being concerned more with 'avarice' than with 'instruction', although he acknowledged the recent emergence of the scientific traveller: 'In the course of the present century . . . a thirst for knowledge, as well as for gold, has led many into different lands' [6–7].

Guthrie in general showed a high regard for the improvement of education, a feature he shared not only with Büsching and the Scottish educators, but also with the whole tradition of the enlightenment in France and in earlier British empiricism from Bacon to Locke. He praised 'the rapid progress, and general diffusion of learning and civility, which, within the present age, have taken place in Great Britain', attributing this to improved government, the extension of prosperity, and especially the availability of books, not only to the learned or the wealthy few, but to the 'great body of the people':

Books have been divested of the terms of the schools, reduced from the size which suited only the purses of the rich, and the avocations of the studious; and adapted to persons of more ordinary fortunes . . . It is to books of this kind, more than to the works of our Bacons, our Lockes, and our Newtons, that the generality of our countrymen owe that superior improvement, which distinguishes them from the lower ranks of men in all other countries.

To promote such an improvement was Guthrie's principal aim, and he pointed out that 'books of geography', by providing 'knowledge of the world, and of its inhabitants', are particularly useful in this [5–6].

The association of geography with such programmes of general education was to remain a significant feature of the following period, when the movement of other sciences more firmly into the hands of specialists tended to cut off scientific inquiry even further from the common man. Both Kant and Humboldt responded vigorously to that challenge in the lectures they offered on physical geography. In Britain, however, this emphasis on popular instruction, combined with the retreat from philosophy, probably retarded the development of geographical thought in the last years of the eighteenth century.

New systems, old theory: the end of the century

A marked characteristic of this period was the absence of any new theoretical statements on geography by British writers, who seem to have produced no more than a series of texts modelled on Guthrie or earlier French and German works. Typical of these was the rather drab *Introduction to Geography* of Richard Gadesby who described himself as a 'Private teacher of Writing, Accounts, Geography etc.'; a cheap text, without maps, it nevertheless reached a third edition by 1787.[46] Another undistinguished book evidently popular at this time was the *Geography for Youth, or, a Plain and easy introduction to the Science of Geography*, which included maps showing recent discoveries by Cook and others; although anonymous, it was probably a development of Demarville's *Young Ladies Geography* since it incorporated several plates from that work.[47]

The history of the discipline too was neglected; indeed, one of the difficulties in writing any historical account of developments in this period is the lack of previous analyses, not only by recent authors but even by those of the eighteenth century. One work from the period, with the title *The History of the Rise and Progress of Geography*, is disappointing in this regard. In it the author, the Reverend John Blair, emphasized the need for a history of all the arts and sciences:

Geography therefore is, in this Respect, like every other Science, whose imperfect Beginnings ought to be traced, and the Time and Manner pointed out in which it received its gradual Improvements.[48]

Blair, however, was obviously one of those who equated geography with cartography, his chief concern being with improvements in map-making from the time of the Greeks, the correction of errors in Ptolemy with a more 'exact' determination of longitude, and the greater precision achieved with improved instruments in the modern period, compared with 'the Coarseness of all the Observations, both in Astronomy and Geography, prior to the days of Tycho' [170]. He concluded with a reference to the way in which advances in the measurement of time have been applied to the *Mensuration of Space* on the surface of the earth [181]. But he offered no account of geography in the eighteenth century: 'It was never my Intention in this Dissertation to enter into the Discussion of the present State of Geography, or to give the comparative Excellence of the Maps of different Countries' [182–3]. Nor was he interested in associating his work with the current textbooks, for he made it clear that although some maps had been added for illustration, 'it is far from my Intention to prefix any System of Geography' [3].

This is indicative of attitudes at that time to geography, which was identified either with cartography and exploration, or else with the so-called *Systems* of regional description produced by textbook writers, often on the model of Guthrie's work. One of these, Joseph Collyer's *New System of Geography*, published in London by John Payne in 1772, was later amplified by Payne in his own more cumbersome *Universal Geography formed into a New and Entire System*, published in Dublin in 1791, while the attempt to link history, geography, and astronomy in the interests of general knowledge was pursued by the Edinburgh headmaster Alexander Adam in his *Summary of Geography and History*, published at Edinburgh in 1794.[49] The title, *A New System of Modern Geography*, was obviously popular at this time, and it can be found prefixed to a later edition of Guthrie's *Grammar*, along with an assurance by the editor on the inclusion of the latest discoveries, notably those from Cook's voyages. This was a larger, more elaborate edition, on good paper, with coloured maps, and probably less suited to the purse of the ordinary person than Guthrie intended.[50] In 1790 Alexander Kincaid produced for a Society in Edinburgh *A New Geographical, Commercial, and Historical Grammar . . . executed on a plan similar to that of W. Guthrie, Esq.*, and in this work recent discoveries in astronomy, particularly those of Herschel, were added to Ferguson's introduction.[51]

Another example of the growing bourgeois tradition in British geography was *The New Royal Geographical Magazine: or A Modern . . . System of Universal Geography*, a lengthy work of over 950 pages with quality maps and illustrations, published at London in 'Periodical Numbers' around 1794 by Michael Adams, who reported in his preface that 'the Study of *Universal Geography* is now become the most *fashionable*, as well as the most *rational Amusement* of the present polite and enlightened Age'. Defining geography simply as 'a Description of the Whole World and what it contains' – a study 'calculated to *please*, as well as to *instruct*' – he claimed that 'it is not at all wonderful that *Geography* should hold a principal Rank among those Sciences, which are the delight of a refined Age, and tend at once to expand the Mind, and enlighten the Understanding'. Moreover, he added, with a reference to public interest in recent voyages of discovery in the South Seas, 'the late Improvements made in this useful Branch of Learning, founded on actual Experience, have greatly contributed to enhance its Worth'. Noting however that 'several Years have elapsed since any New System of Geography has been published', he proceeded to a comprehensive effort at a world description, incorporating information from Cook's maps and reports, as well as from other recent journals, and concluding with an account of developments in the French Revolution to 1794.[52]

The record for the smallest work in this tradition, however, must surely go to W. Peacock's midget pocket-book, *A Compendious Geographical and Historical Grammar*, based on John Mair's 1762 *Survey* and printed at London in 1795. Recommending this as a 'pocket companion' to Peacock's *Compendious Geographical Dictionary*, the author justified this addition to 'the vast number of publications on the science of Geography already put forth' by noting its 'brevity and clearness' and the need to provide information on the continual 'alterations and changes . . . [in] the affairs of the world'.[53]

This problem of constant changes in the world described by geography texts was perceived with even more clarity by the authors of a Scottish compilation of 1796, the *New and Complete System of Universal Geography*, brought out by Robert Heron and others at Edinburgh. In their opinion, 'the object of a *System* of *Geography* is to explain the natural and artificial appearances of the surface of this Terraqueous Orb', but, they pointed out, 'those appearances which it is the business of *Geography* to pourtray, are flitting' – each day 'alters the works of men, and the phoenomena of nature'.[54] The language used here, referring to descriptions not of objects but of appearances or phenomena, suggests the influence of Kantian thought. The plan of the work itself, however, with its historical and astronomical essays, and lengthy descriptions of political divisions, conformed to the established pattern, while the authors' recognition of the inability of geography to aspire to the scientific ideal of perfect knowledge was reminiscent of both Ray and Echard, a century before:

Geography borrows its lights out of every department of human knowledge; and, from the comprehensiveness and complexity of its plan is ever necessarily imperfect. In no part of the World, by no man, by no combination of men, can all the facts falling properly within the circle of Geography, be at once completely known [iii].

This, one of the most intelligent statements in the current geography texts on the limitation of knowledge, might have been the starting point for a criticism of the current ideal of scientific method. Instead, however, the writers proceeded to a traditional defence of geography for its utility and interest, and for its convenience as a repository of what they called a 'medley of information' about nature and society:

Books of *Geography* are multiplied beyond number; and the incessant demands of public curiosity, and the continual fluctuation and increase of materials . . . give daily occasion for new additions . . . Nor is there any more convenient repository than a *System of Geography*, in which the grand facts of *History*, the new discoveries of physical *Science*, the new observations of *navigators* and *travellers*, the communications of our periodical *newswriters*, and all that medley of information concerning nature and society, of which it is necessary for every man, in every condition of life to possess more or less; may be advantageously arranged [iv].

The question of the basis on which selection of such information should be made was not discussed here; this, however, was already becoming an issue in the association of geography with imperialism.

Geography for empire: the American rebellion

In that period the affairs of the world were undergoing significant changes, with the expansion of European empires and at the same time, in the war of

American Independence, the presage of their eventual disintegration. Geography, impeded in the previous century by its encounter with Church antagonism to Copernican theories, now became involved in conflicts over imperialism. Interesting in that regard is a two-volume work on the Guthrie model, *A New and Complete System of Geography*, produced at London in 1778 by 'Charles Theodore Middleton, Esq., . . . Assisted by several Gentlemen eminent for their Knowledge in the Science of Geography'. A large, rather pretentious work, it obviously was popular rather than scientific in outlook, with a motto evidently designed to appeal to the English gentleman who was not contemplating the Grand Tour: *To know the World, from Home you need not stray*. The view of the world that Middleton presented was unashamedly imperialist, a view of Europe not only advanced in culture, but dominant over the world. Geography was used here openly as an *apologia* for conquest. The frontispiece, a copperplate engraving, showed the four races presided over by an angelic Europe (see Fig. 4), and the theme was developed in a short verse:

> *Europe* by Commerce, Arts, and Arms obtains
> The Gold of *Afric*, and her Sons enchains,
> She rules luxurious *Asia's* fertile Shores,
> Wears her bright Gems, and gains her richest Stores.
> While from *America* thro' Seas She brings
> The Wealth of Mines, and various useful things.[55]

Eventually, the European colonists in America who had already achieved political independence became intolerant of relying on such texts for their geography. In 1794 an American edition of Guthrie's *Grammar* was produced at Philadelphia, the editor pointing out the necessity for such a move since Guthrie's account 'was exactly calculated to flatter the grossest prejudices of the English nation, at the expense of every other part of the human species'. Guthrie's article on England, he noted, ocupied 205 pages, one-fifth of the whole in the 1972 edition, while only 42 were given to France, 'thrice as populous, and more than twice as large'. Errors in the English editions were indicated sarcastically, especially those on 'the united states' where even quotations from 'the American Geography of mr. Morse' were found to contain several mistakes; altogether, the editor claimed that alterations and additions, including new maps, in the present volume, justified 'the title of an original work'.[56] Guthrie's title was retained, however, and although his preface was omitted, no alternative theory was offered in its place. Apart from the change of emphasis away from Europe in local descriptions, the American work clearly remained dependent in terms of theory on the British model. This had been equally evident in the first American geography, published five years earlier.

In 1789 Jedidiah Morse (1761–1826) published his famous text, *The American Geography; or A View of the Present Situation of the United States*

Fig. 4. A bourgeois Eurocentric view of the four races: frontispiece from Charles Middleton's *A New and Complete System of Geography* (1778). The link between British geography and imperialism was to persist during the next century.

of America, along with *The American Universal Geography: a View of all the Empires . . . in the known World, and of the United States of America in Particular.* Previously, Morse argued in the preface to his *American Geography,* 'Europeans have been the sole writers of American Geography, and have too often suffered fancy to supply the place of facts.'[57] Methodologically, this statement is of interest in providing an early indication of a positivist approach in American geography – an acceptance of the doctrine of facts, and the opposition of this to imagination. In other respects Morse's work was backward-looking in terms of theory. While offering new descriptions of Kentucky, the Western Territory, and Vermont, he accepted the seventeenth-century division of geography into *universal* and *particular,* and his own brief introduction to the general part retained an account of the Greek climates. Morse defined geography simply as 'a science describing the surface of the earth as divided into land and water' [11] and his work was clearly in the chorographical tradition. It remained, however, a highly respected text in the United States for over half a century.

The long popularity of Morse's work is indicative of the lack of vigour in American geography at this time. As Warntz has pointed out, the association of the subject with detailed regional descriptions was followed by the subsequent decline of geography at Yale and other universities early in the nineteenth century; the study of various aspects of geographical theory was taken over by a number of other disciplines, as eventually 'rising standards in the colleges coupled with declining intellectual content in geography' led to geography being dropped from the curriculum of Harvard in 1816 and Yale in 1825, although it was retained as an entrance requirement. Problems in defining the scope of geography, and the absence of a suitable textbook, may have contributed to that decline: in April 1804 the Report of the Overseers of Harvard recommended preparation of a new textbook on geography, both ancient and modern, dealing with General Geography as well as Particular, this proposal being an outcome of their attempt to determine 'the proper boundary, which divides Geography from the fields of Natural, Civil and Ecclesiastical History, Chemestry, Philosophy and Astronomy'.[58] The use of Bacon's terms by the Overseers in their three-part division of history is not surprising, since this was a time of strong Baconian revival in scientific thought, and similarly their reverting to a seventeenth-century theoretical framework for geography reflected the conservatism of the American approach to the subject in this period. Apparently, no new work resulted from the Harvard Report and by 1825, when geography was excluded from the Yale curriculum, the study of Morse's text was still being used for the entrance examination. It was only after a long hiatus that geography was reintroduced to Yale's Sheffield Scientific School between 1860 and 1863, and to Harvard in 1870. In the interval, it was a Swiss naturalist, Arnold Guyot (1807–84), a student of the current teachings of Humboldt and Ritter, who attempted to change school geography from catalogues of description to a science investigating the relationship of man and nature, and who, with his appointment in 1854 as the first professor of physical geography at Princeton, introduced what Warntz called the second cycle of college geography in the United States.[59]

Guyot, a student of Ritter, was invited in 1848 to give a set of lectures in Boston on the new geography from Europe, and these formed the basis for his book *The Earth and Man: Comparative Physical Geography*, 1849, in which Guyot dismissed the kind of definition favoured by Morse, of geography as 'a simple description of the surface of the globe and of the beings which are found there'. If it was to become a science, he wrote, '*Geography* ought to be something different from a mere description. It should not only *describe*, it should *compare*, . . . *interpret*'. Otherwise, he warned, echoing the challenging ideas of his German teachers, geography will remain a dry collection of 'unmeaning facts':

For what is dryness in a science, except the absence of those principles, of those ideas . . . by which well-constructed minds are nurtured. Physical geography, there-fore, ought to be, not only the description of our earth, but the physical science of the globe, or the science of *the general phenomena of the present life of the globe in reference to their connection and their mutual dependence*. This is the geography of Humboldt and of Ritter.[60]

The theory of science and the general concept of geography proposed here were largely those of Humboldt. It was much more the geography of Ritter, however, with his emphasis on divine design, that Guyot in his own book actually presented. Such teleological views were brought into increasing disfavour as concepts of evolution became widely accepted after 1859, and with it the rest of Guyot's geographical theory itself tended to be rejected. By the time Guyot retired in 1880, as Preston James has observed, 'the "new geography" he preached was not only old but its philosophical basis had been largely discredited'.[61]

This preview, as it were, of nineteenth-century conflicts in geographical theory indicates the kind of problems encountered throughout the first half of that century in efforts towards the reform of geography. American thought on the subject, as Guyot found it in 1848, had remained almost encapsulated, still virtually in the state of Morse's time, and his attempt to apply the ideas developed during the intervening years in Europe draws attention to some of the central issues involved. For geography to escape the encyclopedic role and become a science, as Varenius had argued two cen-turies earlier, it must develop a more adequate intellectual structure, to give coherence to observations. Yet in Guyot's work again, as in that of Ritter or Büsching before him and in the long tradition of natural history, the only conceptualization offered was that of the pre-ordained world of divine providence. It appears likely that the association of this view with the theories of Humboldt, as these were presented by Ritter and his followers, tended to bring them also into disrepute. In any event Humboldt's view of science, itself non-teleological and immensely constructive for geography, seems to have been dismissed in the United States almost without a hearing; if anything, a greater appreciation was sustained for Ritter by his students, as a model for comparative geography.

When in turn the Americans dropped their colonial status with regard to the study of geography and extended their own hegemony in that field into Europe by the mid twentieth century, they brought in many cases an inbuilt antagonism to general ideas, a fear of the speculative, and in general a positivistic empiricism unmellowed by the ideas developed earlier in Humboldt's writings. His theories, no longer enlivened by teaching and discussion, seem to have remained largely unread or uncomprehended in the volumes of his works in American and even European libraries, and the fate of the main attempt of his age to chart a course for geography and science between the twin rocks of Providence and Positivism is recorded today in the uncut pages of his books on library shelves, and in the footnotes to a few secondary sources grudged him in many works by the last generation of geographers – occasional literary genuflections, as it were, to the father-figure. The move by the American positivist school to reject Humboldt's theories as romanticism, indeed, recalled the conflicts in scientific thought during the eighteenth century that had given rise to his work.

CHAPTER 6

SCIENCE AND PHILOSOPHY: ENLIGHTENMENT CONFLICTS IN EUROPE

SCIENTIFIC EMPIRICISM: THE POSITIVIST MODEL

In the course of the eighteenth century, the scientific movement began to take on a more clearly-defined character. Increasingly, with the growth of industrialization and the expansion of European empires, science came to be identified with progress, for it promised the extension both of man's knowledge and of his mastery over nature, twin goals of the enlightenment. Today of course, although it is still customary to attribute this character to the scientific process, the idea of progress is not as confidently acclaimed as it was a century or more ago. Innovation of itself, it has been found, does not always lead to improvement; indeed, some of the most serious of current world problems – overpopulation, industrial blight, pollution – are regarded as an outcome of the scientific age. In the eighteenth century, however, the benefits of science seemed to admit of no questioning. With the adoption of a scientific method based on the apparently solid foundation of sense-experience and experiment, the natural sciences at that time became increasingly identified with materialism and objectivism, while they became progressively more specialized in their study of particulars, so that the achievement of any kind of integrated world view to replace the mechanistic one was made extremely difficult. Questions of aesthetics, values, or social responsibility in relation to scientific activity were rejected: atheists among the scientists generally seem to have regarded those issues as irrelevant, while the proponents of divine management evidently left them to God. In the face of this, a movement of opposition from the romantic school proved insufficient to effect any major change. Meanwhile the separation between science and philosophy, in spite of Hume's efforts, became more pronounced.

In the second half of the eighteenth century Immanuel Kant, the German thinker credited with introducing a modern revolution in philosophy, attempted to construct a reply to the scepticism of Hume and provide a synthesis of the opposing viewpoints of rationalism and empiricism, yet even his contribution evidently had little immediate impact on the science of his day. In England it was not until 1840 that the eminent scientist and philosopher, William Whewell, made what has been called a 'sharp break with the traditions of the British empiricists', by stressing the dependence of perception on ideas and attempting to base a detailed philosophy of science on Kant's theory of knowledge.[1] Significantly, it was in the same period that Alexander von Humboldt moved to extend the Kantian philosophy into the natural sciences, at the same time following Kant's interest in physical geography with a major effort to reinvigorate that subject after its long period of intellectual bankruptcy. His arguments for a modification of the empirical method to accommodate the field of geographical inquiry, and his unprecedented attempt in *Kosmos* to clarify the central concept of geography through a historical study of its development, offered a sound theoretical foundation for such a revival. By that time, however, the commitment of science to a positivist and materialist outlook was already well established. Increasingly, specialism was opposed to universalism, the doctrine of facts and scientific objectivity gave support to claims for the independence of scientific method from history and philosophy alike, while assertions of the amoral status of science were affirmed more strongly as the patent immorality of colonial and industrial society became the target of a movement for reform.

Enlightenment positivism and the doctrine of facts

Positivism, offering at its best some useful guidelines for orderly research in the sciences, at its worst exerting a rigid and restrictive influence on the range of inquiry, received its present name from Comte in the early nineteenth century; its formulation, however, can be discerned much earlier in the enlightenment period. Indeed, the Polish philosopher Leszek Kolakowski, in his study of this movement, pointed to David Hume as 'the real father of positivist philosophy'; in his opinion 'Hume clearly enunciated the basic principles of positivism' and these furthermore 'were one factor in the Enlightenment's struggle against superstition, metaphysics, inequality, and despotism'. These goals of the positivists were important ones and Kolakowski, whose sympathies seem to be indicated by his own antipathy to metaphysics, presented enlightenment positivism in an attractive light, as a movement for the advancement of science and the improvement of human life: 'an attempt to minimize differences among men by a sensationalist theory of knowledge (every human being comes into the world a *tabula rasa*, "blank slate"), an attempt . . . to do away with prejudice and barren speculation'. At the same time he noted that it was the action of Hume, in carrying the premises of empiricism to their ultimate consequences, which led to the destruction 'of all the hopes the Enlightenment had pinned on experience and common sense'; Hume's radical criticism of causality – although not extended, as critics have long pointed out, to his own genetic explanations of ideas – led to consequences evident in later positivism: the effective limitation of knowledge to *individual observations,* and its reduction generally to a pragmatic level. While the internal antinomy in Hume's theory of knowledge left the next generation of positivists with unresolved problems, his achievement was to formulate clearly the fundamental question of the certainty of knowledge, although, as Kolakowski admitted, 'none of the later positivists followed Hume in his rejection of induction'.[2]

On the contrary, indeed, strenuous efforts to defend the legitimacy of induction were a feature of subsequent positivist attempts to prescribe a logical and reliable method of scientific procedure in advancing from 'particulars' to general statements. Basic to all such attempts, as in the work of Hume himself, was an implicit acceptance of what might be called the doctrine of facts, the belief, suggested by Hume and unquestioned in this case even by Kolakowski, that while statements concerning causality or prediction may lead to error, the individual observation of some occurrence is itself free of inference and therefore beyond doubt. In the positivist reliance on facts can be seen the widespread application of a Baconian kind of objectivism with regard to scientific observations.

The word *fact* itself, derived from the Latin *facere*, to make or do, seems to have undergone a transition in meaning in the course of the eighteenth century, so that it came to refer not so much to an action as to some primary observation, a piece of incontrovertible evidence. Resisting this trend, Goethe towards the end of the century, and Whewell in the next, argued that facts must be regarded themselves as theories, while Alfred North Whitehead in 1919, in his famous lectures on the concept of nature, found himself obliged to urge a return to the notion of facts as events.[3] An important part in that transition to the doctrine of facts appears to have been played by the views of John Locke, who argued in his *Essay concerning Human Understanding* that all *matter of fact* involves propositions 'concerning some particular existence' which is accessible to the senses and is therefore capable of human testimony or observation. In discussing the limits of certainty and the degrees of probability in human knowledge Locke accepted the rationalist view that only demonstration from infallible proofs can lead to certainty, and he acknowledged that most of the propositions we reason or act upon therefore involve some element of doubt. Among these, however, he assigned the highest probability to those 'propositions about particular facts' that are confirmed by constant experience and attested in the reports of ourselves and others: 'These probabilities rise so near to certainty, that they govern our thoughts as absolutely, and influence all our actions as fully, as the most evident demonstration.'[4]

Such statements, in his view, represent 'an argument from the nature of things themselves', the force of proof being greater in the original observation than in repetitions of it in what he called *traditional testimonies*, farther removed from original truth: 'The being and existence of the thing itself is what I call the original truth.' Locke denied here any attempt to discredit history – 'it is all the light we have in many cases' – and he drew the parallel, in history itself, of using 'quotations of quotations' where the originals are lacking; nevertheless, his concern to move away from dependence on classical antiquity probably encouraged the objectivism evident in his view on direct observation. With regard to those things that do not fall 'under reach of our senses', Locke assigned an even lower degree of probability, depending chiefly on analogy, and he therefore classed as *speculation* all statements concerning the existence not only of immaterial beings such as 'spirits, angels, devils, etc.', but also of material beings either too small or too remote for observation by the senses, as well as all conjectures about the causes of such effects in nature as the life of animals or the magnetic action of the loadstone [IV, 16, 6–12].

Some of the weaknesses of Locke's theory are indicated here in his attempt to distinguish between fact and speculation. Nevertheless, his ideas exerted considerable influence on scientific thought in the eighteenth century, and his statement that the probability of *propositions about particular facts* can rise *near to certainty* is recalled in the writings of Leclerc de Buffon more than half a century later. Buffon in 1749, in the introduction to his monumental *Histoire naturelle*, went if anything even further than Locke in assigning certainty to observations of 'the facts of nature [les faits de la Nature]'.[5] Like Hume a decade before him, Buffon pointed out the limitations of mathematical proof. In Buffon's view what are called mathematical truths, while being exact and demonstrative, are only truths relative to the initial definitions and assumptions on which all conclusions are based, the truths of subsequent propositions therefore being no more real than the original supposition itself. He attributed a greater reality and certainty to the truths of the physical sciences:

Physical truths, on the contrary, are not at all arbitrary, and are not dependent on us; they rest on facts alone. A succession of similar facts, or if you prefer, a frequent repetition and an uninterrupted succession of the same events, constitutes the essence of physical truth; that which is called physical truth is therefore only a probability, but a probability so great that it is equivalent to a certainty. In mathematics we make suppositions, in physical science we propound and establish: in the first there are definitions, in the second there are facts; we go from definitions to definitions in the abstract sciences; we proceed from observation to observation in the real sciences; in the first we arrive at evidence, in the second, at certainty [I, 67–8].

At the same time Buffon stressed that the study of natural history should not be limited to making exact descriptions and verifying particular facts: it requires wide views (*grandes vues*) as well as laborious attention to detail. Faced with the variety and the *innumerable multitude* of nature's productions, he saw the importance of forming collections and studying individual objects, but in the serious study of nature he also saw the need to 'generalize ideas' and develop a system of explanation, since 'the method of proper inquiry in the sciences is still to be found'. In addition to keen observation, he suggested, 'we need general ideas, . . . and a type of reasoning based more on reflection and study', for the final task is 'to grasp distant relationships, . . . and to form from them a body of systematic ideas, after having carefully estimated and weighed the probabilities' [I, 2–65].

Like practically all the empiricists of his time, Buffon believed that research in science begins with the collection of particulars, or as he had come to call them, *facts*. But beyond this, he reminded his readers, 'we must try to rise to something greater and worthier of our concern: that is, to combine observations, to generalize facts, to link them by analogies' and so eventually to comprehend the more general effects and the grand operations of nature. To this end he expressed regret for the neglect of philosophy in the sciences:

Even in this century in which the sciences seem to be carefully cultivated, I think it is easy to say that their philosophy is neglected and perhaps more so than in any other century. The arts that people call scientific have taken its place: the methods of calculus and geometry, those of botany and natural history, formulas, in a word, and dictionaries occupy nearly everyone.[6]

Buffon's own work reflected this lack of vigorous inquiry in the philosophy of science: while asserting the need in the sciences for general ideas and a type of reasoning based on reflection, he continued to accept a theory of knowledge inherited from the British empiricists of the previous century, and to make even more dogmatic the doctrine of facts outlined in their work, a doctrine which was to provide the basis for the anti-philosophical methodology that took strong hold of science over the next centuries. During Buffon's lifetime the philosophy of Bacon and Locke was widely accepted in France, and the uncritical promotion of their ideas, often in rather simplistic form, is evident also in the extreme sensationist views that became popular among their French followers in this period.

The sensationist approach

The doctrine of facts itself, of course, was closely bound up with the sensationism implicit in the theory of knowledge constructed by the British empiricists. Locke's argument that all ideas are derived initially from sensation, along with his view that the desire for pleasure and the avoidance of pain form the chief motives of all human action, were adopted by Voltaire and many of his French followers, in some cases being taken to extreme limits. Condillac, whose *Traité des sensations* appeared in 1754, claimed that all human experience, even scientific speculation, and imagination, can be traced back to individual sensations recorded by one or other of the senses. This atomic approach to knowledge was widely adopted by the popular writers of the French enlightenment including Voltaire, Diderot, d'Holbach and Condorcet, while more organic views, as suggested occasionally by Diderot or Maupertuis, were not highly developed at this time.[7]

One important conclusion drawn by the French followers of Locke's sensationist approach was that man's conduct can be controlled through education, by what Condillac saw as a process of social conditioning. Locke's theory of the association of ideas, introduced in the fourth edition of his *Essay*, suggested that the operation of the pleasure–pain principle leads to the formation of trains of ideas associated with either pain or pleasure, according to the initial context. In 1823 the editor of a new edition of Locke's works referred to the association of ideas as 'that great and universal law of nature . . . which produces equally remarkable effects in the intellectual, as that of gravitation does in the material world'.[8] Helvétius (1715–71), a close follower of Locke in that respect, argued in his work *On the Mind* in 1758 that inequalities between minds are produced by education, and he there-fore proposed the development of a science of education. Accepting the dualist contrast between the material and intellectual worlds, he looked for comparable laws governing mental operations, and saw such a one in the effect of interest: 'If the physical universe be subject to the laws of motion, the moral universe is equally so to those of interest. Interest . . . changes the appearance of all objects.'[9] Working from Locke's sensationist theories, Helvétius here came close to developing a theory of perception in direct contradiction to the objectivism of Locke himself.

In contrast to Locke also, Berkeley, a consistent sensationist, denied any dualism between subjective sense-impressions and objective properties of physical objects; according to his phenomenalism such objects have no separate existence apart from the sense-perceptions of them, and he di-verged from the earlier British empiricists by rejecting materialism entirely. At the same time he shared their subjectivist outlook and the whole psycho-logical-genetic approach to knowledge which characterized their work. In their lack of concern for the social aspect of knowledge the empiricists of this period were in accord with Descartes, who commenced his inquiry, it will be remembered, with the statement 'I think'. Although Bacon had written of the need for community of effort in the sciences, his own recommendations indeed leading to the later formation of the Royal Society, nevertheless his followers in philosophy were concerned mainly with the thought-processes of the individual in his contact with the external world. They adopted also a genetic approach to knowledge, being concerned more with the origin of ideas in the mind of the individual than with their function in science generally, even though the professed aim of the empirical science for which they attempted to provide a theory of knowledge was to reject the Aristo-telian search for origins or essences in favour of a more functional approach. In all of their work the container theory of the mind was retained, in which, following the mechanical model of current physics, ideas were seen as atomic elements entering the passive mind and subsequently being ex-amined, as Locke suggested, by the inner eye of reflection. Maintaining the Aristotelian view of perception, they attempted also to apply to the mind the atomist and mechanistic views dominant in natural science since the seventeenth century.

Scientific materialism: the mechanist world view

Materialism, and the belief that science is concerned with the investigation of a world composed of material particles, can be traced to Greek thought, with the view of Democritus (*c.* 470–*c.* 400 B.C.), and his follower Epicurus (342–270 B.C.), that reality consists in the motion of indivisible and indestructible atoms in the void. Of the Romans who subsequently adopted the Epicurean philosophy, Lucretius (*c.* 95–55 B.C.) became noted among modern scientists for his work *On the Nature of Things* in which he claimed the action of atoms, rather than any divine creation, to be responsible for the formation of the entire world, including the whole scale of living beings rising to man and the human mind. Opposed by the supporters of natural theology, the atomism of Lucretius was revived during the enlightenment period and modified in the light of recent discoveries in physics and chemistry. This revival itself occurred in the context of a strong commitment by scientists to the mechanist theory of the world advocated by Galileo and Descartes. According to that view, it will be recalled, the entire world, and within it all organisms, even the human body, were believed to function as machines, only the human mind being exempt from the laws controlling the mechanical operations of matter in the divinely ordered universe. Even within the mind, Descartes assumed that sensory experience could be reduced to mechanics and become amenable to a mathematical kind of analysis.

Following Descartes, Hobbes put forward a mechanical theory of sense-perception and extended his materialism to include even the processes of thought. He sought to explain all mental activity as physical responses to matter in motion, following from the impact of sense-impressions. Subsequently, in the case of Berkeley, a thoroughly consistent phenomenalism led to a monistic viewpoint that rejected materialism altogether, but in spite of this the materialist philosophy, or at least the mind–nature dualism, remained the most widely accepted view among scientists for more than a century. Newton had given enormous support to the view of the universe as a machine, with his formulation of a single law governing the operation of all material bodies, and in the course of the eighteenth century further efforts were made to include all aspects of the world, including man, within the materialist framework of explanation. This problem became of extreme importance with the expansion of the biological sciences at this time, and the question of whether life itself is a mechanism capable of scientific analysis preoccupied many thinkers.

In 1747 La Mettrie (1709–50) with his *Man a Machine* developed the ideas of Lucretius and provided an interpretation of materialism which argued not only that matter can give rise to life, but also that organized matter in complex bodies is capable of producing purposive motion and even human thought and feeling. Man differs only in the greater complexity of his structure, and the whole natural order carries evidence of neither an internal moral imperative nor divine order; according to La Mettrie, mind and body form part of one complex and therefore are not governed by separate laws. Along with his vigorous materialism went a degree of moral nihilism that alarmed some of his contemporaries. At the same time, in his theory of knowledge an inconsistency is evident in his approach to the question of experience, since he maintained an anti-theoretical position, advising his followers to use 'the staff of experience, paying no heed to the accounts of all the idle theories of philosophers': one can admire 'all these fine geniuses in their most useless works, such men as Descartes, Malebranche, Leibniz, Wolff', said La Mettrie, 'but what profit . . . has anyone gained from their profound meditations?'[10]

The atheism and materialism of *L'Homme machine* was continued in d'Holbach's *Système de la nature* of 1770, and in a number of works during the intervening years by the great French humanist, Diderot (1713–84). Turning his attention to the problems raised by the materialistic monism of the time with regard to human life and values, Diderot argued that while the universe itself is purposeless, governed by mechanical laws and by chance, the sentience that he believed to be implicit in all matter and active in living organisms gives rise in man to moral order. Man makes his own morality, in accordance with human physiology and needs. Diderot attempted to extend La Mettrie's materialist framework in order to include poetry, philosophy and science in his own system of materialism, although in his *De l'Interprétation de la nature* of 1753 he expressed doubt that experimental science, given the limitations of human perception and understanding, could ever achieve more than to find increasing numbers of links between phenomena; it could never expect, as d'Alembert had hoped, to 'unite them all into a complete system'.[11]

In many respects the work of Diderot and d'Alembert with their great *Encyclopedia* of 1751 to 1766 indicated a more liberal approach to the role of empirical science than that adopted concurrently by many scientists of their time. The enlightenment concern for progress through knowledge, as they expressed it, emphasized man as a social being as well as an individual, and drew attention to the function of science in society as part of what they believed to be a general improvement in education, happiness and virtue for the human race.[12] To this end they showed an interest in the historical development of science itself.

Adam Smith: economics and the historical method

A similar interest in the historical approach was shown by Adam Smith (1723–90), the Scottish professor of moral philosophy who later, with his work *The Wealth of Nations* (1776), was to be acclaimed as the founder of modern economics. As a friend and admirer of David Hume, whose *History of England* had appeared in 1761, Smith now applied the historical method to the science of economics, basing his arguments not on deductions from a set of principles or axioms after the manner of Descartes or Hobbes, but rather on a consideration of previous experiences in different societies with regard to economic problems. At the same time, he brought to economics some of Hume's scepticism with regard to causation, stressing in *The Wealth of Nations* that similar causes can have different effects, and criticizing the mercantilist statesmen of his time who saw trade as the prime cause of national prosperity, rather than as a symptom of it.[13]

In Smith's opinion the chief source of a nation's wealth is the labour on which all production, and furthermore all increase in wealth, must depend, and in this view he took issue with the current French school of 'physiocrats' who argued that economies are governed by nature, since they saw wealth as a constant, dependent on land. Smith at the same time took over many of the reforming ideas of the French economists, their proposals for free enterprise, their dynamic view of the circulation of wealth in the economic system, and their systematic analysis of economic classes, an analysis later developed by Karl Marx. Combining this with the British tradition of 'political arithmetic' with its accumulation of statistical data on the economy, Adam Smith produced a new science of economics, based on the historical method and supported by statistical analysis.

Some aspects of Smith's programme are today open to criticism. His idea of a constantly expanding economy, for instance, is now rejected by leading economists who see the *growth syndrome* as a factor contributing to the present ecological crisis.[14] Furthermore, as the authors of *The Western Intellectual Tradition* pointed out, while Smith's view of society as an immense machine, harmoniously regulated by divine providence, led on the one hand to his argument that individual self-interest and the welfare of society as a whole are necessarily linked together, yet on the other hand the mechanical concept encouraged him to regard workers as mere machines of production, submitting to the brutalizing effects of the division of labour in order to increase productivity and thus promote the general welfare.[15]

The significance of Adam Smith's work, however, was considerable. Appearing in the year of the American Revolution, his *Wealth of Nations* encouraged a change of government policies in Britain towards free trade, a change recommended earlier by Hume in his mid-century writings. In addition, Smith founded the specialized science of economics, alongside the natural sciences of the time, as one of the first of the new sciences dealing with the civil or moral aspects of the world, and for this purpose he indicated the need for a new interpretation of scientific method. Avoiding the limitations not only of the Newtonian form of scientific empiricism as applied to physics, but also the extreme scepticism of Hume with regard to relations of cause and effect, Smith was able to supplement the objectivist 'particularism' of the British empiricists by calling on the experience of the past and incorporating the historical method in his study. As Bronowski and Mazlish saw it, Smith 'preferred to treat economics as a historical rather than as a logical or even as an empirical science'.[16] However, it can be argued that he was in effect expanding the idea of empirical method to include historical inquiry, and therefore was employing a wider notion of experience than that of an immediate encounter with the 'facts'. It is towards Smith's model of inquiry that modern schools of economics in many cases are now returning after a long period of positivistic quantification, a period when Marxist schools of economics were virtually the only ones that continued to assert a close relationship between economic factors and the social situation in which they occur. In the swing back towards this view among Western economists it must be remembered that the historical perspective and a concern for the sociology of knowledge are not in themselves a prerogative of Marxist thought.

Adam Smith's work in economics offers a remarkable parallel to the subsequent attempt by Alexander von Humboldt to develop a general science on the basis of the depressed study of geography. The method that each advanced, combining the historical and the statistical approach, might well have become the standard method for such sciences at that time, when a great expansion of historical research was taking place, with the first *Universal History* in English appearing in seven volumes from 1736 to 1744, Gibbon's great *Decline and Fall of the Roman Empire* from 1776 to 1788, and such distinctive works as Herder's *Philosophy of the History of Mankind*, in 1784. However a strong reaction to this trend was led by the French positivist school of the day, who chose instead to turn back to the empiricism of Francis Bacon.

Lavoisier: the exact science model

Standing in strong contrast to the approach of Adam Smith, or of his own countryman d'Alembert, was the attitude of Antoine Lavoisier (1743–94), whose *Traité élémentaire de chimie* is credited with establishing a revolution in modern chemistry. This work appeared in 1789, at a time when the main philosophic movement of the enlightenment was losing its coherence and a Baconian revival was under way in the natural sciences. Lavoisier's great achievement was to reject the phlogiston theory and introduce the quantitative method into chemistry, applying experimental and mathematical procedures, along with a new chemical nomenclature, to the study of elements. Working on the recent discovery of oxygen and other gases by Priestley and Scheele, Lavoisier was able in this way to transform chemistry virtually into a new science; however, giving little recognition to the priority of their researches, he explicitly rejected any concern in his *Elementary*

Treatise with the history of previous inquiry in chemistry and appears to have been one of the first to incorporate a strong anti-historical approach into the positivist method as it was formulated during the eighteenth century.

In the preface to his *Treatise* Lavoisier explained his programme for chemistry and went on to outline both his scientific method and the theory of knowledge on which it was based, in a way that few elementary texts in the physical sciences would provide today. The method advanced in his proposal, one clearly in the Baconian tradition of empiricism, is to draw conclusions only from experience and experiment, to rely on observation as the source of ideas. Many of Locke's teachings are incorporated in Lavoisier's theory of knowledge, beginning with a concern for linguistic analysis and for more precise nomenclature to improve reasoning in science: 'to designate', as he hoped, 'simple substances by simple words'. In addition he accepted the theory of the association or *enchaînement* of ideas and showed a strong commitment to the sensationist and genetic approach to knowledge. Seeking to explain how learning occurs according to that view, Lavoisier himself turned to the example of the child: basically, he argued that in the child, and in the sciences likewise, 'the idea is an effect of the sensation . . . it is the sensation that gives birth to the idea'. Furthermore, following Hume's application of the pleasure–pain principle, he suggested that, in the child at least, the sensations provide a guide to judgment, since in his view false judgments lead to suffering, correct ones to pleasure.[17]

With regard to the whole problem of classification and generalization in science, Lavoisier followed the objectivist teachings of the British empirical school, as interpreted for him in the extreme sensationism of the Abbé Condillac. It was to Condillac, rather than to Berkeley, that Lavoisier referred in arguing that every abstract idea, such as that of a tree, is based on observation of individual cases in nature, this being in his view an important aspect of the logic of all sciences, including chemistry [9]. Lavoisier's concern to find justification for every generalization not in *metaphysics* alone, but in *nature*, that is, in direct observation, was in itself probably a useful directive for current research in chemistry, and it came to form a central part of the positivist methodology. The real danger of that approach lay in the extension of the anti-metaphysical viewpoint to the point of accepting unquestioningly the doctrine of facts and rejecting all but the narrowest definition of immediate experience. This was the position taken by Lavoisier; in his view, the only means of achieving progress in the physical sciences and preventing the transmission of errors from age to age through the weight of authority, is to rely on facts:

to make continually the test of experience; to preserve only the facts, which are the data of nature, and which cannot deceive us; to search for truth only in the natural sequence of experiences and observations, in the same way that mathematicians reach the solution of a problem by the simple arrangement of data and the reduction of reasoning to operations so simple, to judgments so restricted, that they never lose sight of the evidence that guides them.

'Convinced of these truths', Lavoisier continued,

I have imposed on myself the law of never proceeding except from the known to the unknown, of deducing no consequence that is not derived immediately from experiences and observations [4].

By this means he hoped to enable chemistry to become one day, 'an exact science' [5].

Accepting Condillac's assertion that progress in the sciences will result when better observations lead to more exact language and so to better reasoning, Lavoisier felt confident in eliminating from an elementary treatise such as his own any discussion of the history of the science or acknowledgment of the opinions of those who had preceded him. The sciences already present enough difficulties, without introducing those to which they are strangers, he stated: 'It is neither the history of science, nor that of the human spirit, that is dealt with in an elementary treatise' [12–13]. It was only recent experiences, 'modern experiments', that were of concern to him, and even here he excused his failure to acknowledge the individual contributions of such French contemporaries as Fourcroy, Laplace, Monge, or Berthollet, by explaining that they all formed with himself a community who had adopted the same principles and acquired habits of community life: 'communicating our ideas, our observations, our manner of viewing, to establish between us a kind of communality of opinion' in which it would be difficult to distinguish the particular contribution of each individual [13–14].

This recognition of the social context of scientific inquiry, however, did not lead Lavoisier to question a theory of science that discounted previous learning and placed so much emphasis on the individual, seen as being face to face with the facts. On the contrary, it must be remembered that his *Treatise* reflected the impatience with past traditions that characterized the current revolutionary period in France; indeed, it was published in 1789, at the beginning of the Revolution to which Lavoisier himself was to fall victim five years later: although an avowed revolutionary in science, he was guillotined in 1794 for his association with the oppressive political and economic institutions of the old regime. In some respects his own text of 1789 seems to have shown an equal readiness to condemn the past as a source of error, as well as a similar ruthlessness in attempting to cut off the influence of that past. Turning, then, to the example of Francis Bacon, the deposed Lord Chancellor who had called for a new birth of science during the British century of revolutions, Lavoisier concluded his preliminary discourse with an affirmation of Bacon's approach, as expressed by Condillac:

When things have reached this point, when errors have so accumulated . . . it is a matter of forgetting everything that we have learned, of tracing our ideas to their origin . . . and of remaking, as Bacon said, the human understanding [15].

Commenting on this aspect of Lavoisier's work, J. R. Partington, a distinguished historian of chemistry, traced to his writings the deep hostility evident among contemporary chemists, even in universities, to any studies related to the history of the subject. He pointed out at the same time that Lavoisier's famous error regarding the nature of heat, with which the *Treatise on Chemistry* opens, might have been avoided if, before asserting his own caloric theory, he had studied the history of his science and become familiar with the idea of heat as a form of motion, a theory supported by Newton and proposed earlier in the writings of Francis Bacon.[18]

The importance of Lavoisier as a model in formulating the exact science approach to method seems to have been substantial. Just as the scientific empiricism of the seventeenth century, combining elements of Baconian empiricism and Cartesian rationalism, had been confirmed by the successes of Newton in physics, so the positivism of the eighteenth century was powerfully reinforced by the revolution effected by Lavoisier in chemistry. Meanwhile a series of enthusiastic, although unco-ordinated, efforts to find an alternative to the positivist model of scientific materialism can be traced in Europe during the enlightenment.

RESPONSE TO EMPIRICISM: FROM LEIBNIZ TO KANT

Throughout the eighteenth century, while Newton and Locke remained the two idols of the French enlightenment, the main opposition to their ideas was led by a succession of German thinkers. With the outstanding exception of the Swiss-French author Rousseau in his mid-century outburst against the science of his day, most of the leading French scholars were content to work within the conceptual framework established by the tradition of British empiricism: their suggestions for change, as in the case of Buffon, were often tentative rather than radical, involving contradictions in their own positions that remained unchallenged. Instead, the theories of the British philosophers, translated and popularized in simplistic form by Voltaire and his followers, were converted into the more extreme versions of sensationism and mechanical materialism that came to characterize the positivist model of science in France.

What the scientific empiricism of this time provided was not only a widely-accepted view of the world and its structure, but also some clear directives on the proper manner of studying it. Any major modification of that empiricism involved a confrontation on both these issues; it involved, that is, a new theory of nature and an alternative theory of knowledge. At the beginning of the eighteenth century only one thinker of international stature proved capable of attempting that task: this was the German philosopher Gottfried Wilhelm von Leibniz.

Leibniz: the dynamic world view

A contemporary of Newton, and a competitor for the title of discoverer of the infinitesimal calculus, Leibniz (1646–1716) was a many-sided genius whose wide range of skills, penetrating insights, and vast volume of writings defy any brief assessment of his work. What he offered, however, in his enormous output of ideas, obscure and even contradictory as many of these proved to be, was a major challenge to the mechanistic view of the world. Although attracted in his youth to the mechanical explanation of nature that the scientists of the day had favoured in opposition to scholastic teachings, Leibniz came to realize that an explanation in terms only of *extended mass* was insufficient, and in his article, 'New System of Nature', in 1695, he argued that 'we must again employ the notion of force . . . despite its springing from metaphysics'. He was opposed also, like Ray in the same period, to the Cartesians who 'degrade animals into pure machines', and he criticized the moderns for their lack of broad ideas about the majesty of nature when they equated the mechanisms of nature with those of man, for in his view these differ not only in scale but also in kind. Unity, continuity and harmony form the essential attributes of nature, as he saw it, and he denied that these can come from passive matter alone:

In the beginning when I had freed myself from the yoke of Aristotle, I had taken to the void and the atoms, for they best fill the imagination; but on recovering from that, after many reflections, I realized that it is impossible to find the principles of *a true unity* in matter alone or in that which is only passive, since everything in it is only a collection or mass of parts to infinity.[19]

It is known, said Leibniz, 'that the continuum cannot be composed of points'. Therefore, rejecting the idea of *material atoms*, which in any case he claimed to be contrary to reason, since matter itself must be composed of parts, he proposed instead active centres of force, indivisible and indestructible, and somewhat analogous to souls. In this he showed sympathy with Jacob Boehme (1575–1624) and the school of iatrochemists with their ideas of vital force that had become well known in Germany during the previous century. The nature of substance requires progression or change, in his view, and this requires each unit to contain within itself its own internal force or capacity for action, unless the occasionalist argument of constant divine intervention were to be accepted. Within the chain of being, each unit or soul, according to Leibniz, is independent and capable of spontaneity, yet its actions occur in conformity and harmony with others since each represents in its own way the whole universe. In this manner he explained also what he saw as the *union of soul and body* in man, as a mutual relationship regulated in advance, in accordance with the harmony of the universe [107–16]. Subsequently in his *Monadologie* of 1714 Leibniz gave the name of *monads* to these units, and although his theory remained in many respects obscure, nonetheless in effect it transformed the ancient atomist tradition: with this, as Crocker has pointed out, Leibniz brought into eighteenth-century philosophy the concept of the dynamic, pluralistic universe, continuous but changing, and in addition to that change from a static to a dynamic world view, he introduced a new idea of *organism* as contrasted with the machine, the organism, whether monad or universe, being considered as a *whole*, greater than the sum of its parts.[20]

It has been argued that along with this dynamic view of the universe Leibniz continued to defend the fixity of species, denying the degeneracy or extinction of any species in what he called the best of all possible worlds.[21] Nevertheless, his ideas gave a strong impetus during the eighteenth century to the vitalist school of thought in its conflict with the mechanist views of the Cartesians. Similarly, Leibniz offered a lead to critics of Cartesian and Newtonian physics; arguing that substance cannot be divorced from action, he rejected the Cartesian idea of matter as timeless extension and, as Philip Wiener has observed, 'provided physical nature with a historical dimension'.[22] With this he rejected also the Newtonian ideas of absolute space and time: 'I hold *space* to be something *merely relative*, as *time* is; . . . I hold it to be an *order of co-existences*, as *time* is an *order of successions*.' Therefore, he concluded, 'the fiction of a material finite universe . . . in an infinite empty space, cannot be admitted'.[23]

Apart from Berkeley or Locke, however, few thinkers until the twentieth century shared such relativist views. In general, Leibniz did not attract a wide following among scientists in his own time. To some extent this may have been due to his support of natural theology, his primary aim being to harmonize religion and science through a more satisfactory interpretation of divine providence: a progressive in science, Leibniz remained a conservative in politics and religion. In his theory of knowledge, moreover, where he sought to reconcile the classical philosophies with the new experimental sciences, he took issue with the school of sense-empiricism, and identifying more with the rationalist tradition of Descartes, continued to defend the notion of innate ideas. That view, of course, was anathema to the empiricists of this period, and the more constructive aspects of his criticism were probably rejected along with this.

Leibniz: a rational empiricism

Leibniz himself, however, was by no means completely opposed to the empiricist theory of knowledge, for throughout his career he too stressed the need for an organized programme of experiments and observation in science, arguing as early as 1680 that 'we must distrust reason alone . . . it is important to have some experience or to consult those who have it'; for this purpose he urged also, like Bacon before him, that more of the unwritten knowledge concerning the mechanical arts and practical activities should be recorded. Discussing the need for the rational interpretation of experience, he declared that 'even theory without practice will incomparably be superior to blind practice without theory', but at the same time he pointed out that the distinction between these two is often confused. Any practice can be seen itself as a particular theory if not a general one: 'For if a workman who may not know either Latin or Euclid . . . knows the reasons for what he does, he will possess a genuine theory of his art.' On the other hand, Leibniz added, 'a half-baked scientist puffed up by an imaginary science will project machines and constructions which cannot succeed because he does not possess all the theory required'.[24]

The great hope of Leibniz for progress in the sciences was an improvement in the method of reasoning, and to this end he not only saw the importance of probability theory, but also made an early contribution to the development of symbolic logic. Nevertheless, while defending the importance of effective logic and clear ideas, he took care to point out that concepts without distinct referents are merely symbolic, describing this kind of knowledge as *blind* a century before Kant made famous a similar statement with regard to concepts unsecured by experience.[25]

It was in the philosophy of Kant that the ideas of Leibniz found their strongest expression during the eighteenth century, and subsequently the work of Humboldt gained inspiration from the same source. The strongly theistic teachings of Leibniz, however, provided a framework difficult to reconcile with the growing scepticism among scientists in regard to divine design and determinism generally, and in this his work is to be compared with the natural history of Burnet and Ray in England. Leibniz himself favoured a theory of the two kingdoms of nature, a *kingdom of power* or efficient causes, in which the natural forces of *bodies* are subject to mechanical laws of motion or *mathematical laws of quantity*, and a *kingdom of wisdom* or final causes, in which the natural powers of *spirits* are subject to moral laws. He believed that bodies operate without liberty, while souls which are capable of reason operate with liberty, yet acting in perfect accord with the mechanical kingdom since in the system of pre-established harmony, God has ordained both to function like two clocks regulated together: 'These two kingdoms everywhere interpenetrate without confusing or disturbing each other's laws.'[26] Following from such views, subsequent efforts to distinguish 'social laws' in science from 'exact' mathematical laws were to hamper the development of human geography and the social sciences in the next century.

For Kant, one of the major sources of stimulation provided by the works of Leibniz was the lengthy critique of Locke's theory of knowledge, composed between 1693 and 1709 under the title *New Essays concerning Human Understanding*, but not published until 1765, when it appeared posthumously in a collection of his philosophical works in French and Latin, at Leipzig and Amsterdam. In this work Leibniz mounted a strong attack against Locke's sense-empiricism, denying that in the acquisition of knowledge so much can be attributed to the senses. Instead he stressed the operation of the reason, the function of ideas, and the importance of logical demonstration, as well as the value of comparative and historical studies in science.

Leibniz dismissed Locke's theory that the mind is a *tabula rasa* on which the senses provide the material for all knowledge: intellectual ideas, he argued, do not come from the senses. At the same time he accepted, just as Locke did, the Aristotelian separation between the senses and the reason. His response therefore was to assert again with Descartes the Platonic doctrine of innate ideas, although he differed from the Cartesians to some extent by defining these as principles arising from the original constitution of the soul itself. He argued that we have inborn within us those general conceptions or necessary truths without which the evidence of the senses cannot be united with ideas and so lead, in Platonic terms, to universals: 'The senses, although necessary for all our actual knowledge', in his view, 'never give us anything but examples, *i.e.* particular or individual truths.' Such examples, he pointed out, as Hume was to do later, 'do not suffice to establish the universal necessity of that same truth'; it remained in his terms a contingent truth or 'truth of fact'.[27]

For an assurance of perfect universality and necessity in knowledge Leibniz turned to the innate principles that in his view distinguish human knowledge from that of the brutes and confirm the divine plan of universal harmony. Experience for Leibniz implied no more than induction and examples; principles or necessary truths are innate, and he rejected the possibility that general notions, *koinai ennoiai* in the Greek, on which uniformity of opinion has been reached, can be acquired in any other way.[28] Moving from this position Kant later was to develop a more complex notion of experience in his effort to provide a synthesis of rationalist and empiricist views. Meanwhile in the course of the eighteenth century two further challenges to the mechanistic world view of scientific empiricism were developing in Europe: the vitalist, organic and evolutionary concepts of natural science, and the combination of social and epistemological radicalism that came to be designated as romanticism.

Beyond mechanism: the problem of life

The question of whether living organisms and the existence of life itself can be explained by the mechanistic hypothesis came to the forefront of discussion in France during the enlightenment as the study of plants, animals and social organization became a central concern. On the one hand, the argument of reductionism by the proponents of mechanical materialism claimed that the study of living matter can be reduced to the laws and methods of physics. On the other hand, the vitalists argued that life or vital force requires a changed conceptual structure for its study.

In the context of that debate the study of natural history, meanwhile, was being pursued with great effectiveness. Classification was seen as the prime task of scientific study, and to this end Karl Linnaeus (1707–78) with his *Systema Naturae* of 1735 and his *Philosophia Botanica* of 1751 adopted a method of grouping plants according to the external reproductive organs of the flower, and proposed a system of classifying all plants and animals into classes, orders, genera and species; the system of binomial nomenclature he thus introduced became standard after him. Linnaeus supported the idea of the fixity of species and this, along with his acceptance of discrete classes, accorded quite well with the static and mechanistic view of nature at that time. In contrast, Buffon in his *Natural History* opposed such an artificial classification and favoured the idea of the continuity of nature as advanced by Leibniz. Nature, in Buffon's words, proceeds always by imperceptible *nuances*, so that although the general method of natural history calls for a division into classes, this must be in large measure an artificial imposition of the human mind, as in botany for example a certain number of plants can always be found as anomalies between classes.[29]

Buffon moved his emphasis in natural history from classification to description, and the English translator of his *Histoire naturelle* praised his careful observations of organisms:

Ray, Linnaeus, Rheamur, and other of his cotemporaries, deserve much credit for their classing of animals, vegetables, etc. but it was *Buffon* alone who entered into a description of their nature, habits, uses and properties.[30]

While Ray and the British writers of natural history had actually paid considerable attention to describing living organisms, Buffon extended the scope of his work to attempt a vast synthesis of knowledge on nature: 'Natural history', he claimed in his first volume, *Theory of the Earth* in 1749, and again in his *Epochs of Nature*, 1778, 'embraces all places and all times and has no limits other than those of the universe.'[31] In consequence, like the general geographies of the day, Buffon's work grew to a prodigious size, and although continued by others after his death until it reached 44 volumes in 1804, it remained an unfinished compendium of information. Already the trend was towards the development of specialized sciences – botany, zoology, geology, paleontology – and the comprehensive view that the writers of natural history attempted to achieve seemed for a time to have been lost. In many respects Buffon challenged the static view of the world characteristic of both science and natural theology in his day. Like Leibniz, he supported the idea of continuity and change in nature, and he came to interpret this clearly in terms of changes in the species themselves. Although he saw this chiefly as a process in which degenerate types emerged, his ideas played a significant part in the transition to evolutionary views that became stronger by the end of the century. During this period the concept of evolution was the subject of bitter debate, but it brought to a wide range of researches in the natural sciences a common focus of interest and a measure of coherence that was noticeably lacking in the geography of the time.

The idea of evolution

A number of writers, including Maupertuis (1698–1759) and Diderot, in the mid-century period expressed interest in the idea of evolution, recalling for this purpose the theories put forward in pre-Socratic times by Greek thinkers such as Anaximander and Empedocles. The idea of the great and immutable *chain of being* that had dominated Western thought since the time of Aristotle now came under closer scrutiny, and the attempt to defend this as the embodiment of the principles of plenitude and continuity, and in all as the manifestation of eternal rationality, became, as Lovejoy has pointed out, increasingly difficult. One response, to justify the gaps being recognized in the supposedly perfect chain, was to adopt the historical approach and suggest that, in order to appreciate the completeness of this chain, man would need to know 'the entire sequence of forms in time, past, present, and future'.[32]

This change, then, to viewing the hierarchy of beings not as an instant and final creation, but as a process occurring through time, became more clearly articulated during the eighteenth century, although intimations of it are evident among earlier writers. It represented, moreover, a further step in undermining the ideal of absolute rationality, since for those who previously had sought an explanation of both the world and human knowledge in terms of reason, time had been considered irrelevant. Even in the twentieth century, that ideal continued to exercise a strong influence in the schools of analytic philosophy, where Lovejoy's succinct statement seems particularly appropriate to their approach: 'Rationality has nothing to do with dates' [242]. The ideal of detached and timeless rationality, so important to Descartes and to his followers in the Age of Reason, carried with it many implications for the kind of inquiry pursued in the sciences, where mathematics and logic were highly valued; it influenced also the theory of knowledge held by the rationalists, as evidenced in their quest for eternal truths secured in human reason, and in their approach to education as a logical programme of instruction for the intellect. It is no coincidence that the development of history, the expansion of natural science, the study of the dynamic world in terms of time and place, and the first proposals for radical changes in the schooling of the young, all occurred in the same period during the eighteenth century when the doctrine of feeling and a concern for the moral betterment of man in society challenged the rationalist world view. These, together with the theory of progressive evolution in nature, formed elements of a process that brought a momentous change in outlook in the course of the enlightenment, although not without strong reactionary moves on the part of some groups among scientists and churchmen alike.

To return, rather briefly, to the development of evolutionary ideas in the second half of the eighteenth century, it is sufficient here to note the introduction of the idea of progress into the theories of biological evolution proposed by Jean Baptiste Robinet (1735–1820), who emphasized the progressive development of human knowledge as part of continuous evolution, and Charles Bonnet (1720–93), who postulated a resurrection of higher forms after a series of catastrophes in the earth's history.[33] The search for a mechanism by which such changes could be explained without recourse to divine providence was commenced later in the century by Erasmus Darwin (1731–1802) and Jean Baptiste Lamarck (1744–1829), while in Germany a group of nature-philosophers including Goethe (1749–1832) and Lorenz Oken (1779–1851) became concerned with the morphology of plants and animals, particularly the common features of their structure. In geology also, evolutionary theories produced by Buffon and other researchers in the second half of the century were followed by the doctrine of uniformitarianism expressed in *The Theory of the Earth*, 1795, of James Hutton (1726–97).

Meanwhile in the years prior to the French Revolution a number of writers in different countries contributed to a growing literature on the social and intellectual evolution of man himself. Interesting and important as that literature is, it cannot be examined in detail here, although a significant division of opinion among these writers can be noted. On the one hand was a vocal and articulate group that included Voltaire and the French *philosophes*, who rejected religious orthodoxy and argued that the progress of man will depend not on divine intervention but on the improvement of society, through improved institutions and enlightened programmes of education. Montesquieu (1689–1755), for example, with his *Persian Letters* of 1721, criticized French religious and political institutions, while in his *Spirit of Laws*, 1748, he argued that man's evolution to a developed society depends on the wise legislator, whose influence in promoting the general spirit of the nation can counteract the effects of environment and enable the triumph of morality over physical causes in general.[34] Following Montesquieu, who is credited by geographers with changing the connotation of the term *climate* from the Greek idea of a latitudinal zone towards the modern usage of temperature or weather conditions, the question of the influence of climate and other factors on man was considered further by a number of writers in the following years, among them the English author William Falconer, in a short work in 1781. Noting that this subject, although casually touched on, had never before been particularly considered in any work, Falconer set out to improve on the contribution of Montesquieu, whose inferences concerning the effect of heat and cold on the human body had been made only from observations of 'a dead sheep's tongue'. Falconer himself chose to consider the 'effects on the *living* human body', and adopting a more geographical approach, with a reference to Zimmermann's recent Latin work on the *Geography of Animals*, he argued that while man 'is still liable to be considerably affected, both in his body and mind, by external circumstances, such as climate, situation etc', nevertheless many examples indicate that such influences can be counteracted. In particular he claimed that the universality of the human species, the ability of man alone among the animals 'to subsist in almost every climate and situation' indicates the power of man's reason, of 'his rational faculties, which enable him to supply the defects, and correct the exuberances of particular climates and situations'.[35]

This confidence in the ability of human reason to overcome difficulties in nature and establish an enlightened social order was widely expressed throughout the century. In contrast, however, to that search for a secular morality, a religious interpretation of social evolution was advanced with equal determination, particularly by the German school of nature-philosophy, a school that followed Bruno, Boehme, Spinoza and Leibniz in opposing the mechanists and viewing the world as the outward manifestation of the eternal Spirit. After the time of Herder (1744–1803), the idea of historical development was incorporated more strongly in that conception, and by the early nineteenth century human progress was presented, in the work of Friedrich Schelling (1775–1854) and Oken, as the evolution of the divine idea, a view developed with great force by Hegel in the next decades.

The contrast between the German vitalist interpretation of nature and the mechanist views that had predominated in Britain and France throughout the enlightenment seems to have been accentuated in the years following the French Revolution. If anything, the opposition of German spiritualism to the French atheistic materialism became even more pronounced during the period of Napoleonic conquests, when a movement of German nationalism drew support from romantic doctrines. The revival of natural science, and of geography with it, that took place in Germany during the second half of the eighteenth century occurred, then, at a time of great conflict in thought. For a time under Frederick the Great, French culture and language were promoted enthusiastically in Prussia, the Berlin Academy of Sciences being placed under the presidency of the Frenchman, Maupertuis, from 1746 to 1759. Subsequently a break occurred and it seems that the political and military excesses which followed the Revolution of 1789 prompted a strong reaction against French thought. During that period, a brilliant contribution to evolutionary theory, produced in 1755 by the young Kant in response to the Newtonian teachings of Maupertuis, was largely neglected in his own country.

Kant and cosmic evolution

Immanuel Kant (1724–1804) was one of a number of outstanding German thinkers in this period who attempted to reconcile the scientific and philosophic doctrines of the British and French enlightenment with their own traditions. Prior to his famous contributions to philosophy made in his 'Critical' period of 1781 to 1790, Kant produced twelve philosophical works, two treatises on anthropology, one on education, and at least ten on natural science, including two on the motions of the earth, one on the origin of the winds, three on earthquakes and his *General Natural History and Theory of the Heavens*, published anonymously in 1755. Sub-titled *An Essay on the Constitution and Mechanical Origin of the whole Universe, treated according to Newton's Principles*, this work offered the first clear proposal for the nebular hypothesis of the origin of the solar system, a hypothesis which in effect extended the idea of evolution to the whole universe. Five years earlier, Thomas Wright of Durham, in *An Original Theory or Hypothesis of the Universe*, had attempted to apply Newton's theory outside the solar system, but Kant with his idea of cosmic evolution became, in the opinion at least of his translator, Professor Hastie, 'the great founder of the modern scientific conception of Evolution'.[36]

According to John Goldthwait, another capable translator and critic of Kant, the nebular hypothesis, if published in England or France at that time, would have placed Kant alongside Newton in importance in physics.[37] As it was, however, Kant's work encountered many difficulties in circulation and seems to have been largely ignored; it was the rival nebular hypothesis of the Frenchman, Pierre Simon de Laplace (1749–1827), appearing in his *Exposition of the System of the World* in 1796, that became widely accepted.

Laplace with his research in probability theory and his impressive mathematical exposition of Newton's propositions was convinced that further analysis would enable science to explain all events within the mechanistic system of the Newtonian cosmology. Like Lavoisier's, his philosophy of science followed the tradition of Bacon and Newton in arguing for induction from particular phenomena to general principles, based of course on the determinist assumption of the existence in nature of unchanging and therefore predictable order and regularity:

In the midst of the infinite variety of phenomena which occur in continual succession in the heavens and on the earth one can recognize the small number of general laws which matter follows in its movements.[38]

Like Plato he pointed to the contrast between the apparent and actual movements of the heavenly bodies: 'Guided by induction and analogy we proceed by comparing the appearances, determining their actual movements, and so rising to the laws of these movements' [102]. Subsequently Humboldt, who worked for many years with Laplace in the Paris Observatory and in membership of the private Society of Arcueil, was to acknowledge the *Exposition du Système du Monde* as a model for *Kosmos*. Laplace's *Système*, however, dealt with astronomy and not with geography, discussing the movements, apparent and real, of heavenly bodies, the laws of motion, the theory of gravity, and developments in the history of astronomy. Its significance for Humboldt can be seen in Laplace's effort to provide a comprehensive view of phenomena, and his more liberal view of scientific method:

If man were limited to the collection of facts, the sciences would become nothing more than a sterile catalogue and never come to know the great laws of nature. It is in comparing the facts with each other, in discerning their relations, and in thus linking the most diverse phenomena, that one finally can discover the laws which are always discernible in their most varied effects.

In this way the researcher will find that the chain of causes and their effects in the continuing series of events will be opened to his view: 'It is uniquely in the theory of the system of the world that the human spirit, by a long series of happy efforts, is elevated to this height' [48].

At this time scientific leadership was still being exercised by the British and French, and a full recognition of Kant's contribution to science did not come until well into the nineteenth century. Two editions of Kant's works, commenced in 1838, gave little attention to his scientific theories. In the Rosenkranz and Schubert edition of 1838–42, for example, the interest of the editors was primarily philosophical and Hegelian – Kant's *Cosmogony*, as Hastie noted indignantly, being included there in the section headed 'Physical Geography'. Finally, however, it was another contributor to the study of physical geography who played a large part in the revival of interest in Kant as a scientist: Hastie himself assigned a leading role to Alexander von Humboldt in this revival, suggesting that it was at the instigation of his friend Humboldt that Arago the distinguished French astronomer, in a Paris journal of 1842, initiated a move to acknowledge the scientific work of Kant.[39]

Kant's own work in geography, along with his major philosophical contributions to empiricism and to the idea of moral evolution, will be considered later in this discussion. First, however, it is necessary to return to the period of his *Natural History* when Kant, like other thinkers of his time, was challenged by the iconoclastic assaults of the Swiss writer Rousseau on the scientific, political and educational traditions of the century.

Rousseau and romantic empiricism

A political revolutionary in outlook, arguing vigorously for freedom and equality, Jean-Jacques Rousseau (1712–78) has at the same time been accused of laying the foundations as much for totalitarianism as for democracy, with his doctrine of the subservience of the individual to the general will under the total supremacy of the state.[40] Similarly it can be argued that the effectiveness of his crusade for the reform of social institutions was undermined by his rejection of partial societies, such as religious or political groups, within the unified despotic state. Always the radical –

impatient, daring and impulsive in his proposals for change – Rousseau was not one to emphasize continuity of development or the stability of tradition in his proposals for reform, and the inconsistencies of his position, with its mixture of extremist and reactionary elements, were not worked out in practice. This was the case also with his radical approach to science and the theory of knowledge, an approach which, in view of the stimulus it provided to the romantic movement of the time, has been well termed his *romantic empiricism*.[41]

Rousseau's first publication, in 1749, was an essay, the prize-winning entry in the Academy of Dijon's competition on the topic, 'Has the progress of the arts and sciences contributed to the purification or the corruption of morals?' In response Rousseau launched the first major attack since the time of Francis Bacon on the belief, shared by empiricists and rationalists alike, that advances in science, technology, and the refinements of educated society, must lead to the progress of man. Morality had been given no central position in Bacon's theory of science, and rationalists had assumed that reason, the highest expression of human superiority and dominance over nature, must always lead to what is good. But in his *Discourse on the Moral Effects of the Arts and Sciences* Rousseau accused the sciences, along with the current institutions of society, of corrupting the natural goodness of man. Rousseau, himself of working-class background, rejected the sophisticated life of current French society and with the fervour of the pietist sects of the day, went on in his later works to argue for an ethical renewal of man through a kind of emotional revival.

What must be recognized throughout his work, however, is more than political or religious radicalism – it is a constant, if sometimes ill-guided, attack on the whole theory of knowledge underlying the science and the education of his time. This is particularly evident in his famous treatise on the proper education of a boy: his *Émile* of 1762. Here Rousseau brought a refreshing and humane approach to the teaching of a child, with a concern for meaningful experience that went well beyond Locke's proposals of seventy years before. He defended the importance of emotions and feelings, as well as the child's moral development, giving these priority over intellectual experiences. At the same time, in a piece of extreme reactionism, he turned back to Bacon's 'sweep away all theories' dictum, and argued for the discarding of tradition, the almost complete elimination of book-learning, and an innocent reliance on nature as the guide of youthful development. This was a kind of moral objectivism, comparable with Bacon's objectivism of sense-experience, although it seems that even Rousseau himself was aware of the limitations inherent in his ideal of the noble savage, the good man of nature.

In many cases of course Rousseau was justified in rejecting the literary instruction of his time. Professional pedagogues, he complained, teach nothing but words, and even among the various sciences they boast of

teaching they are careful to choose sciences, not of things, but of *terms*:
'heraldry, geography, chronology, the languages, etc., all studies so remote
from man, and especially from the child, that it would be a marvel if they
ever came to be of use even once in a lifetime'.[42] As a result, in geography
lessons 'the child is always limited to symbols, without ever comprehending
the things which they represent . . . I remember seeing somewhere a
geography which commenced: *What is the world? It is a cardboard globe.*
That is precisely the geography of children' [106]. Instead he suggested that
teaching should begin with the child's observations of his own surroundings:

I would like, for example, to approach geography from both ends and combine the
study of the revolutions of the globe with the examination of its parts, commencing
with the place where he lives. While the child is studying the sphere and being
transported to the skies, bring him back to the regions of the earth and show him first
of all his own home [190].

The child should start with his home town and his father's house, drawing a
map of them for himself and adding rivers or other features as he locates
them: with this proposal Rousseau apparently provided the first popular
model for the later custom in schools of studying the local region.

Rousseau's approach was enormously influential, not so much in effecting
immediate changes in schooling, since his programme was plainly too radical for a conservative society and too elitist, with its requirement of an
individual tutor, for a democratic one. His effect instead was more marked
on the intellectuals of his time, especially in educational thought and in the
literary sphere, where his work stimulated that concern for imagination and
emotions which provided a basis for the romantic movement. Unfortunately
that movement itself degenerated in many instances into the excesses of
sentimentality that brought discredit on romanticism as a whole. Much more
serious issues were involved, however, and it was a limitation of Rousseau's
thought that he was unable to provide coherent leadership with regard to
them.

On the one hand, Rousseau grasped an issue of tremendous importance in
his attack on the science and the education of his day for its lack of morality.
On the other hand, he moved, almost instinctively it seems, to oppose the
exclusion of the feelings and the imagination from the process of learning.
These were all elements, it will be recalled, in the *moral faculties* that Bacon
had adopted from Aristotelian teachings and attempted to exclude from
scientific inquiry. Like Bacon, however, Rousseau was not at all critical of
this aspect of Aristotle's philosophy; it seems he failed to recognize the
classical origin of this division of the mind into separate faculties of sense,
emotion, imagination and reason, and made no forthright effort to question
that division itself. Ignorant of the classical past, he too seems to have been
obliged to perpetuate its weaknesses, and the faculty model of the mind was
continued in his own work.

In effect, Rousseau's theory of knowledge did not advance much past the sense-empiricism of Bacon or Locke; if anything, it extended the sensationism of the French school by giving greater emphasis to the diffuse sensation of emotion than to the specific impressions of the individual senses. But the major elements of romantic thought – a concern for ideas of unity and wholeness, for synthesis rather than analysis, for morality rather than mere intellection, for a total concept of experience, including the feelings and the imagination, instead of a simplistic division between the senses and the reason – these were not directed by Rousseau's work towards an effective programme either for education or for scientific inquiry. Instead, after him, in spite of the efforts during the neo-Platonist revival later in the century by such an outstanding figure as Goethe to effect a link between philosophy, science and art, the contrast widened between a literature of sentimentality and fanciful imagination on the one hand, and a positivistic, 'exact' science of the kind that Lavoisier favoured, on the other. This was the problem encountered by Kant, and he devoted his life to the task of finding a solution.

OBJECTIVE IDEALISM: THE KANTIAN SYNTHESIS

Throughout his life Kant endeavoured in his philosophy to effect a synthesis of the main schools of thought in his time, working always within the Aristotelian framework of the faculty theory of the mind that was common to them all. Strongly influenced by the rationalism of the Cartesian school, he joined with Leibniz, Spinoza and Wolff in asserting the power of human reason. At the same time, he was to a large extent in sympathy with the tradition of British empiricism, being in particular a close reader of the works of Hume, whose scepticism provided a major challenge for Kant. Writing on metaphysics in 1783, he stated that no event 'since *Locke's* and *Leibniz's* Essays' had been 'more decisive in respect of the fate of this science than the attack which *David Hume* made on it'; he admitted readily that it was Hume's remarks, with his questions on cause and effect, and his challenge to reason, 'that first, many years ago, interrupted my dogmatic slumber and gave a completely different direction to my enquiries in the field of speculative philosophy'.[43]

Kant: experience and morality

Moreover, the enlightenment search for a secular morality concerned him deeply, as well as the problem – already considered to some extent by Locke and Hume but raised more aggressively by Rousseau – of the role of the feelings in human knowledge. One of Kant's earliest writings on these issues was stimulated by Rousseau's work and in it he seems to have come as close as at any point in his career to challenging seriously the division between sense and reason. Two years after the appearance of *Émile*, Kant in 1764 published his *Observations on the Feeling of the Beautiful and Sublime*, a rather youthful and uneven work but one of considerable interest as an early statement of his philosophical views and as an indication of his response to the romantic movement. In this respect it must be considered as more than a treatise on aesthetics, and Joseph May has already drawn attention to the interest of this book for geographers, suggesting that the mid 1760s marked 'the years of Kant's greatest flirtation with empiricism' and also the time when geography, as the empirical study of nature, occupied a position of greatest importance in his thought.[44]

Kant's special concern in the *Observations* was with the role of the feelings in moral education and in his concluding statement, reflecting Rousseau's concern for the moral betterment of the masses, he expressed the hope that

the as yet undiscovered secret of education be rescued from the old illusions, in order early to elevate the moral feeling in . . . every young world-citizen to a lively sensitivity, so that all delicacy of feeling may not amount to merely the fleeting and idle enjoyment of judging, with more or less taste, what goes on around us.[45]

It seems that a central aim of this book was to develop for its readers what might be called a sociology of taste, and Kant in that respect was something of a pioneer in the sociology of knowledge. Departing from the accepted rationalist or neo-classicist view that beauty, as unity, order and harmony, is inherent in objects themselves, he argued that the subjective response of the viewer also plays an important part in aesthetic judgments and he considered these to be closely linked with feelings of pleasure: 'the various feelings of enjoyment' in his view 'rest not so much upon the nature of the external things that arouse them as upon each person's own disposition' [45].

The aesthetic experience, involving, in Aristotelian terms, the imagination and the feelings as well as the reason, was of particular interest to writers of the romantic movement. Kant himself moved to introduce a moral component as well, by developing the concept of the sublime or noble as an aesthetic category. Sublimity, advanced as a literary style in Roman times, had been revived in the seventeenth and eighteenth centuries when British writers, and later Rousseau also, praised the sublime as the wild element in nature and in art, in contrast to the neo-classical ideal of order and regularity in both. Whereas Shaftesbury, for example, associated sublimity with the infinite in nature and considered it to be unattainable for man, Kant on the other hand argued in 1764 that a consciousness of the sublime in man is the basis of all moral principles and he described this, not as a purely rational response, but as a feeling – '*the feeling of the beauty and dignity of human nature*' [60]. Commenting on this, Goldthwait claimed that Kant's conception here was quite new: 'man himself, by his own nature, and universally, exhibits the sublime' [25].

It must be pointed out, however, that a universal application can scarcely be defended in Kant's case, since the sublime in his view was limited, quite literally, to 'man himself', and to civilized white man at that. Women, he explained in some detail, exhibit beauty rather than nobility and although 'the fair sex has just as much understanding as the male' their philosophy 'is not to reason, but to sense'; with regard to their instruction, taking the example of geography, he advised that a map of the globe might be a pleasant diversion, similarly 'they will need to know nothing more of the cosmos than is necessary to make the appearance of the heavens . . . a stimulating sight to them' [78–80]. Moreover, in his discussion of *National Characteristics*, based on Hume's 1748 essay 'Of National Characters', Kant refused to commit himself on the dependence of such characteristics on climate or on type of government but in contrast to Büsching he expressed a strongly racist view with regard to the *Negroes of Africa*: according to Hume, he wrote, 'not a single one was ever found who presented anything great in art or science or any other praiseworthy quality' [97–111]. At another point Kant could express a grand concept: within the 'design of nature', he stated, 'the different groups unite into a picture of splendid expression, where amidst the multiplicity unity shines forth, and the whole of moral nature exhibits beauty and dignity' [73–5].

The concept of nature as order and harmony, as unity in multiplicity, already discussed by Ray and Leibniz, was central to Kant's thought, receiving in his works a strong emphasis on an internally sustained moral order, through his concern for the intellectual and ethical development of man. In the *Observations* he attempted to find a principle of human conduct, something akin to Hume's *moral sense*, in man's immediate aesthetic and intuitive response to the evidence of the sublime in nature. Subsequently, however, as Goldthwait has noted, Kant turned to a reliance on the reason rather than the feelings as a guide to correct judgment and in his final work on aesthetics, the *Critique of Judgement*, 1790, no longer attempted to base the supreme rule of conduct on experience [31]. Two years earlier, in his major work on ethics, the *Critique of Practical Reason*, 1788, Kant had asserted a kind of moral rationalism, arguing, in opposition to the social utilitarianism and even nihilism of his day, that in man as a rational being, freedom of the will implies a conception of laws and duty, leading to the advancement of humanity. In this work he argued that an unfailing guide to right conduct, a *categorical imperative*, is grounded in human reason, the faculty by which man distinguishes himself from the rest of nature. Following Leibniz to a large extent, he made a distinction between the *kingdom of nature* – a machine, acting under necessity from without – and the *kingdom of ends* – a world of rational beings (*mundus intelligibilis*) acting according to self-imposed rules under the moral imperative. In any attempt to understand the universe, he claimed, only the intellect can provide man with a consciousness of the enormity and wonder of the whole – the external world of sense, and the internal world of his own personality, linked in a vast system – 'the Starry Heavens *above*, the Moral Law *within!*'[46]

Phenomena and noumena: Kant's theory of knowledge

Kant's theory of how man comes to know the world, although modified considerably in the course of his long career, involved at each stage some kind of dualism. In his *Inaugural Dissertation* of 1770, published in the year he became Professor of Logic and Metaphysics in the University of Königsberg, he expounded the concept of infinity, and also 'the *intellectual* concept of a whole', as a world or totality in contrast to an aggregate, and he went on to argue that the limitations of the human mind should not require the rejection of these two concepts: the limits of human knowledge are not the limits of the universe. In terms of human knowledge he made a division at that stage between the *sensible* world of phenomena, or representations of things as they appear, and an *intelligible* world of noumena, the rational knowledge of things as they are. With regard to empirical knowledge concerning phenomena, he argued that perceiving is associated closely with the logical use of the intellect: intellectual cognition of a number of appearances leads to reflective cognition or *experience*, and from this, empirical concepts can be formed. Kant's definition of experience, by effectively involving the intellect in perception, represented a strong break from earlier sense-empiricism. Already he was concerned with the role of concepts in knowledge, allowing in his theory for the formation of empirical concepts on the basis of experience, although he stipulated that these do not pass out of the class of sensitive knowledge by being brought to a greater universality: only pure rationality, he claimed at this point, leads to knowledge of *things as they are*.[47]

Subsequently Kant rejected the kind of dualism proposed in the *Inaugural*, and in letters written between 1770 and 1780 – the years of his long silence in terms of publications – he came to argue for the subjectivity of both sensibility and intellect. Even through rationality, he admitted, we cannot know the thing in itself. In 1771, referring to his plans for a book to be titled *The Limits of Sensibility and Reason*, he indicated his growing concern with 'the fundamental concepts and laws' that provide the conditions for knowledge of the sensible world, commenting that 'a fairly lengthy period of time is necessary for a single concept to be examined intermittently, in all kinds of relationship and in contexts as extensive as possible'.[48]

A decade later, Kant's famous reply to the subjective idealism of Locke and Berkeley was expressed in his *Critique of Pure Reason*, where he emphasized the objective aspect of human knowledge, in what has been called his *objective idealism*. Arguing that all phenomena are perceived in the relations of space and time, Kant stressed that all perception is interpreted in terms of concepts in the mind, and these for Kant, although he continued to recognize empirical concepts, were primarily what he termed the *categories*, pure concepts or principles of understanding, inborn in the mind, which make possible the objective reference of experience. Among these he specified quantity, quality, matter and causality. These concepts themselves, he claimed, remain useless, however, unless they can be applied in experience: 'Thoughts without content are empty, intuitions without concepts are blind.'[49] He maintained therefore that the categories do not enable man to arrive by pure thought at valid conclusions about objects as they are in themselves: such knowledge cannot in effect pass beyond *the limits of sensibility*. Accordingly, Kant proposed a new dualism, between a natural world of *phenomena*, or appearances in space and time, perceived by man's senses and providing the basis for empirical knowledge; and a transcendental reality that gives rise to these phenomena – a world of *noumena* which affect the mind yet are inaccessible to sensible experience. At the same time, in attempting to define the limits of rationality, Kant claimed in his theory of transcendental idealism that knowledge concerning ideas can be achieved by pure reason, independently of sense-experience. He divided such rational or non-empirical knowledge into two kinds: first, *mathematical* knowledge in general, and secondly, *philosophical* knowledge or metaphysics, based on pure concepts, in which he included, along with ethics and his own *Critique*, a large number of studies under the heading of the *metaphysics of nature*, among them ontology, rational physics and psychology, theology, and cosmology, the last including *a priori* laws of nature and such cosmological ideas as the idea of the absolute whole.

Kant's basic division of sciences, then, by 1781 distinguished these *rational sciences* from the *empirical sciences* which were themselves divided into *empirical physiology* or general natural science, considering nature as the object of the outer sense, and *empirical anthropology*, considering man or soul as the object of the inner sense. Physical geography, associated with the idea of *space* (the form of intuition of appearances or occurrences in the external world), was included by Kant in natural science, while history, which he associated with the idea of *time* (the form of intuition of events in a sequence), seems to have been grouped with anthropology. In this classification, then, Kant in effect confirmed with a more sophisticated justification the old division of geography–space, history–time, a division that geographers even in his own period were finding untenable. For all its brilliance, indeed, Kant's philosophy has been found unsatisfactory on many counts. Norman Kemp Smith, for example, drew attention to the contradiction arising from Kant's inclusion of empirical studies in philosophy,[50] while a number of writers have joined him in criticizing the obscurity of Kant's doctrine of inner sense.[51] P. F. Strawson, moreover, pointed to what he saw as the incoherence in Kant's dualism of appearance and reality, and condemned the distinction of human faculties – cognitive, sensitive and imaginative – in what he called the 'Kantian model'.[52] For scientists in particular it appears that some of Kant's most valuable contributions – including his emphasis on the potential objectivity of ideas and his useful teaching that the data of the empirical studies are perceptions and ideas, not the objects themselves – were obscured by the more difficult aspects of his philosophy. With regard to the classification of sciences, the deficiencies of Kant's own theories were, if anything, accentuated by his concern to promote specialization and his insistence that the boundaries of a science must be clearly defined:

> If a field of knowledge is to be exhibited as a *science*, its differentia, which it has in common with no other science and which is thus *peculiar* to it, must first be capable of being determined exactly; otherwise the boundaries of all the sciences run into one another and none of them can be treated soundly according to its own nature.[53]

Kant repeated that belief three years later in his *Metaphysical Foundations of Natural Science* (1786), where he attempted to define the limits of the metaphysics of nature, commencing with what he called the general or transcendental part, the field of pure philosophy, which in his view discovers natural laws through rational cognition from concepts and so provides the necessary foundation for all natural science. Any doctrine of nature, he argued, 'deserves the name of natural science only when the natural laws that underlie it are cognized a priori and are not mere laws of experience'; accepting that such empirical principles remain merely contingent, he claimed that reason alone, without reference to experience, can demonstrate the absolute necessity or 'apodeictic certainty' of inherent laws of nature:

> Since the word 'nature' already carries with it the concept of laws . . . so every doctrine of nature must according to the demands of reason ultimately aim at natural science and terminate in it, inasmuch as the necessity of laws attaches inseparably to the concept of nature.[54]

The term 'nature', he noted, can signify either a quality, the primal internal principle belonging to the existence of a thing, or it can refer to 'the sum total of all things insofar as they can be objects of our senses and hence also objects of experience, under which is therefore to be understood the whole of all appearances'. Taking it in the second or 'material signification' he divided nature again into two parts, one containing objects of the external senses, the other the object of the internal sense: 'Therefore, a twofold doctrine of nature is possible: a *doctrine of body* and a *doctrine of soul*. The first considers extended nature, and the second, thinking nature' [3]. Without a foundation in general metaphysics and in addition the application of mathematics, which itself depends on pure philosophy, a doctrine of nature, however, cannot be called a science: it remains in his view merely a historical doctrine, either a description of nature as an ordered system of classes, or else 'the history of nature as a systematic presentation of natural things in different times and in different places' [4].

According to Kant – writing before the publication of Lavoisier's *Treatise on Chemistry* – no such a priori foundation had been demonstrated for chemical phenomena; therefore, 'chemistry can become nothing more than a systematic art or experimental doctrine, but never science proper; for the principles of chemistry are merely empirical . . . they are incapable of the application of mathematics'. For the same reason he rejected the possibility of a science of the soul:

the empirical doctrine of the soul must always remain yet even further removed than chemistry from the rank of what may be called a natural science proper. This is because mathematics is inapplicable to the phenomena of the internal sense and their laws [7–8].

The main difficulty in his opinion is that 'the pure internal intuition in which the soul's phenomena are to be constructed is time, which has only one dimension'; moreover, a thinking subject is not amenable to experiment, 'and even the observation itself alters and distorts the state of the object observed'. He still believed, however, that such limitations did not apply to the study of *extended nature* and the concept of matter:

This is the reason why in the title of this work, which, properly speaking, contains the principles of the doctrine of body, we have employed, in accordance with the usual practice, the general name of natural science; for this designation in the strict sense belongs to the doctrine of body alone [8–9].

His main aim in *The Metaphysical Foundations of Natural Science*, then, was to explore 'the limitations of the whole faculty of pure reason and therefore of all metaphysics' by attempting 'a complete analysis of the concept of matter in general'. All mathematical physicists, he stated, even those who 'solemnly repudiated any claim of metaphysics on their science', must make use of such metaphysical principles, and he claimed furthermore that while other sciences such as pure mathematics and the empirical natural sciences can never attain absolute completeness, this can be confidently expected in the metaphysics of corporeal nature: 'I believe that I have completely exhausted this metaphysical doctrine of body' [9–11]. All deductions regarding the universal concept of matter can be logically derived, he explained, from the pure concepts of the understanding, namely quantity, quality, relation, and modality, as defined – with complete certainty, in his view – in the table of categories of his *Critique of Pure Reason*. Therefore, arguing that 'natural science throughout is either a pure or applied doctrine of motion', he offered his analysis in four sections: *Phoronomy* considering matter as pure quantum, *Dynamics* taking account of the quality of the matter, *Mechanics* considering relation, and *Phenomenology* referring to 'the mode of representation, or modality, i.e. as an appearance of the external senses'. The essential function of a separate metaphysics of corporeal nature, he emphasized, is to provide instances which give meaning to the concepts and propositions of transcendental philosophy or general metaphysics. In this context he made the interesting comment that it is possible 'to delineate the boundaries of a science not merely according to the constitution of its object and the specific mode of cognition of its object, but also according to the aim that is kept in view as to the further use of the science itself' [12–15]. Generally, however, Kant's claims to completeness and certainty in his system remained unconvincing, while the problems inherent in his definition of science and the division between inner and outer sense were to become evident in his approach to geography.

Kant's physical geography

Kant's concern with geography dated from early in his career, his first lectures on this topic at Königsberg being recorded in 1756, the year after his own *General Natural History*, and two years after Büsching's *Erdbeschreibung*. In his *Outline and Prospectus for a Course of Lectures in Physical Geography*, 1757, Kant did not feature a distinction between geography and history. He divided the study of the earth, much as Büsching had done, into political, mathematical and physical aspects, and devoted his own introductory lectures to physical geography, for which, as he pointed out, it was agreed that no suitable textbook existed.[55] Noting that this study was inadequately treated, Kant affirmed its limitation to a concern with the surface of the earth, defining it closely in accordance with Büsching's outline:

Physical geography considers only the natural conditions of the earth and what is contained on it: seas, continents, mountains, rivers, the atmosphere, man, animals, plants, and minerals.[56]

The German approach to physical geography in this period clearly included man in its scope, in contrast to the subsequent British interpretation, which restricted the study to the non-human world.

Although over the next four decades manuscript notes of Kant's lectures were circulated and widely discussed, he resisted publishing them until, after an unauthorized version appeared in 1801, he requested his assistant, Rink, to produce from lecture notes an official edition which appeared the following year under the title *Physische Geographie*. Several sections of this appear to have been added by Rink; however, according to the researches of Erich Adickes the introduction, composed evidently around 1772 when he began to lecture separately on anthropology, may be regarded as a genuine statement of Kant's mature theory of geography.[57] In this introduction Kant departed from the plan of Büsching by defining physical geography as a description of nature, or the world as the object of 'outer sense', in contrast to anthropology, concerned with mental phenomena or experience of man's consciousness as provided by the 'inner sense' through perceptions rather than through sense-experience.[58] At the same time Kant's definition allowed for man, as himself a phenomenon of nature, or 'one of the appearances of the sensible world',[59] to be included in the study of physical geography; moreover, although much of what Büsching referred to as civilization or learning was now assigned by Kant to anthropology, he regarded a concern with customs or *mores* to be a part of moral geography, itself based on physical geography.[60] There seems insufficient justification for Roger Minshull's claim that Kant restricted his physical geography to 'all the visible, concrete, tangible things on the earth's surface. The intangibles and imponderables are left out.'[61] It is clear, however, that Kant's attempt to distinguish sciences on the basis of his rather unsatisfactory distinction between inner and outer sense, raised considerable difficulties for his followers in geography.

In the *Physical Geography* introduction Kant also distinguished between geography as a description of simultaneous occurrences in the present, perceived under the aspect of space, and history as 'a continuous geography', an account of such events in the past, in relation to time; he therefore dismissed the term 'natural history' since a correct history of nature would have to describe the occurrences of the whole of nature through all time [260]. Much attention has been given to a further distinction that Kant made, in this introduction, between a logical or artificial classification of nature, such as the Linnaean *Systema Naturae,* and the physical classification provided by a geographical description of nature, in which 'things are observed according to the places which they occupy on the earth' [259]. This was in many respects a useful distinction, and one developed further by Humboldt. However, the accompanying statement in the introduction, that 'the classification by concepts is the logical, that by time and space the physical classification', must be rejected as meaningless in terms of Kant's own philosophy, since geography also is based on concepts. As Joseph May indicated, in drawing attention to that discrepancy, too much has been made of this statement by some geographers this century who have imputed to Kant a threefold distinction, contrasting geography, as a spatial or chorographical study, with logical or systematic sciences on the one hand, and historical or temporal sciences on the other.[62] More significant was Kant's argument that in his day the so-called systems of nature based on logical order were in every case no more than an *aggregate*, 'because a system presupposes the idea of a whole out of which the manifold character of things is being derived' [260]. He emphasized that such an idea is provided by the geography of nature, to which, in accordance with his own theory of knowledge, he assigned an important role.

Moving beyond Locke's notion of geography as a preparative or propaedeutic to further studies for the child, Kant argued for its function as a 'propaedeutic to understanding our knowledge of the world', while admitting, like Büsching before him, that 'instruction in it still appears to be very defective'. Since man's experience occurs in the world, a preliminary idea of that world, gained through instruction, is a necessary first step to knowledge: with such an idea, gained from physical geography and anthropology, 'we are in a position to classify and arrange, within the framework, every one of the experiences we have had'. Constantly, Kant stressed that for knowledge to form a system, the idea of the whole must precede the parts, and in this he applied to the sciences themselves Aristotle's notion of the prior concept:

The idea (*Idee*) is architectonic; it creates the sciences. For example, he who wants to build a house first creates for himself an idea for the whole, from which all the parts will be derived. So our present preparation is an idea of the knowledge of the world. Here we make for ourselves in a similar way an architectonic concept, which is a concept wherein the manifold is derived from the whole [256–7].

Looking beyond the inadequate schoolbook geography of his day, Kant from as early as 1765 suggested a more significant position for the study in unifying learning: the revival of geography, in his view, should bring about 'that unity of knowledge, without which all learning is only piece-meal (*Stückwerk*)'.[63] It would appear that Kant intended assigning to geography the function of Bacon's universal science.

Kant himself was not explicit on the procedure that should be followed in developing the subject, and the descriptive content of his own *Physical Geography* was by no means distinctive, showing little departure from the secondary sources he used. It was in his attempt to provide a conceptual basis for the study that his original contribution lay, and here his concept of experience was of considerable importance. While declaring with the sense-empiricists that knowledge of the world begins with the senses, he pointed out that an individual's personal observations are limited to the portions of time and space in which he lives: 'therefore, we have to avail ourselves of other people's experiences'; these, he added, should be reliable, and preferably recorded in writing. This concern with vicarious experience, and with what Bacon had tended to deprecate as *tradition*, or learning from instruction, is a feature of the kind of empiricism advanced by Kant, and his departure from the Baconian view on this issue is evident in his advice to the scientific traveller:

more is needed for knowledge of the world than just seeing it. He who wants to profit from his journey must have a plan beforehand, and must not merely regard the world as an object of the outer senses [256–7].

Again, Kant did not himself engage in travel or in geographical research: that project, along with the task of developing physical geography as a science, was taken up by Alexander von Humboldt at the end of the eighteenth century.

While it is important to avoid ascribing too much originality to Kant with regard to the new programme for geography, his works evidently provided a fruitful source of inspiration for Humboldt, offering not only some challenging developments in epistemology but also a set of more elevated aims for the study in providing a concept of the system of nature. Kant's contribution to the enlightenment ideal of the moral betterment of man was a conception of nature as a dynamic system, developing through man himself towards its final end in the full expression of human potential, an end that would be embodied in improved social organization and a perfect constitution. The idea that the whole history of man could be considered from this point of view was outlined in Kant's essay, 'Idea for a Universal History with Cosmopolitan Intent', published in the *Berlin Monatschrift* in 1784. Subsequently, after the French Revolution, rejecting the notion of happiness as a goal for man, Kant reaffirmed the ideals of peace and freedom within a *civil community* in his *Critique of Judgement*.[64] In that work the idea of inherent forces in matter, already elaborated in his *Metaphysical Foundations of Natural Science (Metaphysische Anfangsgründe der Naturwissenschaft)*, of 1787, was combined with a notion of the organism as a natural system, to provide a dynamic concept of the unity of nature, one that Humboldt found extremely meaningful in his own research.[65]

A century, then, after Leibniz had commenced the task of constructing an alternative to Baconian empiricism and the mechanist world view, one of the best intellects of the age joined those who continued Kant's great work in that regard. For Humboldt the Kantian system was not a definitive framework but a foundation for his own response to the problems of empirical science. In particular, the unformed study of *physical geography*, on which Kant himself had offered introductory lectures without attempting a serious publication during his years of vigour, came to be chosen by Humboldt as a focus for his own contribution.

CHAPTER 7

GEOGRAPHY REVIVED: THE AGE OF HUMBOLDT

Towards a new empiricism: 1790–1859

The final decades of the eighteenth century can be seen as a time of considerable ferment in scientific thought, particularly among German thinkers for whom the teachings of Kant provided an additional challenge in regard to empiricism, as well as contributing to a revival of interest in geography. In Britain and France, where the full impact of Kantian ideas was delayed for many years, the positivist empiricism of Hume and Lavoisier continued to find wide acceptance, and in those countries, although significant advances were made in many aspects of natural science, the position of geography in general remained relatively depressed. It was among German writers that both the strongest opposition to positivism and the most interesting developments in geographical theory took place over the next half-century, and in this a leading part was played by Alexander von Humboldt, from the 1790s until his death in 1859. To call that period *the age of Humboldt* is therefore appropriate for the present inquiry, and while this may not meet with the same general agreement as is the case for the *age of Newton* a century earlier, it can be argued that Humboldt in his own day was acknowledged in the leading scientific societies of Europe as an outstanding figure, even being widely regarded at the height of his career as second in fame only to Napoleon.[1]

In considering the encounter between empirical science and geography in Humboldt's lifetime, particular attention needs to be given to the period before 1800, when the essential outlines of his own position were formulated. The later years, encompassing the first half of the nineteenth century, offer in themselves a vast field for research and in dealing with that period the present study must be confined rather strictly to the main themes in tracing the development of Humboldt's theory of science, and considering aspects of his contribution to geography. In both of these he provided some of the most significant innovations of his time. Operating, as he did, constantly in the centre of scientific activity in his day, Humboldt provides in effect an excellent focus for a consideration of an important formative period for the natural sciences.

Born in 1769 at the time of James Cook's first voyage, Humboldt died in the year that saw the publication of Darwin's *Origin of Species* and Marx's

Contribution to the Critique of Political Economy. During those ninety years most of the present sciences were not only defined in terms of scope and methodology but also, quite literally, *institutionalized* as their teaching and practice became established in various institutions. Humboldt himself was closely involved in those developments, proving himself a leader in a number of fields, including geology, mineralogy, botany, vulcanology, earth magnetism and meteorology, as well as geography. His commitment to scientific inquiry began in his youth, during a time of experiment and debate in the natural sciences, before a rigid structure was acquired by many of these studies and while vigorous controversy centred on new discoveries and techniques, on aims and limits for new sciences, and on the nature of the scientific process itself. In that context his concept of geography emerged in conjunction with his theory of scientific method.

THE FORMATIVE YEARS: 1790–1800

Humboldt and geography

Humboldt's interest in geography was stimulated early in his career. After an upper-class education which he received, along with his elder brother Wilhelm, from a number of tutors at the family home of Tegel and in nearby Berlin, he proceeded to periods of study at several different institutions, commencing with six months in the winter of 1787–8 at the small university of Frankfurt-on-Oder. Later, in 1789, he joined his brother at the University of Göttingen, at that time a leading centre of intellectual activity in Germany. There he met the geographer George Forster (1754–94), the celebrated author of *A Voyage Round the World*, a record of Forster's travels with his father, Reinhold, who had replaced Banks as naturalist on Cook's second voyage from 1772 to 1775.

In that work, published in English in 1777 and translated into German the following year, Forster was outspoken in his comments on the problem of observation in science, and in his condemnation of mere fact-gathering. Faced with 'the seeming contradictions in the accounts of different travellers', and what he called the illusory systems based on these by 'the philosophers of the present age', he suggested that,

The learned, at last . . . raised a general cry after a simple collections of *facts*. They had their wish; facts were collected in all parts of the world, and yet knowledge was not increased. They received a confused heap of disjointed limbs, which no art could reunite into a whole; and the rage of hunting after facts soon rendered them incapable of forming and resolving a single proposition . . . Besides this, two travellers seldom saw the same object in the same manner, and each reported the fact differently, according to his sensations, and his peculiar mode of thinking.

Aware that all such observations must be considered in relation to the particular observer, Forster admitted that for his own part,

I have sometimes obeyed the powerful dictates of my heart, and given voice to my feelings; for as I do not pretend to be free from the weaknesses common to my fellow-creatures, it was necessary for every reader to know the colour of the glass through which I looked.

Above all, in the manner of Buffon, he stressed the need for the observer 'to combine different facts, and to form general views from thence, which might . . . guide him to new discoveries', and he declared his own aim 'to throw more light upon the nature of the human mind', and so to contribute 'to our general improvement and welfare'.[2]

Forster's scepticism with regard to the claims of sense-empiricism, his interest in geographical description and his concern for a clear and readable style in scientific writings evidently provided a significant model for Humboldt. In 1790 he accompanied Forster on a journey down the Rhine and across to England, and Forster's account of that journey in his *Ansichten vom Niederrhein* (*Views of the Lower Rhine*), with its lively concern with all aspects of the landscape, from geology to the cultural life of the people, again invites comparison with Humboldt's later travel accounts. On many issues, of course, the two men disagreed, Forster for example questioning Humboldt's refusal at that stage to admit a volcanic origin for basalt, while at the same time stressing the current lack of knowledge on the form and structure of the earth-surface, or even the plants and animals on it.[3]

Returning across France, the travellers passed through Paris at the time of the Bastille Day celebrations of 1790, and the idealism of the French Revolution in that early period made a lasting impression on Humboldt. Altogether, this journey with Forster seems to have provided one of the important formative influences on Humboldt's ideas and goals. When his first book, *Mineralogical Observations of certain Basalts on the Rhine*, was published anonymously the same year, it was dedicated to George Forster.[4] Many years later, Humboldt in *Kosmos* referred warmly to him as a friend and teacher, and acknowledged Forster's contribution to geographical inquiry: 'Through him began a new era of scientific voyages, whose aim was the comparative study of peoples and regions [vergleichende Völker- und Länderkunde].'[5]

At Göttingen, Humboldt continued his studies in natural science and classical antiquities, under professors as distinguished as Christian Gottlob Heyne (1729–1812), professor of classics (and, incidentally, father-in-law of Forster), and Johann Friedrich Blumenbach (1752–1840), professor of anatomy, author of a *Handbook of Natural History* and acknowledged as a founder of the science of anthropology. His brother Wilhelm von Humboldt (1767–1835), who showed a keen interest in both Kantian and Platonic thought, became more closely involved with the classical revival in literature. Alexander however left Göttingen in 1790 to pursue research in the natural sciences, attending first – in accordance with his mother's plans for him to have a career in finance – an Academy of Commerce in Hamburg, where his studies included commerce, botany, mineralogy, and some geography.[6] At that time such practical or empirical subjects were taught in separate technical institutes, and the question of whether studies of this kind should have a place in the general or liberal education which universities claimed to offer was to be widely discussed in Europe.

After a year in Hamburg, Alexander spent eight intensive months, until February 1792, at the Freiberg Academy of Mines, which had earned distinction under the direction of Abraham Gottlob Werner (1749–1817), a pioneer in the study of chronological succession in rock strata and leader of the Neptunist school in geology. Along with mineralogy and geology, Humboldt maintained at Freiberg his early interest in botany, a study in which he had been encouraged previously by the capable botanist Karl Ludwig Willdenow (1765–1812) in Berlin. From Humboldt's researches and experiments on underground plant life in the mines came his first major publication, a Latin work, *Florae Fribergensis*, in which he showed a characteristic concern, not only with the plants themselves but also with the relation of these as organisms to their environment. Indeed, in his introduction he suggested that the *geography of plants* should form an essential part of a subject that he called in Latin *Geognosia*, in German, *Erdkunde*. Arguing that geognosy should be concerned with animate as well as inanimate nature, he proposed three divisions: the geography of rocks, also called geognosy and 'industriously studied by Werner', the geography of animals, 'for which the foundations have been laid by Zimmermann', and the geography of plants 'which our colleagues have left untouched'. Evidently following Kant to a considerable extent, Humboldt rejected the term 'natural history' and stressed the way in which plant geography, along with geognosy as a whole, differed from other natural sciences and from a further study that he called earth history:

Plant geography [Geographia plantarum] traces the connections and relations by which all plants are bound together among themselves, designates in what lands they are found . . . This is what distinguishes geognosy [Geognosiam] from *physiography* [*Physiographiam* (Naturbeschreibung)], falsely called natural history [historia naturalis]; zoology, botany, and geology [Zoognosia, Phytognosia et Oryctognosia] all form parts of the investigation of nature, but they study only the forms, anatomy, processes, etc., of individual animals, plants, metallic things, or fossils. *Earth history* [*Historia Telluris* (Erdgeschichte)], more closely related to geognosy than to physiography, but as yet attempted by no one, discusses the kinds of plants and animals that inhabited the primeval globe, their migrations and the annihilation of many of them, the origin of mountains, valleys, rock strata and ore veins, . . . Thus the history of animals, the history of plants, and the history of rocks, which indicate only the past state of the earth, are to be clearly distinguished from geognosy.[7]

In scope and in method Humboldt's *Geognosia* of 1793 can be regarded as providing an important model for modern geography. His idea of plant

communities, extended to a study of the distribution and relation of rocks
and animals, suggested the basis for a new science, one concerned with the
interrelationships of organic and inorganic phenomena on earth. At this
stage, significantly, he gave no specific indication of including man in this
study, although his footnote to Büsching's *Neue Erdbeschreibung* at the
same point shows that he was familiar with that work, and was probably
aware of Büsching's outline for physical geography. Humboldt himself, it
must be noted, did not here call his general study 'geography', although he
used the term *Geographia* in connection with the three composite studies of
rocks, animals, and plants. In part, very likely, he wished to dissociate his
proposed science from the amorphous geographies of his day; at that time,
moreover, many of the studies now established as sciences were still in the
process of becoming more precisely defined and formalized, so that there
was no firm agreement on names. Referring, for example, to Werner's study
of rock formations as the 'geography of rocks [Geographia oryctologica]',
Humboldt noted that it was also called simply 'geognosy', to which in
Kosmos he later added 'or even geology'.[8] In his own *Geognostical Essay* of
1823, furthermore, Humboldt himself restricted the term *geognosy* to a
study of rock strata in the earth's crust,[9] a change that itself draws attention
to the problems associated with variations in terminology and concepts in
the history of science.

The challenge of empiricism: Humboldt's theory of science

Published with the *Florae Fribergensis* in 1793 was Humboldt's work *Aphorisms on the Chemical Physiology of Plants*, in which his concern with the
analogies between plants and animals, and the contrast between inanimate
matter and living organisms, was clearly expressed. These remained central
issues for him throughout that decade and he undertook a series of experiments on plant physiology, organic chemistry, and animal electricity. From
the outset, however, although his use of the term 'Aphorisms' suggests a
Baconian model for this work, he showed also a strong preoccupation with
man's effort to consider all of nature as a totality, and with the relation
between universal ideas and the study of particular examples.[10] In these
years Lavoisier's anti-phlogiston theory of chemistry was just beginning to
find a reluctant acceptance in Germany, Humboldt being one of the leaders
in adopting the new theory, and his reference in the *Aphorisms* to the *Traité
élémentaire de chimie* [134] indicates that by this time he must have encountered not only Lavoisier's teachings on analytic chemistry, but his theory of
'exact science' as well. Humboldt's own efforts in this period to develop a
modus operandi with regard to scientific inquiry are of considerable interest.

From 1792 to 1796 Humboldt held an appointment with the Prussian
Ministry of Mines, rising rapidly to the position of inspector and impressing
officials in the bureaucracy with his effectiveness in handling records, his

indefatigable activity in the mines themselves, and his ability to increase production. Fieldwork, data collection and experiment occupied Humboldt at this time: it seemed he was determined to excel in the empirical programme of the Bacon-Lavoisier tradition. At the same time his concern for the social and moral betterment of man, a legacy of the enlightenment, found expression in his energetic efforts to improve or invent safety devices for the miners, to establish a pension fund for them, and to offer them some education in connection with their work. By 1794, on his own initiative, he had opened the Free Royal Mining School at Steben, which although limited in the classes it offered, was operating effectively some years before the free schools of Pestalozzi in Switzerland and the workers' schools of the philanthropist Robert Owen in Britain. It was a gesture, but it became a prototype for similar institutes in Germany. With his practical work among the miners of Bayreuth, Humboldt was emerging as a social activist, an unusual role for a Prussian nobleman in his day. In letters at that time he spoke also of his depth of feeling for his work and his passionate involvement with these activities: his own version of the scientific empiricist began to appear more in line with the radical tradition of Rousseau than that of Lavoisier.

These were exploratory years for Humboldt, when the foundations for his future scientific procedure were nevertheless established, and at this stage it seems that he was already moving to modify and extend the Baconian tradition in science. This was actually a goal being proposed for him at that time by his brother Wilhelm, who saw Alexander's genius surpassing that of Francis Bacon. Writing to the Swedish diplomat Karl Gustav von Brinkmann in March 1793, Wilhelm remarked on the outstanding quality of Alexander's intelligence, vision, energy and scholarship: 'He seems made to connect ideas, to see links and combinations.' Alexander's special genius, in his opinion, would be equal to the great task facing their age – 'a complete restoration of learning, and more than that, of all human endeavours'. The necessity for this, Wilhelm continued, had been perceived by Bacon, many of whose ideas had been of great value:

But where he wanted to indicate the whole, where he sought to assemble the particulars into a unity, there he lacked genius. With true genius he would not have tolerated so much fiddling and confusion in his plan . . . Of all the great minds known to me from history or from personal experience, in all times, my brother is in my opinion the only one capable of combining the study of physical nature with that of moral nature, so bringing for the first time true harmony into the universe as we know it, or if this should prove too much for the powers of one man, then of preparing the study of physical nature in such a way that the second step is made easier. Moreover, it is almost immaterial how he pursues his studies and in what way he arranges them. Whatever he deals with leads him automatically, as I have often noted, to the viewpoint just mentioned . . . I hope and indeed am sure that he will devote his entire life to these studies, and will not enter into any personal relationships that, however fine they may be in themselves, always prevent the giving of all energies completely to one purpose, and as he is in a position that enables him to pursue his aims in whatever parts of the world his studies require, so I have the greatest assurance in expecting something great from him.[11]

In the hands of the Humboldt brothers it appeared that the Baconian revival was to be given a new direction, and throughout his life Alexander seems to have maintained a total dedication to the task outlined by his brother, pursuing always his own course in this effort.

German *Naturphilosophie*: Goethe and Herder

Early in 1794 a visit to join Wilhelm at Jena brought Alexander in close contact with Goethe and the famous Weimar society of writers and idealist philosophers, many of them belonging to the nobility, and a number holding distinguished appointments at the University of Jena. Prominent among them was J. C. F. von Schiller (1759–1805), Professor of History at Jena, along with J. G. Fichte (1762–1814), Professor of Philosophy from 1794 until a charge of atheism obliged him to leave in 1789, and F. W. J. von Schelling (1775–1854), Fichte's successor at Jena in Philosophy and a leader of the German romantic school. In their *Naturphilosophie* this group favoured the neo-Platonic idea of polar forces in the universe; following Leibniz, Spinoza, and Kant, they opposed the mechanistic and materialist science of the French and stressed the need to comprehend the unity and harmony of nature among the phenomena.

Goethe, whose work on the *Metamorphoses of Plants* had appeared in 1790, welcomed the visits of Alexander over the next years as a stimulus to turn again to natural science and record his studies in comparative anatomy; writing from Jena to Duke Carl August in March 1797, he praised Alexander as 'a positive cornucopia of natural sciences'.[12] His own approach opposed the limitation of scientific inquiry to rigid experiments, mathematical analysis and strict tabulation of results. Concerned in his own plant and animal morphology with a search for essential forms, he objected to the increasing specialization of modern science and its disregard of subjectivity and symbolism in nature.[13]

Both Goethe and Schiller were suspicious of experimental science as threatening to destroy the charm of nature, a view that Humboldt always denied strenuously.[14] He was sympathetic, however, with many of their views, adopting for a time the theory of vital force favoured by the nature-philosophers. In August 1794, accepting Schiller's invitation to compose a contribution for his journal *Die Horen*, Alexander wrote the essay *The Life Force or the Genius of Rhodes*, in which his speculations on life were presented in allegorical form, in the accepted literary style of the time.[15] It was his only attempt at this method of expression – subsequently he turned entirely to science for his productive output, confining any personal sentiment, along with his lively humour, largely to private letters.

Although he wished to avoid the sentimental literary approach of his time, Humboldt continued to assert that it is not necessary to exclude from scientific writing all that is aesthetic and creative: frequently, in his later publications, he attempted to demonstrate that a scientific work can be also a contribution to literature, and to the end of his life he held that the essential inspiration of poetry, the wonder of the boundless universe, in Kant's phrase, is also the foundation of a science of nature. In this he found a model in Goethe, and later in *Kosmos* recalled that Goethe, both scientist and poet, 'stimulated his contemporaries "to solve the profound mystery of the universe", and renew the bond which in the primitive age of mankind united philosophy, physics, and poetry'.[16] Humboldt never lost sight of the possibility of such a synthesis, although he realized that it involved a different interpretation of all three elements – a poetry more meaningful and less sentimental, a philosophy not entirely divorced from scientific inquiry, and an empirical science less narrowly conceived and more productive of generalizations. All of these lead to comprehension of the world; it was to the sciences and the problems of empirical method that Humboldt directed his energies.

In that task a fertile source of ideas in this period was the work of the Protestant pastor Johann Gottfried von Herder (1744–1803), a former student of Kant and for some time an influential member of Goethe's circle. Herder's *Ideas on the Philosophy of the History of Man*, 1784, based on a small tract he had written a decade earlier, and called by Herder 'my grand work', gave an important role to geography in his effort to provide the foundations for future studies in this field. Commencing his own philosophy with what he called a general view of the earth as the home of man, and an examination of the various organized beings, Herder argued that 'mere metaphysical speculations . . . unconnected with experience and the analogy of nature, appear to me aerial flights, that seldom lead to any end'.[17] At the same time he stressed the need to generalize from the data and for this purpose suggested the collection of observations on chemistry, heat and electricity, to produce 'a geographical aerology . . . a comprehensive view of geography and the history of man', in order to complete for all of nature, including man, 'the picture, of which we have at present but a few, though clear, outlines' [14–15].

In every case Herder believed that further research would lead to proof of divine design in nature, and his own accounts of curiosities often showed much of the credulity evident in earlier works of natural theology. Nevertheless, many of his suggestions for future projects appear to have been carried out in Humboldt's later work. With regard to the earth's surface, for example, Herder called for the collection and extension of information concerning mountains to supplement the works of Varenius and others, and he drew attention to the Peruvian Mountains, as 'perhaps the most interesting tract in the world in regard to the higher branches of natural history' [25]. Conscious also of the significant function of plants in the system of decay and regeneration that he saw in nature, Herder praised both the *Philosophia botanica* of Linnaeus, 'which arranges plants according to the elevation and quality of the land, air, water, and temperature', and the *Natural History of Southern France* in which Soulavie 'has given a sketch of a general physical geography of the vegetable kingdom, and promised to extend it to animals, and to man'. 'We must', said Herder, 'recommend to some one, particularly skilled in the science, our wish for a *universal botanical geography for the history of man*' [32–4].

The history of man, in Herder's view, must be considered in relation to the creatures that occupied the earth before mankind. He introduced his own discussion of the animal kingdom by pointing to 'an evident equilibrium, not only over the whole Earth, but in particular regions and countries', an equilibrium maintained by nature through the constant 'adjustment of opposing species' [35]. With regard to the distribution of animals, he acknowledged 'a profound work, compiled with scientific industry', Zimmermann's *Geographical History of Man, and universally-distributed Quadrupeds.*[18] In turning to man, Herder emphasized the significance of tradition in human culture. Man is a teachable animal; even man's reason, in Herder's view, is *learned*:

This is not innate in man, but acquired: and according to the impressions he has received, the ideas he has formed, and the internal power and energy, with which he has assimilated these various impressions with his mental faculties, his reason is . . . stunted or well-grown, as is his body [91].

Furthermore, he noted, each society or nation is able to comprehend only 'according to the circle of their own conceptions'. Comparing the mythologies of various nations, in what he referred to as 'a complete geography of the inventing mind', Herder pointed out that many of the ideas so held are not derived solely from visual objects, they are *inherited*: 'What one nation holds indispensable to the circle of its thoughts, has never entered the mind of a second.' Among primitive peoples, he believed, reliance on verbal communications encourages the passing on of national fictions, and in that process he noted the significance of the imagination:

Of all the powers of the human mind the imagination has been least explored . . . it seems to be not only the band and basis of all the finer mental powers, but the knot, that ties body and mind together [195–201].

Efforts to oppose both the faculty theory of the mind and the mind–body dualism, on which that theory has been based, are evident in Herder's work.

Arguing that language, and especially writing, must provide the chief means for the 'progressive improvement of the human mind', Herder called for another Leibniz to fulfil the wishes of Bacon, Leibniz himself and others for a thorough study of the languages of various nations. While Herder like Bacon remained critical of the written tradition, expressing concern that 'this means of perpetuating our thought' should leave 'our best thoughts . . . crippled by dead written characters', nevertheless he considered writing to be 'the most durable . . . institution of God, by means of which nation acts upon nation, age upon age, and through which probably the whole human species will in time find itself encircled in one chain of fraternal tradition' [238–9]. Meanwhile, he ascribed Europe's leadership in civilization to *invention* and *united emulative exertions* in the arts and sciences, and from this he looked forward to a new universal association 'of all ranks and nations, by means of education, laws, and a political constitution' [631–2].

Herder's liberalism and concern for moral progress draw comparison with the views of Kant in the same period. In contrast to Kant, however, Herder's respect for other cultures and his concern for the immediate welfare of all humanity led him to express sympathy for peoples such as the Indians of America who had suffered degradation as a result of foreign destruction of their traditional way of life. He criticized all violent change in clearing forests for cultivation in those regions: 'Nature is every where a living whole, and will be gently followed and improved, not mastered by force' [186–7]. Moreover he spoke out vigorously against European dominance and the brutalizing effects of great wealth:

What are the objects of our luxury? for what does it disturb the whole World, and plunder every quarter of the Globe? . . . Why do the poor suffer hunger, and with benumbed senses drag on a wretched life of toil and labour. That the rich and great may deaden their senses in a more delicate manner, without taste, and probably to the eternal nourishment of their brutality [191].

In many respects it seems that Alexander von Humboldt found in Herder's work a stimulus for his own ideas and activities.

The idea of a universal science

During 1794 Humboldt referred in correspondence to his plans for a work already outlined in his *Florae Fribergensis*, a book to be completed 'in twenty years', under the title: 'Ideas on a future History and Geography of Plants or Historical Account of the gradual Extension of Vegetation over the Earth-surface and its general Relation to Geognosy'. It was to consider plant-life 'in connection with the whole of the rest of nature, along with its influence on sentient mankind'.[19] By 1796, however, he appeared to be moving beyond plant geography to a plan for a more comprehensive science. Writing to Pictet in Geneva in January 1796, he reported that in the six years since his voyage to England with George Forster he had travelled widely in the mountains of Europe and had not ceased to be occupied with physical observations, studying nature from the most varied points of view: 'I have conceived the idea of a universal science [physique du monde]; but the more I feel its need, the more I see how slight the foundations still are for such a vast edifice.' Already he was concerned with the problem of developing such a study on an empirical basis, yet in accordance with the concept of harmony in nature: 'to reduce experiments to general laws, to establish harmony among the phenomena'.[20]

While Humboldt's ideas at that point were by no means clearly formulated, this letter evidently marks a significant stage in his movement towards the project that was to become *Kosmos* fifty years later. It seems to have been only in the later months of 1796, however, as Humboldt was concluding his experiments on the problems of animal sensitivity and the chemical changes involved in life processes, that he was able to articulate more clearly the dynamic conception of nature that became central to *Kosmos*. Of considerable importance to him in this development was his encounter at that time with Kant's ideas on organisms and on the explanation of material effects – ideas expressed in the *Critique of Judgement* and in Kant's writings on chemistry, especially his *Metaphysical Foundations of Natural Science*.[21] For Humboldt it appears that later developments in Kantian theory enabled him to give a more precise formulation to a concept previously expressed in more general terms by Herder and the nature-philosophy school, and subsequently explored in his own research. His work on plant associations in 1793 suggests that Humboldt was already beginning to perceive the continuity of nature not just as an ideal harmony, but in terms of functioning interdependence.

In 1797 the results of Humboldt's experiments with electricity, stimulating with electrodes the nerves of various animals, including himself, were published in the two-volume German work, *Versuche über die gereizte Muskel- und Nervenfaser* (Experiments with irritated Muscle and Nerve Fibres), subtitled 'with conjectures on the chemical process of life in the animal and vegetable worlds'. Here, rejecting the notion of a single vital force, he stressed the complexity of the changes involved in the processes he had studied, and – as if extending Herder's concept of a dynamic equilibrium maintained through conflict in the animal world – he pointed to the reciprocal action on each other of all other forces in nature: all elements exist in association with others, and none remains isolated.[22] In place of speculations on life-force, Humboldt saw this work as providing the foundations for a new science of vital chemistry. For this purpose, like Lavoisier, he gave particular attention to improving scientific procedures: in expressing his results he emphasized the convenience of mathematics for analysis, enabling propositions to be represented by a few signs or a formula, and he introduced a symbolic notation of a kind still used today in physics and chemistry. This interest in the graphic presentation of data led him subsequently to similar innovations in the course of his work on climatology and geology, notably with his use of isothermal lines and geological profiles.

While over the next years Humboldt continued to publish his researches on chemistry and magnetism, he did not persist with attempting through physiological experiments to find a solution to the problem of life and the relation between the inner and the external world. Accepting neither the vitalist nor the mechanist terms in which such problems had been conceived throughout the last century, he pursued a different approach and resumed his preparation for a general science. By the time his *Versuche* appeared he had already declined promotion and resigned his position with the Department of Mines, the death of his widowed mother the previous year having left him with financial independence and freedom to travel. During 1797 he set off with his brother for Vienna, on the way to Italy, and considered preparing for an expedition to the West Indies. In Vienna he studied astronomy and geodetic measurement, mastering the use of various instruments and carrying out intensive field research in the Alps with the geologist Leopold von Buch. When in 1798 Napoleon's military advance caused a change in their plans, Alexander followed Wilhelm to Paris where he met the great scientists of the Institut de France and obtained the latest in accurate equipment. Finally, after plans to join other expeditions collapsed because of the war, Humboldt with a French botanist, Aimé Bonpland, travelled through Spain to Madrid where they secured the exceptional favour of royal approval for a voyage to carry out scientific research in the Spanish domains of America. Writing to his friends, immediately before their departure from Corunna, Alexander expressed his elation at this opportunity: 'What a treasure of observations I will be able to make to enrich my work on the construction of the earth! . . . Man must strive for the good and the great!'[23] In a letter to the mineralogist Karl-Marie von Moll (1760–1838), he reaffirmed the general purpose guiding his research:

I shall collect plants and fossils, and I shall be able to make astronomical observations with some excellent instruments; I shall analyse the air by chemical means . . . but all of this is not the principal object of my voyage. My attention will never lose sight of the harmony of concurrent forces, the influence of the inanimate world on the animal and vegetable kingdom.[24]

The extent to which Humboldt achieved these goals was remarkable; researches he carried out during his travels in America over the next five years established him as an expert in each of the fields he mentioned. Those years, forming a colourful and significant period in Humboldt's life – one given considerable, even undue, prominence in many accounts of his work – can be mentioned only briefly here. Throughout the voyage from Spain, during a visit to the volcanic island of Tenerife, and from his arrival at Cumana in July 1799, until his return to France in August 1804, he carried out a strenuous programme of research in the best tradition of scientific empiricism, recording observations, collecting specimens, correcting existing records of latitude, longitude, and altitude, as well as completing measurements for new maps and profiles of a vast area of Spanish America.

At the same time, he was gathering important data on social and economic conditions in those regions; subsequently his appreciative studies of native cultures in the New World, along with his strong indictment of slavery and exploitation in colonial society, helped bring about something of a revolution in European attitudes on these questions. Constantly his work emphasized the moral dimension of scientific inquiry. Furthermore, this exposure to an apparently unlimited range of new phenomena in the tropics evidently reinforced for him the warnings of both Kant and Forster on the indiscriminate collection of facts, and convinced Humboldt that he would be unable to complete his scientific work in America. Early in his second year in the New World, and not long after completing a successful expedition with Bonpland to demonstrate the interconnection of the great river systems of the Orinoco and Amazon, he acknowledged the importance of a more stimulating intellectual milieu in which to operate. Writing to his brother in October 1800, Alexander spoke of his happiness in the Spanish colonies. He found the climate healthy and praised the richness and variety of nature there, its 'immensity and majesty beyond all expression'. The inhabitants were pleasant, the whole situation profitable for his researches, without 'the distractions which occur in civilized countries'; he wished to spend 'some more years' there, especially in 'the study of the various human races'. But he realized the deficiencies of his present location:

The only thing that one could regret in this solitude is to remain stranger to the progress of civilization and of science in Europe and to be deprived of the advantages which result from the exchange of ideas.[25]

It is evident that he was testing, in isolation, and to an extent experienced by neither Bacon nor Lavoisier, the limitations of the individual-centred theory of empiricism.

EARLY NINETEENTH CENTURY: 1800–1825

The American works: a picture of nature

Returning to France in 1804, Humboldt commenced the publication of his American works, an enormous task that continued over the next three decades and consumed most of his fortune. The Paris *Grande Edition* of 1805 to 1834, produced in conjunction with other editors and written almost entirely in French, extended to thirty elaborate volumes,[26] while his own output in that period included a number of other publications in both French and German.

Throughout these works Humboldt held to his original plan. Significantly enough, the first volume to appear in the *Voyage* edition was his *Essai sur la Géographie des plantes* (1805), in which Humboldt not only established plant geography firmly as a separate science but also expressed the central aim of his American researches, to provide a coherent picture of nature from his observations. In the preface to his *Geography of Plants*, introducing the *Voyage* works, Humboldt emphasized that his concern was not only with the 'great phenomena that nature presents in these regions', but with their *ensemble*. He went on:

It was in my early youth that I conceived the idea of this work. I communicated the first sketch of a geography of plants, in 1790, to the celebrated companion of Cook, Mr. George Forster . . . The study that I have since made of many branches of physical science has served to extend my first ideas. My voyage to the tropics has provided me with valuable materials for the physical history of the globe [histoire physique du globe].

His research, he pointed out, 'encompassed all the phenomena of nature [physique] that are observed both on the surface of the globe and in the atmosphere which surrounds it'.[27]

Following closely his outline of 1793, Humboldt took up again in his *Geography of Plants* the question of the general study of nature. An essential part of *physique générale*, he argued, is the geography of plants: 'This science considers vegetation in relation to local associations in different climates' [13–14]. In considering plant communities and plant migration, he pointed to an important role for geographical inquiry, in studying the actual groupings and distribution of living organisms on the earth, in relation to their environment. Although he devoted a number of volumes in his American works to the painstaking description and classification of individual plants, Humboldt realized that the study of plant geography was essential also: it could provide a synthesis of much botanical research and so contribute to a fuller understanding of the functioning system of nature.

In the *Tableau physique* which accompanied the *Géographie des plantes*, Humboldt included an elaborate and original profile of the Andes, illustrating the changes in plant life with altitude (see Fig. 5), and in the text he continued his effort to provide a comprehensive view. In this work, he explained, 'I have attempted to unite into a single picture the whole complex [ensemble] of physical phenomena in the equatorial regions.' His discussion of these *phénomènes physiques* included such aspects as *vegetation, animals, geological relations, temperature,* and *the chemical construction of the atmosphere, its electrical tension, its barometric pressure, and the decrease of gravitation* [41–2]. With these, it is important to note, he included a reference to human culture and the *cultivation of the soil:*

We have analysed up to now the physical phenomena which are presented in the equatorial regions; we have examined the modifications of the atmosphere, the vegetable products of the soil, the animals which live at different altitudes, and the nature of the rocks that form the Cordillera. Now we turn our eyes to man and the effects of his industry [139].

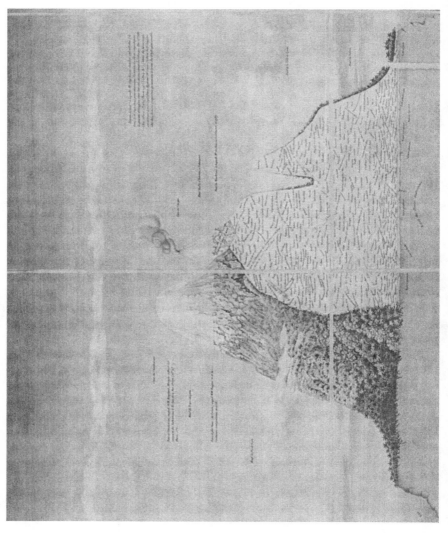

Fig. 5 'Géographie des Plantes équinoxiales: Tableau physique des Andes et Pays voisins' (Geography of equatorial plants: physical profile of the Andes), from Humboldt's *Voyage aux régions équinoxiales du Nouveau Continent,* vol. 1 (1807). His studies of plant communities, shown here in their relationship to altitude, provided a pioneer contribution to biogeography and ecology.

Clearly he now regarded man as part of the ensemble of nature. In considering the response of different societies to their surroundings he expressed support for the theory that adversity acts as a stimulus to human culture:

The civilization of the people is almost always in inverse relation to the fertility of the land which they inhabit. The more that nature presents difficulties to be surmounted, the more rapid is the development of their moral faculties [139].

Physical science, according to Humboldt's approach, was not to be dissociated from the study of man. The task of establishing a link between the physical and moral spheres had remained an important one for him since his days at Jena; indeed, the German edition of this work, which appeared in 1807 under the title, *Ideen zu einer Geographie der Pflanzen, nebst einem Naturgemälde der Tropenländer*, was dedicated to Goethe, and Humboldt at that time openly acknowledged the insights provided for him by Goethe's views of nature.

During those years the romantic school was gaining strength in Germany, and its leaders challenged the assumptions underlying the mechanistic science of France, not only with their own concept of the world as an organic whole but also with their argument that the aim of science is not utility and control but wisdom and understanding – not merely the domination, but the interpretation of nature. Humboldt, working more in Paris than in Berlin for over twenty years after his return from America, faced the need to reconcile these two points of view. His own approach to scientific research favoured careful observation and experiment; throughout his career he showed little patience with attempts, of the kind associated with the school of Schelling and Hegel, to construct a natural philosophy solely through intuition and speculation, although he was by no means committed to the Baconian view of eliminating theory from scientific method. Schelling himself protested that it was quite possible for investigators to conduct experiments under the guidance of philosophical ideas, and in 1805 he wrote to Humboldt asking him to test this new theory, that 'reason and experience can never be more than apparently opposed'.[28] In reply Humboldt said he welcomed the revolution produced in science by natural philosophy, and in the following years he continued to maintain that the empiricist should not necessarily be opposed to the philosopher. Increasingly, it seems, he came to realize that such a position implied a reassessment of the empirical method as it was then conceived in the exact sciences. Writing in the first volume of *Kosmos*, forty years after his earlier exchange with Schelling, he expressed a more critical viewpoint:

The totality of empirical knowledge and a fully developed *philosophy of nature* . . . cannot be in conflict as long as the philosophy of nature, according to its promises, is a reasoned comprehension of the actual phenomena in the universe. Where any contradiction appears, the fault must lie either in the hollowness of the speculation or in the arrogance of empiricism, which believes more to be proved through experience than can be justified [I, 69].

In 1807, acting again in sympathy with the ideals of Schiller and the Weimar circle, Humboldt moved also to establish closer links between science and literature. His *Ansichten der Natur* (1808), written during a period of residence in Berlin at the time of Napoleon's invasion of Germany, is a series of essays providing evocative descriptions from his South American journey. These views of nature in the tropics of the New World, showing a constant order maintained through 'the concurrent action of various forces', were presented to his people at a time of trouble and disorder in their own country, to renew their enjoyment of such scenery. It was an attempt, he explained in the preface, to bring an artistic and literary treatment to subjects of natural science. Humboldt always regarded this work with much affection, and in the preface to the second and third editions, of 1826 and 1849, he was even more explicit on its significance: 'the combination of a literary and of a purely scientific object – the endeavour at once to interest and occupy the imagination, and to enrich the mind with new ideas by the augmentation of knowledge'. Without at this point questioning the division between imagination and reason, he suggested that works of science can appeal to both of these at once. Far from rejecting the methodical approach of scientific procedure, however, in entering the field of literature, he explained that his aim was to promote 'a love for the study of Nature, by bringing together in a small space the results of careful observation on the most varied subjects, by showing the importance of exact numerical data, and the use to be made of them by well-considered arrangement and comparison'.[29]

Humboldt's stand on these issues was a significant one, especially in the years immediately following the death of Kant, when the romantic movement seemed to offer the main source of resistance to the positivist conception of science. Already in scientific circles it was becoming customary to distinguish the natural sciences from the cultural or intellectual studies on the basis of method, while efforts to extend the exact science model into such fields as sociology, history, and linguistics made the issue a more serious one. Humboldt was distinctive as a leading scientist who joined with members of the literary circles of the time in objecting to the growing identification of science with technological advance and its consequent divorce from both philosophy and art.

Commenting on the striking affinity between the early nineteenth century and the mid twentieth in this respect, an affinity to which many writers have drawn attention, A. Gode- von Aesch has suggested that in the struggle against the kind of science favoured in the era of positivism, the arguments previously advanced by those known as the romantics are now being reinforced with the recent revival of criticism in regard to the anti-metaphysical viewpoint in science and its claims to objectivity:

This dissatisfaction with the metaphysical aphasia of our sciences is one of the clearest manifestations of a very general crisis . . . The sciences are accused of having ceased to be human sciences . . . Their results are called meaningless, at least in so far as they are concerned with the construction of an objective world detached from man as a center of vision.[30]

For Humboldt, at least, the sciences always remained human in their focus, their aim being in his view the betterment of mankind through the extension of knowledge. In this he was carrying forward the ideas developed at the end of the previous century by Kant and Herder in Germany, and by A. N. de Condorcet in France.[31]

Human geography: the people of the New World

A number of Humboldt's works, published between 1810 and 1834 in the *Voyage* series, were concerned in effect with human geography. His *Vues des Cordillères et monumens des peuples indigènes de l'Amérique* (1810) dealt with the origins and culture of the indigenous peoples of America, pointing to evidence for ancient links with Asia and with the civilization of the Old World.[32] Working methodically from documentary sources and architectural remains, as well as from his own observations, Humboldt again, as in his *Views of Nature*, applied the procedures of empirical inquiry to the human sphere, this time to historical and sociological research. At the same time, his scientific work incorporated a clear moral purpose and the view that he presented of the New World cultures was in distinct contrast to current European notions of savage life. Pursuing his theme of man's relation to the natural order, as he discussed the agriculture, arts and political organization of the Mexican and Inca civilizations, he viewed each society as a functioning system, evolving in the context of local conditions while also sharing in characteristics common to all human societies (see Figs. 6 and 7). He emphasized in effect that the development of American cultures, along with the obstacles to their progress, must be considered as part of the general progress of the human mind.[33]

The sympathetic consideration of native civilizations – a viewpoint in which Humboldt joined more with his brother Wilhelm and the tradition of Herder than with that of Kant – was continued in his political essay on Mexico, *Essai politique sur le Royaume de la Nouvelle-Espagne*, of 1811. Here his thorough training in finance, his experience in mining, and his growing competence in the field of social geography, were combined to produce both a classic regional study and a skilful exposure of the effects on a native society of European colonialism. Commenting on the civilization of the Indians before the conquest, he pointed out that their records and monuments were comparable with those of the most civilized people; later, however, under the oppression of what he called European ferocity and Christian fanaticism, the destruction of their hieroglyphics deprived them of the means of communicating their knowledge, leaving a race ignorant and indigent, with 'porters, who were used like beasts of burden', and 'crowds of

Fig. 6 Raft on the River Guayaquil, after a sketch by Humboldt. From Humboldt's *Vues des Cordillères et monumens des peuples indigènes de l'Amérique* (1810)

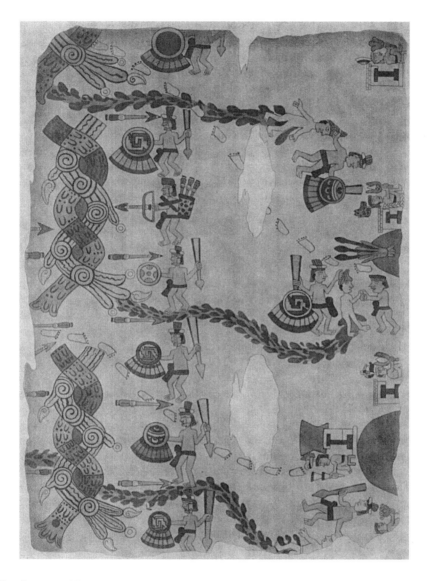

Fig. 7 Aztec Hieroglyphics, showing the 'Migrations of the Aztec people', from Humboldt's *Vues des Cordillères* (1810). Respect for the pre-conquest cultures of the native Americans was a strong theme in Humboldt's works.

beggars, who bore witness to the imperfection of the social institutions, and the existence of feudal oppression'. He brought the lesson home to Europe:

How shall we judge, then, from these miserable remains of a powerful people, of the degree of cultivation to which it had risen from the twelfth to the sixteenth century, and of the intellectual development of which it is susceptible? If all that remained of the French or German nation were a few poor agriculturists, could we read in their features that they belonged to nations which had produced a Descartes and Clairaut, a Kepler and a Leibnitz? [34]

Civil liberty, he added, remained rare even in Europe and is not a necessary product of advanced civilization: 'There are countries there, where, notwithstanding the boasted civilization of the higher classes of society, the peasant still lives in the same degradation under which he groaned three or four centuries ago.' Reports of the social state of such peoples were not to be found however in the current histories:

We shall seek in vain this relation in the annals of history. They transmit to us the memory of the great political revolutions, wars, conquests . . . but they inform us nothing of the more or less deplorable lot of the poorest and most numerous class of society [I, 177–8].

Although remaining to some extent tactful towards his Spanish hosts, Humboldt made this study of a colony a biting indictment of oppression in any society. Quantities of statistics, presented in tables, graphs and maps, along with historical records of Mexico, were assembled in his report, always to support what he mildly called a 'general view' but providing together a strikingly succinct picture of the effects of unequal distribution of wealth and productivity: 'Mexico is the country of inequality. No where does there exist such a fearful difference in the distribution of fortunes, civilization, cultivation of the soil, and population' [I, 184]. At every point, it seems, he used his thorough examination of the available data as the basis for proposals for future planning and reform, turning his attention to the administrative divisions, the legal system, agriculture and mining, while including also his own recommendations for a canal across the isthmus. In conclusion he stressed again the central theme of his work, his concern to examine the sources of public prosperity, and to impress upon administrators that

the prosperity of the whites is intimately connected with that of the copper coloured race, and that there can be no durable prosperity for the two Americas till this unfortunate race, humiliated but not degraded by long oppression, shall participate in all the advantages resulting from the progress of civilization and the improvement of the social order! [IV, 282].

Humboldt's study of Mexico stands as a significant contribution to the field of human geography, and even today in the new era of social activism and advocacy-research it remains an outstanding model. It is important to note again, however, that Humboldt himself referred to this work, and to his later studies of Venezuela and Cuba, not as a geography but as a political essay. The problem of geography, and in particular the place of the civil aspect, continued to occupy him at that time.

A new science: *Physique du monde*

In the preface to the *Relation historique du Voyage* (1814), his personal account of the American expedition written in French, Humboldt explained that while he wished to make known the countries he had visited, his main aim was 'to throw light on a science which has scarcely been outlined and which is called vaguely enough by the names of *Physique du monde, Théorie de la Terre* or *Géographie physique*'.[35] This preface was omitted from the 1818 English version by Helen Maria Williams, but the translation of Thomasina Ross in 1851 interpreted these three names as 'Natural History of the World, Theory of the Earth, or Physical Geography'.[36] It is evident that while an English equivalent for *Theorie de la Terre* was readily available, the French *Physique du monde* offered greater difficulty to the translator, and the choice here of 'Natural History' was unfortunate in identifying Humboldt's new science with the old composite study that it was intended to replace. Moreover, even the literal translation of *Géographie physique* as 'Physical Geography' remains to some extent misleading, since in English usage by the nineteenth century, physical geography implied a study of surface features of the earth such as mountains and rivers, whereas Humboldt followed Büsching and Kant in ascribing a wider meaning to it as the geography of the whole natural world, living as well as non-living. The term *physique*, used by Francis Bacon in proposing a study of dynamic processes in nature, came into regular use in France during the eighteenth century to indicate the general study of nature, and in particular to distinguish what the English called 'general' or 'universal' geography from topographies or local descriptions. But its connotations are not always clear, for it must be related to the current conception of nature, and especially to the debate on the place of man.

In his 1814 preface Humboldt expressed his own intention to include a study of man in physical geography; his aim, he explained, was

to comprise in one view the climate and its influence on organized beings, the aspect of the country, varied according to the nature of the soil and its vegetable cover, the directions of the mountains and rivers which separate races of men as well as tribes of plants; and finally the modifications observable in the conditions of the people living in different latitudes, and in circumstances more or less favourable to the development of their faculties.

A central problem for such a study of the earth, he suggested, is to consider 'the eternal ties which link the phenomena of life, and those of inanimate nature'. With regard to the division between the different branches of science, he pointed out that 'the natural sciences are connected by the same ties which link together all the phenomena of nature'. Noting that the geography of plants, which deals with the relations plants have to each other, to the soil and to the air, is dependent to a large extent for its progress on that of descriptive botany, he acknowledged that it would be detrimental to the advancement of science to attempt rising to general ideas, while neglecting the study of particular facts. At the same time he expressed regret that travellers highly trained in isolated branches of natural history rarely united their observations effectively, so that in this respect the results achieved did not seem to keep pace with the immense progress which had been made by the end of the eighteenth century in many sciences, particularly geology, the study of the atmosphere, and the physiology of animals and plants. Moreover, he directed attention to an aspect not explored within the natural sciences of his day, the relation between mind and nature:

one of the noblest characteristics which distinguish modern civilization from that of remoter times is that it has enlarged the mass of our conceptions and rendered us more capable of perceiving the connection between the physical and intellectual worlds [x–xiv].

The problem of reconciling the epistemology of Kant and Herder on the one hand, with the scientific empiricism of the Lavoisier school on the other, clearly preoccupied Humboldt at this time as he resumed the task of constructing a general science. In considering the phenomena of nature he returned frequently in the *Relation historique* to the Kantian emphasis on the need for a prior concept of the whole to guide research:

The ties which unite these phenomena, the relations which exist between the various forms of organized beings, are discovered only when we have acquired the habit of viewing the globe as a great whole [105].

Humboldt moved away from Lavoisier also in developing historical and comparative studies, although in his lengthy *Examen critique*, a study of the development of knowledge concerning the geography of the New World in the fifteenth and sixteenth centuries, he emphasized again the importance in historical research itself of a methodical approach, the scrupulous comparison of dates and the study of contemporary documents.[37]

Pursuing further the question of scientific method, Humboldt in his *Essai géognostique* of 1823 indicated his own concern to go beyond the limits of a narrow empiricism. In that work he introduced a comparison, based on his own researches in America, between the rocks of Europe and the New World, drawing attention, in some cases for the first time, to similarities of structure and form. There, in the preface, he outlined his own procedure:

In this *geognostical essay*, as well as in my researches on the *isothermal lines*, on the *geography of plants*, and on the laws which have been observed in the *distribution of organic bodies*, I have endeavoured, at the same time that I presented the detail of phenomena in different zones, to generalize the ideas respecting them, and to connect them with the great questions in natural philosophy . . . These subjects, I believe, are not mere vague theoretical speculations; far from being useless, they lead us to the knowledge of the laws of nature. It would degrade the sciences to make their progress depend solely on the accumulation and study of particular phenomena.[38]

The marked similarity with Forster's views is interesting. In the following years Humboldt himself became more closely involved with the study of geography, which at that time was attracting the attention in Europe of a growing number of authors, whose various titles for their works – 'Modern Geography', 'Scientific Geography', 'Earth Science' – reflected their effort to raise the subject to the status of a modern science. A brief consideration of several such works from the first quarter of the nineteenth century will indicate the trends occurring in the subject.[39]

Modern geography

In the preface to his *Modern Geography* of 1802 the English author John Pinkerton (1758–1826) pointed to the need for a new system of geography at the beginning of a new century. The eighteenth, he noted, had brought not only 'the gigantic progress of every science, and in particular of geographical information', but also drastic changes in most countries of Europe:

Whole kingdoms have been annihilated . . . and such a general alteration has taken place in states and boundaries, that a geographical work published five years ago may be pronounced to be already antiquated.

He criticized the latest popular texts as 'not only abounding with numerous and gross mistakes, but being so imperfect in their original plans, that the chief geographical topics have been sacrificed to long details of history, chronology, and commercial regulations'. Moreover recent discoveries were omitted or poorly treated in them, even though 'more important books of travel, and other sources of geographical information, have appeared within these few years, than at any period whatever of literary history'. Claiming to have extracted the essence of these sources in his own book, Pinkerton recalled that such works, combining 'poetical and romantic narration, with the study of man, and the benefits of practical instruction' had appealed to Montesquieu and Locke, as well as to 'my late friend Gibbon' – and he added, 'Why did he not write geography! Why has a Strabo been denied to modern times!'[40]

Some years later, in the 1807 edition of his *Modern Geography*, Pinkerton was pleased to note that his own work had been called by its admirers 'the English Strabo'.[41] Meanwhile in 1802 Pinkerton claimed his system of geography to be more complete than similar works in any other language, and his comments on these were scathing:

the Spaniards and Italians have been dormant in this science, the French works of La Croix and others are too brief, while the German compilations of Busching, Fabri, Ebeling, etc. etc. are of a most tremendous prolixity, arranged in the most tasteless manner, and exceeding in dry names, and trifling details, even the minuteness of our Gazetteers [I, viii].

In particular he rejected as absurd some recent attempts by French authors to revive the inclusion of ancient names on maps. Geography in his view must be concerned with the description of modern countries, on the basis of recent observations. Arguing, in the tradition of Baconian empiricism, that 'the grounds of any branch of science are to be found in modern precision', he condemned what he called 'theoretic geography, always useless, because it cannot alter the face of nature, and often blameable, as by suppositions of knowledge, it impedes the progress of genuine observation, and patient discovery' [I, xi–xii]. Instead, directing his own work to 'the advancement of science', he proposed to acknowledge doubts or ignorance and to welcome the correction of errors, while in contrast to former works, 'blindly copied from preceding systems', he made a feature of thorough documentation with 'regular references to the authorities, here observed for the first time in any geographical system'. In the interests of severe accuracy also, he noted at the end of his discussion on method, his aim was for a style concise rather than elegant [I, x–xiv].

Pinkerton's book itself, a rather ambitious work in two volumes, well bound and printed, incorporated recent maps and references to some of the latest research, including a report by Humboldt from South America in 1801. However, beyond attempting to bring geography into line with some of the better aspects of early-seventeenth-century empiricism, his work offered few innovations in terms of theory. He still saw geography, like chronology, as an aid to history [I, x], and his preliminary comments, distinguishing geography from cosmography, and contrasting it with chorography and topography in the Ptolemaic tradition, showed little change from geographical texts of the Baconian period. Similarly, in referring to the divisions of geography he mentioned first hydrography and, in more detail, general geography:

What is called General Geography embraces a wide view of the subject, regarding the earth astronomically as a planet, the grand divisions of land and water, the winds, tides, meteorology, etc. and may extend to what is called the mechanical part of geography, in directions for the construction of globes, maps, and charts [I, 1].

Following the custom of the previous century Pinkerton arranged for these topics to be treated in a lengthy 'Astronomical Introduction' written by the Reverend S. Vince, Professor of Astronomy and Experimental Philosophy at Cambridge. Other divisions suggested by Pinkerton included 'Sacred' and 'Ecclesiastic' geography, concerned with the Scriptures and Church administration, and also what he called 'Physical Geography, or Geology, which investigates the interior of the earth, so far only as real discoveries can be made'; in his opinion the so-called 'systems of the earth' are 'cosmogonies' and 'have no connection with the solid science of geology' [I, 1–2].

For what he saw as the popular interpretation of geography, as 'the description of the various regions of the globe, chiefly as being divided among various nations, and improved by human art and industry', Pinkerton did not use the term 'Special Geography' but suggested instead 'a scientific term' – *Historical Geography* – not only from 'its professed subservience to history', but because of the example of Herodotus or Tacitus. Proposing three divisions of historical geography – Ancient or Classical, to A.D. 500, Middle Ages, to the fifteenth century, and Modern Geography, 'the sole subject of the present work' – he explained that in describing each country he gave priority to the human aspect, discussing 'its political state, including most of the topics which recent German writers, by a term of dubious purity, call statistic', and 'the civil geography' – the government, chief cities, towns, and so on – before going on to the *Natural Geography*.[42]

Altogether it would probably be fair to say that while the English Strabo did not advance the theory of geography much beyond his Hellenistic namesake, his work represented a serious attempt to upgrade the geography texts of his time and introduce the current standards of scientific method. On that issue he was ready to criticize the procedure of Humboldt, whose account he used extensively in the South American section. Referring to Humboldt as 'a French naturalist, who has lately visited a considerable part of South America', Pinkerton commented that 'he is too fond of bending nature to his theories, while he ought to have been content with the observation of facts' [II, 675]. In response perhaps to those comments, Humboldt himself later made a point of criticizing Pinkerton's *Modern Geography* for its inaccuracies in regard to Mexico.[43] Moreover, the two authors diverged also in their views on Western imperialism. In strong contrast to Humboldt's subsequent defence of native cultures in America, Pinkerton concluded his geography with a recommendation, in effect, for a military invasion of Africa:

the wrongs of Africa can only be terminated by a powerful European colony, an enterprize worthy of any great European nation, a scene of new and vast ambition, and among the few warfares which would essentially contribute to the eventual interests of humanity, and raise a degraded continent to its due rank in the civilized world [II, 777].

A work evidently more in sympathy with the ideas of Humboldt at this time was the *Géographie Universelle* of Conrad Malte-Brun (1775–1826) which appeared at Paris in a number of editions, with successive modifications, from 1810 to 1865. Acknowledging in his preface that this was 'an immense undertaking' and a rash one, Malte-Brun explained that the aim of the work was to bring together the whole of Ancient and Modern Geography, in order, as he said,

> to generate in the mind of an attentive reader a lively picture [image vivante] of the entire earth, with all its different countries, . . . and the peoples who have inhabited them or inhabit them now.

These are subjects, he explained, 'which have usually been consigned by the moderns to erudite rather than elegant pens, and have been regarded as susceptible of no brilliancy of literary composition, or depth of philosophical meditation'. But he expressed his own conviction, 'that the science of geography admits of being made very different from what it now is'. He hoped to raise geography to equal history, and in phrases reminiscent of Kant, stressed the importance of considering place as well as time, 'delineating to the mind the permanent theatre' for men's actions, and describing the unchanging course of nature. At the same time he suggested that such a description of the globe must be 'intimately connected with the study of man, of human manners and institutions'.[44]

In the first two volumes of this work, as he explained in the preface, Malte-Brun provided first a historical introduction, dealing with the progress of geography to the time that the discovery of the New World 'opened to the view of science', in his terms, 'the vast ensemble of the globe'. He followed this with some discussion of the general theory of geography, before presenting what he called, using the term featured by Humboldt in 1807, a 'physical picture [tableau physique] of the globe'. Here he described the general features of the earth and atmosphere, as well as the distribution of vegetation and animals, while considering finally, in a way that recalls Büsching's plan for physical geography, the races of man, their appearance, languages, beliefs and 'the progress of their civilization' [3–4]. The remaining six volumes offered 'a successive description of all the parts of the world' and here he decided, after long meditation, to vary the order of description according to the significant features of each country or group of states, in order to improve on previous methods:

> After having examined all the so-called classifications of special geography, we have found that it is the too rigorous adherence to these abstract methods that gives so much aridity to these books of geography. Due to this empty parade of science, geography, this living image of the universe, seems to be nothing more than a cold and sad anatomy: the young dread it, the learned neglect it, the world scorns it [I, 5].

In presenting his own special geography Malte-Brun did not deny the merits of other approaches: he welcomed a new *Varenius* to produce a purely mathematical geography, or, as he continued,

> let another *Bergmann* discuss, in the language of chemistry and natural history, the elements of a new *physical geography*; let the naturalists subdivide the same physical geography into many particular sciences, such as the geography of plants, mineral-

ogical geography, etc.; let the pupils and successors of *Busching* collect with indefatigable patience the materials of *chorography* and *topography* . . . let them display, in immense columns of numbers, the details of that branch of *political geog-*
raphy which the Germans call *statistik*; let other savants consider . . . the critical
comparison of ancient geographers, or the history of voyages and discoveries.
Nothing is more useful to science [I, 7].

Furthermore, noting that 'the mathematical and physical principles of
geography are immutable, but the state of human knowledge varies', he
favoured also a series of geographies for different epochs, suggesting, as
Pinkerton had done, the ancient period, the middle ages, and 'modern
geography' from the discovery of America. Although he acknowledged the
need to restrict modern geography within limits, to avoid confusing it with
other sciences, Malte-Brun saw the wide extent of geography as providing
an opportunity 'to reunite under the same point of view the results of the
most different sciences' [I, 7–8].

On this concept of geography as on other issues, Malte-Brun's ideas show
a close resemblance to those expressed earlier by Kant and Humboldt, and
without entering here into questions of priority it is clear that the *Géo-*
graphie Universelle marks an important attempt to incorporate recent developments in theory into a major geography text. Linking Malte-Brun with
the earlier tradition of Rousseau and Voltaire, and with the later work of
Whewell with his interest in the imaginative process of conceiving the nature
of the world, W. R. Mead noted that Malte-Brun 'is claimed to be the first of
the new philosophers of geography'.[45] Later philosophers of geography
were to encounter serious opposition in what might be called the positivist
backlash to the movements of romanticism and idealism. That reaction
became prominent in the second quarter of the century, when a more
restrictive definition of scientific method by the schools of Comte and Mill
left the followers of Malte-Brun apparently with only two clear alternatives:
either they could defend the concepts and procedures of the new geography
and accept what is known as an exceptionalist position in regard to other
sciences; or they could identify with positivist methodology and reject any
conception of geography that failed to conform with it. A third alternative
seems to have been attempted seriously in that period by no geographer
apart from Alexander von Humboldt. That alternative was to assert the
values of geography and to challenge the very basis on which an anti-
metaphysical and fact-oriented empiricism rested.

It is evident, however, that most geographers of the time were unwilling
to become involved in any philosophical debate. August Zeune (1778–1853)
in his *Gea*, which appeared in three editions in 1808, 1811 and 1830,
commenced with the aim of formulating a scientific geography but in the end
confined himself to producing a work of *earth-science (Erdkunde)*, concerned with the description of natural rather than political units and therefore, as he claimed, based on 'a more secure foundation' than the changing

boundaries of states in the period of the Napoleonic wars.[46] Following the traditional approach he divided his book into a general and a special part, although his short general geography with its frequent acknowledgment of Humboldt's researches showed a distinct change of format, including, for example, a section on earth magnetism.

A more interesting German work, in the field of human geography, at the time was *Die Erde und ihre Bewohner* (The Earth and its Inhabitants) published at Leipzig (1810–11) by Eberhard August Wilhelm von Zimmermann (1743–1815), whose earlier studies on the geography of animals had been commended by Herder. His regional descriptions, although traditional in many respects, incorporated discussion of recent statistics and publications, while in his concern for *Völkerkunde* (ethnology) he maintained a strong social commentary, featuring in volume I on Africa a study of the slave trade which included examples of harsh treatment of the blacks and a condemnation of an epoch that in his estimate had already produced forty million slaves.[47] The book remained unfinished at Zimmerman's death; the final volume in the 1816–20 edition was completed by others, and it did not deal with Europe. A central theme, however, in the existing volumes was a concern to show the effects of European colonization throughout the world, a theme developed forcefully in the same period by Alexander von Humboldt.

During the early nineteenth century indeed, while relative freedom of expression existed in western Europe, the strengthening of liberal and humanitarian ideals seemed to promise the emergence of a new climate of thought, in which the ideas of the *Naturphilosophie* school, with their rejection of mechanist and dualist views and their affirmation of the unity and interdependence of nature, could provide the foundation for a new social philosophy and theory of knowledge. Humboldt's commitment to science was in this direction, and the constant effort of his work in geography was to develop such views, though even among his supporters there appeared no movement strong enough to effect any radical change.

Notable among those who promoted his concept of geography in the following years was Carl Ritter (1779–1859), who became professor of geography in 1820 at the University of Berlin, after the appearance of the first volume of his major work *Erdkunde* in 1817. Ritter had been an admirer of Humboldt since their meeting in 1807 and in his subsequent lectures and publications continued to acknowledge the leading contribution of the older geographer: 'Alexander von Humboldt has become, by his thorough studies of nature in Europe, Asia and America, the founder of Comparative Geography.'[48] In stressing the importance of the comparative method, Ritter adopted Humboldt's argument that geography as a science should provide, more than a catalogue of places, a study of the earth in terms of dynamic relationships:

There have always been detailed descriptions of the different parts of the earth, many of them remarkable for their accuracy, yet there has been lacking a knowledge of the principle of organic unity which pervades the whole . . . The whole subject of *relations* was unstudied. And it is a knowledge of the relations of things that leads to a scientific interpretation [10].

He returned frequently to this theme: 'When Geography ceases to be a lifeless aggregate of unorganized facts, and becomes the science which deals with the earth as a true organization, a world capable of constant development, . . . it first attains the unity and wholeness of a science' [13]. Science, he believed, would discern order in the apparent chaos of phenomena and here he turned, as Humboldt had done earlier, to the Kantian view of the world, not as a mechanism but as an organic whole:

There is, above all this thought of parts, of features, of phenomena, the conception of the Earth as a whole, existing in itself, for itself, an organic thing, advancing by growth, and becoming more . . . perfect [18].

In contrast to Humboldt, however, Ritter did not pursue the epistemological issue which was to become central to *Kosmos* – that it is only *by means* of such concepts, built up in relation to experience, that the human mind can apprehend the world. Although he defended the importance of historical understanding and showed strong sympathies with the romantics and the *Naturphilosophie* school, Ritter in terms of method seems to have accepted the Baconian approach to induction, supporting the procedure of commencing with observations rather than with opinions or hypotheses. Like the great Swiss educator Johann Heinrich Pestalozzi (1746–1827), whom he visited at Yverdon in September 1807, and whose ideas of *Anschauung* as man's recognition of an inherent world order formed the basis of his concept of a new Christian geography, Ritter did not overtly challenge Bacon's theory of knowledge. His work, including the massive *Erdkunde* or *General Comparative Geography* which in his lifetime expanded to nineteen volumes of descriptions on Africa and Asia without reaching to Europe, represented largely an attempt to reconcile standard empirical methodology with his own teleological view of the world.[49] The design theory, with the doctrine of nature's subservience to man, and the Hegelian-inspired notion of the divine will working itself out in the history of the continents, received great prominence in Ritter's geography and in this also the contrast with Humboldt was significant. Whereas Humboldt, who declined any university chair during his career, left no organized school behind him, Ritter's theory of geography was continued by some of his more famous students – Guyot in the United States, Reclus in France, Semenov in Russia – and the outcome in Guyot's case has already been noted.

For Humboldt himself the appearance of the elaborate special geographies of Malte-Brun and Ritter seems to have confirmed his dedication to a different task and with it the role, in Malte-Brun's phrase, of a new Varenius. In the second quarter of the nineteenth century he turned with growing concentration to the revitalizing of general or physical geography. This was one area of reform left open to him, for the years following the defeat of Napoleon were marked by increasing political repression, with the restoration of the French monarchy in 1814 under Louis XVIII and the reassertion of absolute rule in both France and Prussia. Censorship and persecution of republicans or liberal sympathizers placed greater restraints on intellectuals, even in Paris, and research in the colonies too was restricted. In 1814, on one of his many diplomatic missions, Alexander accompanied Friedrich Wilhelm III to London and with the Prussian King's assurance of financial support outlined to the British government his long-held plan for an expedition to the Himalayas and southern Asia. Permission for him to enter India, however, was refused by the British, despite repeated applications in the following years; presumably the East India Company opposed such a visit, especially after the appearance of the first part of the *Personal Narrative* with its trenchant criticism of conditions in the Spanish empire as Bolivar led the liberation of South America (1810–26).

THE IDEA OF KOSMOS: 1825–1859

Empiricism and reaction

In the period after 1825 Humboldt took a closer interest in the progress of German geography, assisting Heinrich Berghaus (1797–1884) and Karl Hoffmann (1796–1842) with a journal for *Erd-, Völker- und Staatenkunde* which they had founded that year under the title of *Hertha*. Over the following years he maintained a regular correspondence with Berghaus, contributing letters and research reports from other authors, as well as his own articles, to the journal.[50] From 1825 to 1827, although he remained in Paris, Humboldt came under increasing pressure from Prussian authorities to return to Berlin. At this time the third volume of his *Relation historique*, containing in its final chapters the political essays on Venezuela and Cuba, was published with evident haste and in an incomplete form in a Paris edition dated 1825. With the apparent loss before publication of the fourth volume, this marked his last substantial work in the field later called human geography. The vigour of his attacks on slavery and his support for reform in these essays suggests that, in the period of social and political reaction then occurring in Europe, few of the ruling conservatives must have regretted the loss of his final volume. In his *Essai politique sur l'île de Cuba* for example, having discussed the strategic location of Havana and examined 'the extent, the climate, and the geologic constitution of a country which opens a vast field to human civilization', he presented a table of population figures which he said 'might give rise to the most grave reflections' for it indicated that 83 per cent of the population were coloured – more than two million people, and over a million of them slaves. It was Humboldt's method to use statistics to support a moral purpose, and he warned:

> If the legislation of the West Indies and the condition of the coloured peoples does not soon undergo some salutary changes, if there continues to be discussion without action, political power will pass into the hands of those who have the strength to work, the will to be free, and the courage to endure long privations. This bloody catastrophe will occur as a necessary consequence of circumstances . . .

While noting that the fear of such consequences was likely 'to act more powerfully on minds than the principles of humanity and justice', he added that meanwhile 'the whites believe their power to be unassailable'.[51]

To appreciate the significance of Humboldt's position it must be remembered that he was writing as one of the most prominent scientists of his day; the issue at stake was the right, and the responsibility, of scientists in general and the political economist in particular, to engage in the crucial social issues of the time. To realize the extent and the effectiveness of the reactionary movement that followed it must be recalled that exactly a century and a half separated the *Essay on Cuba* from the iconoclastic Presidential Address of Zelinsky to the Association of American Geographers. Wilbur Zelinsky himself drew attention to the fate of radical thought among Western intellectuals in the intervening period: 'We must confess that the conspiracy of silence within the academy concerning Marx and Marxism is perhaps the greatest scandal of North American scholarship.'[52] It was in 1845, the year the first volume of *Kosmos* was published, that Karl Marx, already in exile, was deprived of his citizenship by the Prussian government, and in the preceding decades all liberal thought, including much that was far less extreme than the later communist doctrines, was under constant attack in Germany following the Carlsbad Decrees of 1820, as moves towards democracy were resisted by the ruling classes. Wilhelm von Humboldt, a leading humanist and reformer who had advocated a constitutional monarchy, was forced to leave the Prussian ministry in 1820 and retired from political life.

Alexander himself, in his post as Chamberlain to the Prussian King – a post on which he was now financially dependent after spending his own fortune on the publication of the American works – was finally recalled from Paris in late 1826, probably in an effort to control his reforming activities, and he returned with some reluctance to Berlin in May 1827. In November of the same year he commenced a series of public lectures which continued, along with a second series offered in response to public demand, until April 1828. With these he was continuing a tradition introduced in Berlin by the romantic school, with Schlegel's lectures on aesthetics from 1801 to 1804, and Fichte's lectures on politics and his famous *Discourses to the German Nation* of 1804–5 and 1807–8, followed by other courses over the years on science and philosophy. At this time some of the most significant lectures of the century were being delivered in Europe: Hegel (1770–1831), the dominant figure in German idealist philosophy, gave his lectures on the *History of Philosophy* from 1822 to 1831 at Berlin, while in Paris from 1826 to 1829 Auguste Comte offered his *Course of Positive Philosophy* in which, like the British utilitarians and philosophical radicals, he moved to link scientific empiricism with social reform. Humboldt's subject was *Physical Geography* and man's effort to understand the cosmos. Avoiding contentious social problems he nevertheless continued his aim of the improvement of man through the contributions of science, by directing attention to a problem that concerned all mankind, and the task implied in this was the reshaping of science itself.

The success of Humboldt's lectures was an outstanding event in Berlin at the time. Writing to Berghaus in December 1827 of the unexpectedly enthusiastic reception of '*my lectures on physical geography*', he added that he had declined Cotta's offer to publish transcripts of the lectures, since he wished to give much more thorough preparation to 'a book on *physical geography*' (*physische Geographie*), and he expressed the wish that Berghaus would prepare an *Atlas of physical Geography* (*physischen Erdkunde*) to accompany it. In this letter Berghaus saw being clearly expressed 'the idea for the book that was to become celebrated under the name *Kosmos*'.[53] That project, for which by April 1828 Humboldt had already planned the title 'Entwurf einer Physischen Weltbeschreibung' (Outline of a Physical Description of the World),[54] continued to occupy his attention for the last three decades of his life, the first volume alone taking another eighteen years to produce.

Meanwhile, within the limits imposed by the conservative reaction, he maintained an impressive output of publications and other contributions to scientific research from the time of his return to Berlin – adding his prestige in 1828 to the organization of the first major scientific conference in Germany, and initiating an international programme of magnetic observations. During 1829 he completed a journey across Russia, travelling at the Tsar's invitation although with the stipulation that he avoid all commentary on social conditions. Humboldt observed that restriction in his subsequent report, limiting himself to a study of mountains and comparative climatology, with a few introductory comments on an environmentalist theme that had interested him for a long time: the influence of mountain systems on human societies, on their migrations and 'the progress of intellectual culture'. Moreover, even his comments on method at this stage were in conformity with the standard theory of induction as a procedure commencing with facts. Suggesting that each century is marked by 'a new direction of ideas' he indicated the aim of his own work, to bring together different branches of the physical sciences, to lead to their mutual enrichment through a method of 'assembling a very great number of facts, grouping them, and raising them by means of induction to general ideas'.[55] Over the next years he apparently became more sceptical of that theory of induction.

Already Humboldt had emphasized throughout his career the importance of generalizations, the grouping of data, and the comprehensive view, as outcomes of scientific inquiry. In the course of his preparation for *Kosmos*, and even during the writing of it, he seems to have become more critical of the doctrine of facts. While the details of that development cannot be traced here, it can be noted that in the period after 1825 those issues were being explored by others in a number of works, which Humboldt himself later acknowledged as sources. The publication of Hegel's *Lectures on the History of Philosophy*, soon after the author's death in 1831, gave Humboldt access to the incisive Hegelian attacks on positivist empiricism and the whole objectivist fallacy. Describing Aristotle as a 'thinking empiricist', Hegel in that work pointed out that Aristotle's *Physics* included what was later called the 'Metaphysics of Nature', whereas contemporary physicists merely 'devote their attention to what they call experience, for they think that here they come across genuine truth, unspoiled by thought, fresh from the hand of nature'.[56] Humboldt himself in his Berlin lectures of 1827 had taken pains not to give any public denial of Hegel's philosophy although he was opposed to many aspects of it.[57] Writing to his friend Varnhagen von Ense in May 1837 he agreed to read Hegel's historical studies carefully, despite 'a wild prejudice against the theory that every nation must be the representative of some particular idea', and reported in July from his reading of the *Philosophy of History*, that despite its many false assertions, 'there is indeed a forest of ideas for me in this Hegel'.[58] Along with the philosophy of Kant, the epistemology of Hegel evidently became meaningful to Humboldt in his later years.

A further stimulus in that direction came in the same period from the work of William Whewell (1794–1866), whose studies in the history and philosophy of science received a number of acknowledgments in *Kosmos*. Whewell, Professor of Moral Philosophy at Cambridge, and distinguished also for his contributions to the natural sciences, was the originator of the term 'scientist'. In his influential philosophical works of 1837 and 1840 he used a historical study of the inductive sciences to reveal the methods scientists had actually employed, in order to provide a theory of science that would avoid 'the fallacies of the ultra-Lockian school', especially those of the French sensationist and anti-metaphysical doctrines.[59] He traced these fallacies to the incomplete model provided by Francis Bacon as 'the supreme Legislator of the Modern Republic of Science', and pointed to the difficulty of extending his 'Philosophy of Induction' beyond the physical sciences. Bacon himself had planned 'the extension of the new methods', he noted, 'to intellectual, to moral, to political, as well as to physical science . . . But the philosophy which deals with mind, with manners, with morals, with polity, is conscious still of much obscurity and perplexity'. Whewell criticized Bacon for an emphasis on facts rather than ideas, and a neglect of 'the Conceptions which the intellect itself supplies' [II, 389–99]. Moreover he drew attention to the rejection of past philosophy by both the Baconians and Cartesians, with their overemphasis on the superiority of modern science and the 'new philosophy'. With regard to the rationalist–empiricist conflict he noted that Descartes himself depended on observation in his own method, while Newton's contrary doctrine of the rejection of all unfounded hypotheses would be disastrous to inquiry if applied strictly [II, 456].

At the same time Whewell did not draw attention to the acceptance by both schools of the faculty division of the mind. Instead he stressed the importance of the intellect in providing general ideas 'which govern the synthesis of our sensations'. Applying this to the sciences themselves he went on to advance what he claimed to be his own doctrine of 'Fundamental Ideas belonging to each science, and manifesting themselves in the axioms of the science'. This, along with his emphasis on the activity of the mind in all observation and experience, was an important contribution of Whewell's thought. He concluded his 1840 study with an acknowledgment of Kant's achievement in transferring attention to the subject, the knower, as the foundation of reasoning, rather than the object known, and his own work continued that movement [II, 463–79].

In a later edition of his *Philosophy of the Inductive Sciences* Whewell expanded his ideas in a volume entitled *Novum Organon Renovatum*, with the obvious aim of reforming Bacon's own 'new organon'. Here he was even more explicit in relating facts to concepts: 'facts cannot be observed as Facts, except in virtue of the Conceptions which the observer himself unconsciously supplies'. He stressed furthermore the futility of attempting to exclude ideas and inference from the process of perception: 'A certain activity of the mind is involved, not only in seeing objects erroneously, but in seeing them at all.'[60] At the same time he continued to maintain a distinction between sensations and ideas: 'the antithesis of *Sense* and *Ideas* is the foundation of the Philosophy of Science' [5–6]. Moreover he believed it essential to separate the intellectual faculty from the emotions and imagination:

Facts, for the purposes of material science, must involve Conceptions of the Intellect only, and not Emotions . . . the Facts which we assume as the basis of Science are to be freed from all the mists which imagination and passion throw around them [50–57].

By contrast, although he did not directly question the faculty theory itself, Humboldt in *Kosmos* defended the contribution of imagination and interest along with intellect in the process of scientific inquiry [I, 40; V, 8]. While he constantly made it clear, as in his Berlin lectures of 1827, that he condemned the excesses of some romantics and idealist philosophers who made ill-founded or extravagant speculations on the philosophy of nature, he continued to support the holist views of the nature-philosophy school and their belief that reason cannot be divorced from imaginative and affective processes in experience and hence in science.

An open rejection of the faculty model of the mind actually appeared at this time in the work of Johann Friedrich Herbart (1776–1841), a former student of Fichte at Jena from 1794 to 1797 and later an exponent of Pestalozzian theories, who was appointed to Kant's former chair at Königsberg in 1809 during Wilhelm von Humboldt's term as Minister for Public Instruction. For many years Herbart's philosophy was overshadowed by that of Hegel, who remained professor from 1818 to 1831 at the new University of Berlin, Prussia's third university, founded in 1810 largely through Wilhelm's efforts. After Hegel's death, having failed to gain the Berlin chair and disturbed by political events in Prussia, Herbart returned to Göttingen for the rest of his life, publishing there his final statement on education, *Outlines of Educational Doctrine*, 1835. In this, repeating his view that it is an error 'to look upon the human soul as an aggregate of all sorts of faculties', he argued that 'faculties are after all at bottom one and the same principle'.[61] Throughout his work he had attempted to provide a unitary theory of knowledge to reconcile what he saw as inconsistencies in both the sense-empiricism of the Rousseau–Pestalozzi view and the Kantian approach with its dualism of phenomena and noumena and its emphasis on the role of innate concepts or categories which themselves are determined by the basic structure of the mind. Instead, Herbart suggested that the mind itself is continually in process of being formed by the organization of experiences or presentations (*Vorstellungen*), derived from observation of the external world, which in association create what he termed an *apperception mass*. Learning or creative thinking occurs as new presentations are related to those previous experiences, ordered as concepts, which can be raised above what he called the *threshold of consciousness*, and the effectiveness of that process depends upon the range and clarity of such concepts. Following Herder rather than Kant he thus explained differences between various cultures, a theme developed also by Humboldt in describing Indian societies in colonial America.

Herbart's model of the mind, a brilliant contribution to the thought of his time, opened the way for subsequent Freudian research on the subconscious, while his suggestions for teaching procedures were widely adopted by educators later in the century, although in their form of Herbartianism his own concern with moral and metaphysical issues was generally rejected in favour of a Baconian framework with its implied acceptance of mental faculties. With the strengthening of positivism the significance of Herbart's theory was not explored, and even Humboldt seems to have had little direct contact with his philosophy, although they shared a belief in the importance of concept formation in knowledge, and a belief also in the integrative role of geography in the field of learning. In his major work of 1806 *Allgemeine*

Pädagogik (*General Pedagogy*, later translated rather inaccurately as *The Science of Education*), Herbart assigned geography a central position linking what he saw as the two main groups of subjects – those like literature, history and religion, dealing with mankind and society, which provide social experience, and those like natural science, mathematics, and crafts, which provide knowledge of the external world or objective experience.

As far as Humboldt was concerned, however, judging by the absence of any reference to Herbart in *Kosmos*, the writings of such 'speculative' philosophers were less likely to attract his attention than the work of a scientist such as Whewell. As already noted, like other thinkers whose works were available to Humboldt at that time, Whewell seems to have advanced some, although not all, of the themes developed in Humboldt's theory of science. Whewell's own attempt, for example, to classify sciences according to the basic concepts of each was not extended by him to the social sciences and it gave a place neither to geography nor to the concept of cosmos. That task became a central one for Humboldt.

Kosmos: The knowledge of nature

In October 1834 Humboldt wrote jubilantly to his old confidant Varnhagen: 'I am going to press with my work, – the work of my life.' He had taken the crazy idea, he went on, 'of representing in a single work the whole material world, – all that is known to us of the phenomena of heavenly space and terrestrial life', and this, too, with a lively style to retain interest while instructing the mind. The whole was not to be what had commonly been called 'Physical Geography', he explained, since it would comprise both heavens and earth, everything existing: 'I began it in French fifteen years ago, and called it "Essai sur la Physique du Monde".' The title *Kosmos* was chosen after some hesitation, he added, to indicate in a striking way the scope of the work and to stand 'in contrast to the "Gäa" [*sic*] (that rather indifferent earthy book of Professor Zeune, a true *Erdbeschreibung*)'. His own work involved a much wider conception: 'It must represent an epoch in the mental development of man as regards his knowledge of nature.'[62]

Outlining the contents to Varnhagen, Humboldt explained that it would commence with a general section in which the rewritten introductory lecture on physical geography from 1827 would be followed by 'the picture of Nature.' Next, a discussion of incentives to the appreciation of nature in modern times from literature, art and the study of plant life would precede 'the history of the physical description of the World; how the idea of the Universe – of the connection between all phenomena has been becoming clear to different nations in the course of centuries'. He stressed the significance of those sections: 'Those Prolegomena form the most important part of the work.' They were to be followed 'by the special part, comprehending the detail' and he included a list:

Space – the physics of Astronomy – the solid portion of the globe – its interior and exterior – the electro-magnetism of the interior – Vulcanism, *i.e.*, the reaction of the interior of a planet upon its surface – the arrangement of matter – a short Geognosy – sea – atmosphere – climate – organic life – distribution of plants – distribution of animals – races of man – languages, – and so on, to show that their physical organisation (the articulation of sound) is governed by intelligence.

As projected, the scope was clearly beyond any previous work, extending to a concern not only with man and society but with human intellect as a part of nature. At that stage, while preferring the effect of a single volume, he hoped that 'two volumes will include the whole'.[63]

In *Kosmos* as it finally appeared the detailed part, in which Humboldt attempted to gather the results of the latest research, was never completed. After the Prolegomena, which comprised the first two volumes in 1845 and 1847, the third volume, dated 1850, was devoted to the astronomical section, while the discussion of the earth itself in the fourth volume of 1858 and the unfinished fifth volume had not reached beyond aspects of vulcanism and geology at the time of his death in 1859; the majority of volume V, published posthumously in 1862, consisted of a detailed index to *Kosmos* prepared by an editor. Humboldt's original outline of the detailed part, with its specific inclusion of plant and animal geography as well as a study of man, was surveyed completely however in the first volume, in what he called '*a general picture of nature [Naturgemälde] or comprehensive view of phenomena in the cosmos*' [I, xii]. Eduard Buschmann, editor of the last volume, affirmed that in December 1856 Humboldt's plan had remained unchanged[64] and in February 1857 Humboldt wrote to Varnhagen that 'the last (fourth) volume of "Kosmos" will consist of two parts', of which the first was already printed, but both were to appear together to avoid spoiling the effect of a continuous description from the temperature of the earth's interior to the 'races of man'.[65] In 1858, with the second part still delayed, he expressed his regret at being unable to fulfil his original plan to produce a single work on 'what is usually called *physical geography*', and explained his slow progress by pointing to the expanding horizons of knowledge in his time and 'the decreasing vitality of an almost ninety year old man'.[66] The place of human geography in his final work remains an unresolved question.

Subsequently, the *Physical Geography* of Mary Somerville in 1848 and Arnold Guyot's comparative physical geography, *The Earth and Man* of 1849, both followed Humboldt's plan and included a discussion of man, although the pronounced imperialist views of Somerville's work and the emphasis on divine design in Guyot's geography represented a break with Humboldt's views.[67] Later nineteenth-century texts, however, tended to revert to a dualist approach and did not persist with his conception of geography. Instead, after Darwin, with an increasing emphasis on the adjustment of all organisms, including man, to the environment, the question of determinism came to be interpreted even more in terms of the opposition of man's will to immutable laws of nature, and at its worst the issue degenerated into an argument over whether man is controlled by the brute world, or whether he exploits it.

Humboldt's attempt to include a study of man and the human intellect in a science of the earth remained an isolated one in his own century. His concept of nature as unity in diversity, constancy in change, was neither mechanist nor dualist and he saw the improvement of human culture as part of the functioning of an ordered cosmos. Moreover, since he believed the gradual understanding of nature to be part of the intellectual development of mankind, he gave prominence to that aspect in *Kosmos*, devoting almost the entire second volume to a history of that process in Western civilization. In this section, using the term *Weltanschauung*, world view, he set out to study 'the history of thought on the unity of phenomena and the inter-relationship of forces in the universe' [II, 536]. His aim was to trace the 'main stages in the gradual development and expansion of the idea of the cosmos as a natural whole', and his work was a significant contribution to the history of ideas:

The history of the philosophy of nature [physischen Weltanschauung] is the history of the recognition of a natural whole, the description of the striving of mankind to comprehend the cooperation of forces in the earth and in space; it shows accordingly the epochs of development in the generalization of views, it is a part of the history of our world of ideas [Gedankenwelt] [II, 135].

In discussing '*the limits and scientific treatment of a description of the physical world* [*physischen Weltbeschreibung*]'[68] Humboldt seems to have been conscious of the constraints exercised by earlier models. Characteristically he saw his own work issuing from a historical context, and he made it clear that he linked *Kosmos* with a geographical tradition extending from the *Geographia generalis*: 'That important work of Varenius is in the true sense of the word a *physical geography* [*physische Erdbeschreibung*]', he wrote [I, 74]. He was prepared to break with that tradition in extending his own study to the universe instead of confining the astronomical part to an introduction [I, 59–60], although he acknowledged from the beginning the problem involved in taking 'the hitherto vaguely conceived idea of a *physical geography* [*physischen Erdbeschreibung*]' and expanding it, by means of 'a perhaps too enthusiastic plan', to a physical *Weltbeschreibung* [I, viii].

Throughout the following century this was regarded as a retrogressive step by many geographers, and *Kosmos* was labelled as the last of the great cosmographies rather than as a book pointing the way to the future. In view of the present development of space exploration there could well be a resurgence of interest in Humboldt's cosmography, with the possibility that the study of satellites and planets – as well as selenography – might be linked with geography in the late twentieth century.[69] In 1963 A. N. Strahler, for instance, followed his *Physical Geography* with a textbook, *The Earth Sciences*, which despite its title extended to the solar system and was intended to meet the growing enthusiasm for the field of space research.[70] It is probably a reflection on the restricted interests of recent geography that Strahler's new project was carried out in a department of geology.

With regard to Humboldt's other major divergence from Varenius, in planning to include in general geography a closer study of plants, animals and aspects of human geography, he already had the support of the model provided by the physical geography of Büsching and Kant, both of whom saw these as part of a *Weltbeschreibung*. Although these sections were never completed in *Kosmos*, Humboldt's principles in this respect evidently remained unaltered. In sketching with Berghaus the outline of a geography textbook for Indian schools, after a request from Joseph Dalton Hooker in October 1848 for such a work, he proposed that it should commence with a general section which was to include plant and animal geography, along with 'the races of man', and he referred to this as *physical geography*, in contrast to the second part which he envisaged as *special geography* or *Länderbeschreibung* (description of regions).[71] This project, unfortunately, lost English support and remained incomplete after four years.

An enormous field for research in human geography had been opened by Humboldt's American work; however, following the successful revolutions in Spanish America, what seems to have been a virtual embargo on his writings on social issues after 1825, with the strengthening of political reaction in Europe, must have inhibited his later moves in that direction. On this account it is interesting to note that in addition to the silencing of Humboldt's social criticism during his Russian visit of 1829, two other works which involved similar commentary were terminated abruptly before completion between 1830 and 1834. One of these was the *Examen critique* of 1834, which despite its rather disarming title – *Critical examination of the history of the geography of the New Continent and the progress of nautical astronomy in the 15th and 16th centuries* – was more than a thorough compilation of historical records and a clarification of Vespucci's role in the naming of America. It provided not only a brilliant demonstration of the continuity of ideas through history and their influence on action, as shown in the Hellenistic concepts of the earth that inspired the voyages of Columbus; it examined also the impact of European conquest on the New World. In the second section of the work Humboldt drew attention to the effects of colonial repression and the suffering of the Indians, faced more often with inhumane legislation and persecution than with humanity or kindness, and he noted the sad picture of human misery confronting the reformer in Europe and the United States as well as in the rest of the New World – the struggles for the abolition of slavery, the enfranchisement of the serfs and the general amelioration of the condition of the workers. The final preface, however, written in November 1833 in Berlin, indicated that the last two sections, which were to deal with early charts of the New World and the progress of navigation, would not be completed:

The work which I intended to prepare on the history of the geography of the two Americas and the progressive correction of astronomical locations, has been abandoned since my voyage to Northern Asia and the Caspian Sea. A new set of ideas has been presented to my mind, and has diminished the interest which I had for the kind of work with which I have been occupied since my first return to Europe. I have decided to bring to an end my works on America . . .[72]

To some extent his explanation seems plausible, the new set of ideas obviously referring to his work on *Kosmos*. Humboldt went on to explain that his decision to end the American works caused him fewer regrets now that the French explorer Joseph Boussingalt had returned after twelve years in the New World, with the latest data on magnetic and meteorological phenomena, geology, topography and chemistry; it is significant, however, that no reference was made to a continuing study of economic, social or political conditions. His next statement is also interesting:

I hope to publish soon the fourth and final volume of the *Relation historique*, the only work of this long series of publications on America which remains to be completed [v].

As it turned out, the fourth volume of the *Relation historique* was never published, and its mysterious disappearance is an intriguing subject for research. The account given by Julius Löwenberg in the standard nineteenth-century biography of Humboldt, namely that the fourth volume was largely finished by 1810 and printing had already begun when the volume was destroyed, has been widely accepted although his statement does not seem to have been substantiated.[73] Commenting on that account, Anne Macpherson expressed doubt that the fourth volume was actually nearing completion in 1810, suggesting instead that perhaps another historical work was being referred to in the correspondence of that time, and that despite Humboldt's frequent references to his progress on the last volume in letters to Arago and others after his return to Berlin, he had little more to say on the journey and actually lost all interest in the project as he turned to the writing of *Kosmos*.[74] In reviewing the Löwenberg account, however, Hanno Beck noted that Humboldt as an old man had thought of finishing the report of his travels, even long after 1833: 'Eduard Buschmann wrote to the publisher Georg von Cotta, Berlin 14.12.1859 that Humboldt had told him personally on 20.5.1855 and at the end of October that he still wanted the fourth volume of his *Relation historique* to be published.' A proposal by Buschmann that the work might be concluded from the manuscript of Humboldt's travel journal, and a plan by Cotta to publish the *Tagebuch* itself in two volumes never seem to have materialized.[75] Indeed the standard German translation of the *Relation historique* by Hermann Hauff, published by Cotta in 1859–60, ended with Humboldt's arrival in

Cuba in December 1800, thus omitting the more contentious final essays in the third volume of the French edition.

The missing fourth volume, as Beck has noted [II, 69], was to deal with the period from May 1801 to August 1804, covering his travels in the modern states of Colombia, Ecuador, Peru, and Mexico, the second visit to Cuba and the last weeks in the United States. Already the earlier volumes had made a significant impact: Charles Darwin, for instance, wrote to J. D. Hooker in February 1845 of Humboldt's influence on his own research in natural science: 'I shall never forget that my whole course of life is due to having read and re-read as a youth his *Personal Narrative*.'[76] The last volume was likely to have been of immense interest in Europe, especially following the revolutions there of 1830 and 1848, and the suspicion must remain that some form of conservative intervention led to its suppression.

Humboldt himself clearly maintained his humanitarian outlook: as late as 1856 he protested in a liberal Berlin paper against the omission from a recent New York translation of the *Political Essay on New Spain* of his chapter on slavery in Mexico, and he saw as a personal achievement the passing of Prussian legislation freeing any slaves brought to that country.[77] However, official opposition to any renewal of his analysis of contemporary social conditions was apparently effective in curtailing further publications.

In a number of letters to liberal-minded friends during these years, Humboldt complained of his 'moral and mental isolation' in the intellectually barren environment of Berlin, where there were constant accusations against him of impiety, republicanism or revolutionary sympathies. He wrote to Varnhagen in June 1845 describing one such attack in a German newspaper:

In the 'Rhein-und Mosel-Zeitung', No. 122, May 29th, I am accused of Voltairianism; of denying all Revelation; of conspiring with Marheineke, Bruno Bauer, and Feuerbach . . . and all on account of 'Kosmos', p. 381.[78]

In the first volume of *Kosmos*, published earlier in 1845, the page in question dealt with explanations of the origin of mankind, Humboldt quoting from the work of his brother Wilhelm on language, and the anatomist Johannes Müller in his *Physiologie des Menschen*, to the effect that in the absence of evidence from experience the notion of human descent from a single couple could well be mythical. This was a sensitive issue at a time when the young Hegelians, including Phillipp Karl Marheineke (1780–1846), Bruno Bauer (1809–82) and Ludwig Feuerbach (1804–72) were questioning many orthodox Christian beliefs, Bauer and Feuerbach in particular arguing for an end to all religion since in Feuerbach's terms, man by creating the idea of God was alienating himself from his own divinity.

For his own part, without expressly denying a religious interpretation, Humboldt had followed the earlier example of Keckermann and Varenius in detaching his science from Christianity and in this, as Hanno Beck has noted, he developed the more recent work of Kant and Forster, to become the first great exponent of a secular geography.[79] Although Humboldt actually never allied himself with the radical Hegelians, nevertheless he was seen as a threat by the conservatives. Towards the end of 1845 Varnhagen recorded in his diary:

He assures me that but for his connection with the Court he should not be able to live here. So much was he hated by the Ultras and the Pietists that he would be exiled [132].

Writing to his patron, Friedrich Wilhelm IV of Prussia, in March 1846, Humboldt noted rather wryly that many royalists would like to see him buried in the family home of Tegel, or else safely back in France:

Certainly there are very estimable people who, from mere affection towards Your Majesty, would be glad to see me either under the Column in Tegel, or once more on the other side of the Rhine.[80]

At this time, while Humboldt in Prussia was being attacked for his radical tendencies, Karl Marx (1818–83) and the group of German revolutionaries then in exile in Paris were accusing him of complicity as an envoy of the Prussian King in securing an order for their expulsion from France during his visit to Paris in January 1845.[81] Although that charge was denied, the radicals continued to express their hostility. As late as 1856 Marx reported to Friedrich Engels (1820–95) that Bruno Bauer, leader of the young Hegelians in Berlin, had spoken quite deprecatingly of Humboldt during a recent visit to London.[82]

Little is actually known of any personal contact between Marx and Alexander von Humboldt, although obviously they knew of each other, and noting this Hanno Beck has suggested that the question of their relations needs to be clarified through further research.[83] Although both were concerned with social reform, Humboldt's position was more that of a liberal conservative, advocating freedom of communication and an improved constitution through democratic processes, and he evidently hoped to achieve this initially with the support of an enlightened monarch. His meetings earlier in the century with South American revolutionaries, especially those with Simon Bolivar in 1804 and 1805, have been given some prominence, but it seems likely that Humboldt, with his sense of the historical continuity of social patterns, was wary of radical promises of dramatic change through revolution. Moreover, the kind of materialism favoured by Marxists and other left-wing Hegelians was diametrically opposed to the more organic and holist world view that he was developing. Marx saw nature primarily as the object of human activity – either actual or potential – and although he recognized society as a part of nature, as a natural environment in itself, his main interest was in the dominance of nature, the extension of man's control over all productive forces in nature, including therefore society itself.[84]

Rejecting Hegel's conception of history as the development of the World Spirit, finding its expression in the mass of mankind, in nation states and in the world-theatre, Marx and Engels substituted a materialist conception of history in which they argued that the character of societies is derived, not from an abstract spirit but from the material conditions of existence, especially the modes of production and other economic factors which in their view determine political and cultural conditions. Their approach was anti-theological, anti-metaphysical, socialist and of course anti-capitalist; they claimed that the industrial revolution had created a new class, the proletariat, and that the emancipation of these workers from oppression would lead to the emancipation of all society through socialist organization. In that process they assigned a revolutionary role to the masses, outlining this for the first time in their joint work *The Holy Family* of 1845, a satirical attack on Bruno Bauer and the speculative left-Hegelians. With the aim of providing a scientific account of history which would give a justification of communism, Marx and Engels described here a line of continuity leading from the materialism of Bacon, Hobbes and Locke through to their disciples in France – including La Mettrie, Helvétius and d'Holbach – while from there the ideas of the bourgeois French revolution were taken back to Britain in the socialist programmes of Bentham and Owen, the founder of English communism. Tracing their own position back to earlier French materialism, they explained that movement as the product of two distinct trends: on the one hand the mechanical materialism of Descartes whose *physics* provided the basis for the advances of French mechanical natural sciences in the eighteenth century, and on the other hand the materialism of the British empiricists, introduced to France through the works of Locke: 'Materialism', they wrote, 'is the *natural-born* son of *Great Britain*.' A synthesis of those traditions, they believed, would lead to a positive science of society, rejecting both metaphysics and romanticism, and continuing the work of Owen and the scientific French communists who 'developed the teaching of *materialism* as the teaching of *real humanism* and the *logical* basis of *communism*'. In many respects the views of Marx and Engels coincided with the positivist ethos which dominated nineteenth-century science, with its commitment to progress and reliance on Bacon's sense-empiricism as a theory of knowledge. They argued further that the humanist aim, the development of society, would be achieved by changing the material conditions that govern individual experience in the social context:

If man draws all his knowledge, sensation, etc., from the world of the senses and the experience gained in it, then what has to be done is to arrange the empirical world in such a way that man experiences and becomes accustomed to what is truly human in it and that he becomes aware of himself as man . . . each man must be given social scope for the vital manifestations of his being. If man is shaped by environment his environment must be made human. If man is social by nature, he will develop his true nature only in society . . .[85]

A concern with what later came to be called the sociology of knowledge was an important contribution of Marxist thought to the more individual-centred tradition of Baconian empiricism, although the concept itself did not originate with Marx – it can be traced in many earlier writers, including Herder, Kant and the Humboldt brothers. In that respect the work of both Marx and Alexander von Humboldt represented a departure from the kind of positivism that was currently finding reinforcement in the philosophy of John Stuart Mill (1806–73).

Educated by his father to participate actively in the utilitarian reforming programme that James Mill shared with Jeremy Bentham, John Stuart Mill turned, in his *System of Logic*, to producing an interpretation of scientific method based on logical induction from facts and rejecting Whewell's emphasis on the priority of concepts.[86] Mill himself steered positivism firmly away from the extreme utopianism of Auguste Comte's later works, in which the French founder of positivist sociology gave renewed stress to the establishment of a 'religion of humanity', a new science of society leading to the moral regeneration of the Western nations and extending 'to the rest of the white race, and finally to the other two great races of man'.[87] Positivism, as a substitute for Catholicism, according to Comte, would provide an 'objective basis' for social reform by revealing 'the laws or Order of phenomena by which Humanity is regulated', for in his view 'the economy of the external world . . . embraces not merely the inorganic world, but also the phenomena of our own existence'; and although he added, 'Our conception of this External Order has been gradually growing from the earliest times', his epistemology shared more with Bacon than with Pestalozzi or Humboldt [25–7]. Accepting the faculty theory, he used it to justify both a scientific method and a social hierarchy. His ideal society was to be ruled by an intellectual elite from which he evidently excluded, as deficient in rationality, not only the working classes but also all women, in whom he claimed that a natural 'subordination of the intellect to the social feeling' determined the position he assigned them – an important one, in his view – as guardians of morals and the finer sensibilities [3–5].

In that context Humboldt's efforts in *Kosmos* to articulate a dynamic concept of nature and an alternative theory of scientific method indicate a clear divergence from the work of Comte, Marx and Mill. Meanwhile in 1848, deeply disappointed with the failure of the liberal revolutions in Europe, Humboldt wrote in August to Berghaus of his concern to search for order in the wider relationships of the universe:

In the disorder of our days and with the discord of our times . . . I seek refuge . . . as often as my position allows, in the unending cosmos, seeking and finding, in the investigation of its phenomena and laws, the peace that is so necessary to me in the evening of an extremely eventful life.[88]

Again in March 1853, writing to Varnhagen of 'the perplexities of my desolate life and the morally disgraceful condition of the times', he expressed regret at the recent death of Leopold von Buch, seeing it as a prelude to his own:

And in what condition do I leave the world? – I, who remember 1789, and have shared in its emotions! However, centuries are but seconds in the great process of the development of advancing humanity. Yet the rising curve has small bendings in it, and it is very inconvenient to find one's self on such a segment of its descending portion.[89]

With this he turned in the next six years to his survey of research in physical geography and the questions this raised on the need for a more coherent theory of science.

Mind and nature

A central problem that came to occupy Humboldt in the writing of *Kosmos* was the relationship of mind and nature, for he accepted neither the materialism of the Marxist and positivist schools nor the extreme idealism of the orthodox Hegelians. Initially in writing this work Humboldt seems to have been reluctant to overstep the limits already prescribed for physical geography by Kant, who had stipulated the external world as the proper sphere for geography, and in the first volume of *Kosmos* he actually affirmed – although in rather vague terms – the current belief that the laws which apply to the higher intellectual sphere are different from other natural laws [I, 386]. Late in his life, however, he retained doubts as to the limits to be drawn between the physical and intellectual spheres. In the third volume he acknowledged the place in nature of the products of thought:

Man elaborates within himself the material which the senses present to him. The products of such mental labour belong as essentially to the domain of the cosmos, as the appearances that are themselves reflected in the mind [III, 7].

The study of man's ideas assumed an important place in his work as he realized that since phenomena are interpreted as mental perceptions, a science is concerned with the organization of these. Following Kant and Hegel he regarded as objective, rather than subjective, all knowledge which does not depend on an individual mind for its existence. Concepts, which enable personal experience to be objectified, have therefore an important role in scientific inquiry; moving beyond Kant he saw these as being developed by man himself through history. In particular he regarded a science as a means by which such concepts are generated, tested and incorporated in the sphere of ideas. Thus a significant aspect of his *Weltbeschreibung* consisted in his effort to trace the concept of nature into the modern period. The scope of his entire project led him to express frequent misgivings in *Kosmos* and in 1850 he concluded, 'the criticism of incompleteness must apply particularly to the part of this work that concerns the intellectual life in the cosmos, external nature reflected in the world of thought and feeling' [III, 7]. At the outset he had stated his belief that he should limit himself to the area in which he felt most secure after a scientific career occupied with 'measurements, experiments, and the investigation of facts' [I, 68]. However, as his introduction to the final volume indicated, he became increasingly alert to the dependence of facts on concepts themselves.

Humboldt's own philosophy emerges from his writing. Accepting at the outset a distinction between two spheres – the external world on the one hand and the 'intellectual world [Sphäre der Ideen]' on the other – Humboldt proceeded to establish a relationship between them. 'Freedom and order, nature and mind [Freiheit und Maass, Natur und Geist]' [II, 288] were dualisms that his belief in the unity of the world led him to question. Freedom of the human intellect need not be inimical to the notion of an ordered world; indeed, he believed that the world of thought (*Gedankenwelt*) can be understood as part of the process of nature. His interpretation of the ideal world, then, is a dynamic one, in harmony with his concept of change and motion as part of the necessity of nature. Avoiding the impasse encountered by Kant in attempting to relate the perception of the individual subject to the noumenal sphere, Humboldt viewed the world of ideas in historical perspective. Ideas generated by mankind are passed on through education, with each generation borrowing what they judge to be applicable to their own needs; through the input of highly gifted individuals this store of ideas is increased and social improvement, accompanied by further investigation, expands man's mental world. Intellectual development is a process in which the essential function of science is the ordering of experience. In any given period, man can apply to the study of objects only the concepts available in the human intellectual heritage. The free intellect is the means by which the sphere of ideas is enriched through new discoveries or the explanation of connection in the world, but the successful application of reason to the understanding of the world requires both ordered observation and knowledge of relevant concepts.

Humboldt was aware that the phenomena of science are indeed 'appearances' which are not necessarily a true reflection of external objects. The method of induction, then, involves efforts to establish validity through experiment, direct observation, enumeration of data and the refinement of instruments and methods, including analysis as an intellectual method; classification and comparison enable the practical application of ideas to develop theories and produce new concepts. Where in the past, 'fantastic creations' [I, 16] or errors have been incorporated in the mental world and perpetuated in language, science becomes a means of making the inner world of ideas a more adequate reflection of the external world. Empirical investigation is a continuous process, leading to progress in knowledge, consonant with ordered change in the cosmos: observation of the diversity of nature leads to the development of theories and ideas of interrelation, and so to general views of the unified whole.

The concept of the pre-existing universe is essential to Humboldt's philosophy, and although like Kant he accepted 'the whole' as preceding human experience, he made this the basis, not of a static classification but of an evolutionary process. Intuition of 'the wonder of the boundless [Zauber des Unbegrenzten]' is evident in primitive peoples; with intellectual development and the operation of thought on experience, mankind achieves an understanding of order in nature. Imagination acts as a creative stimulus in this process, arising from 'the connection of the sensible with the intellectual [Zusammenhange des Sinnlichen mit dem Intellektuellen]' [II, 74]. Empirical science is the means by which the contact of man's mind with the external world can lead to intellectual development: the ordering of experience contributes to the control and increase of ideas; thought and its products are incorporated in the world totality. In Humboldt's view the two spheres of the one cosmos are in dynamic interaction.

In this way 'the apparently impassable gulf between thought and being, the relationship between the knowing mind and the perceived object' [II, 282] was seen by Humboldt as the locus of the sciences. Science is mind applied to nature:

One may regard *nature* as opposed to the realm of the *mental*, as it would be if mind were not already contained in the entirety of nature: or one may oppose *nature* to *art* . . . but these contrasts must not lead to such a separation of the physical from the intellectual that the *physical science of the world* is reduced to a mere aggregation of empirical specialties. Science begins where the mind itself takes hold of matter and attempts to subject the mass of experiences to a rational understanding; it is mind directed towards nature [I, 69].

Humboldt's view of science remains a challenging one in the context of the twentieth-century debate. While working ostensibly within the framework of the doctrine of a created universe – for despite accusations against him of atheism, he was careful to avoid any evidence of impiety – nevertheless he emphasized that all future progress for mankind depends on the coherent development of knowledge through man's own efforts. For the sciences to become 'a mere aggregation of empirical specialties' was inimical to such development. In contrast he believed that a comprehension of nature would emerge from the natural sciences, encompassing 'both spheres of the one cosmos – the external world, perceived by the senses, and the inner, reflected intellectual world' [III, 8]. Combining elements from the scientific and philosophical traditions of his time, Humboldt himself made significant steps towards providing a viable methodology for scientific inquiry and at the same time a consistent theoretical basis for geography.[90]

It is important to note the significance of Humboldt's references to these issues in some of the last pages he wrote at the end of his life. In the introduction to the final volume of *Kosmos* in 1858 he reasserted his aim to provide in that work a general *world picture* and to contribute to the generalization of views with regard to 'continuous, active *natural processes*' by showing the way that dynamic elements form a natural whole. He contrasted this with an 'attempt to classify and examine the individual elements' – a procedure, he noted, with an obvious reference to Baconian empiricism, 'on which, up to a certain period of time in our scientific knowledge, it was believed that results are based'. Describing his own method as 'a thoughtful consideration of empirical phenomena', that is, of phenomena given through experience [V, 5], he went on to refer in some detail to the growing criticism of Bacon's interpretation, along with a more searching reassessment of Aristotle's view of experience, in recent German works on the history of philosophy. He quoted with approval the argument of C. A. Brandis that in this respect Aristotle was not only Bacon's predecessor but his superior, 'because he was convinced that the human mind can apprehend the world of actuality . . . only *by means* of the concept [Begriff], and indeed, only to the extent that the latter is developed in its correlation with the data of experience'.[91]

Brandis himself, who opposed the extreme idealism and the 'immanent dialectic' of Hegel as well as the sense-empiricism of Bacon, had argued in the same context that while Aristotle cannot be seen as supporting any theory of the dialectical unfolding of a divine idea, it would be equally a mistake to regard him 'as the founder of a sensualistic empiricism'. Aristotle, he continued, 'is the most decided advocate of the claims of experience, but of an experience dependent on the recognition of productive ideas' and he suggested that Bacon would not have become misled on that issue if he had been acquainted with the Aristotelian system 'more from the actual writings of the author than from inadequate reports and distorted practices'.[92]

For his own part Humboldt, who could respect the useful aspects of Hegelian philosophy, went on to point out that Hegel too had referred to Aristotle as 'a complete but at the same time a *thinking empiricist [denkenden Empiriker]*'.[93] The term is one that can well be applied to Humboldt himself, and the development of his own *thinking empiricism* is an epic story, of which only the outlines can be sketched here. The search for an adequate theory, commencing at least from his early days at Göttingen, extended over seventy years, and it accompanied at every stage his efforts to construct, through a series of geographical works, what he saw as a necessary counterpart to the science of his time. The way of the geographer, he finally came to realize, led through the paths of philosophy: the study of epistemology and the re-evaluation of Aristotle and Bacon were necessary tasks on the way. At the age of 65 he even attended university lectures for two years on the history of Greek literature, as he became aware of the importance of the history of ideas in tracing the development in different societies of the concepts that give meaning to human experience. In Humboldt's last months he was still engaged in the effort to articulate his new concept of empiricism:

For it does not become the free spirit of our time to reject as groundless hypothesis every philosophical attempt to enter deeper into the linkage of natural phenomena, when such attempts are also based on induction and analogy. Considering these noble talents which nature has granted to man, it does not become him to condemn either reason meditating on causal relationships or the lively power of imagination, indispensable for stimulating all discovery and creation [V, 8].

Humboldt himself made no claims to the possibility of achieving complete knowledge, either by strict observation or by reasoning and speculation:

The boundless spheres of observation, which through newly discovered means (instruments) daily become wider, and indeed the *incompleteness* of perception for every single moment of speculation, make the task of a *theoretical philosophy of nature* to some extent an *uncertain* one [V, 6–7].

Defending the philosophy of nature – 'it is not necessary to *deny* everything that cannot be *explained*' [V, 13] – he continued to explore the concept of the global ecosystem or, in his terms, 'the great and complex community that we call nature and world'. He saw this community as 'the combination into a *natural whole* of elements that have the capacity for evolution', including in it 'the *intellectual life in the cosmos* . . . the world of thought and feeling' [V, 5–7]. For ideas on a universe of change he recalled again the theory of Empedocles (*c.* 494–434 B.C.) and Anaxagoras (*c.* 500–428 B.C.) that in all phases of *being*, in life and death, the elements merely change their relationships [IV, 12; V, 20–1]; he noted also the earlier teachings of Heraclitus (*fl. c.* 500 B.C.) that in a continuous process of transformation everything *has its being in its becoming* – referring here to a German work of 1858 on this topic, and quoting Hegel's view that 'it is a great conception of Heraclitus, to turn from *being* to *becoming*' [V, 19–20; pp. 287–8 below]. Finally, advising that research should focus on natural processes, searching for laws and causal relationships, he praised recent advances linking theoretical chemistry and physics, research on electro-chemical processes and the nature of heat and light, and studies in electro-magnetism which provided evidence of dynamic interaction in the cosmos [V, 10–22].

Humboldt's *Kosmos* remained unfinished; his personal contributions to the sciences, the method of thoughtful empiricism he explored, and the constant moral basis underlying his own scientific work, represent his outstanding achievements. In this brief survey the aim remains, however, not so much to offer an evaluation of his work, its successes or failures, but rather, by considering the traditions he inherited and the goals that he set himself, to appreciate the immensity and, in his own time, the loneliness of the task.

CHAPTER 8
EPILOGUE: THE WAY AHEAD

With the death of both Alexander von Humboldt and Carl Ritter in 1859 an era in the history of geography came to an end and a distinctive period in the development of the sciences commenced, when the so-called romantic views of the nature-philosophers in the first half of the century were displaced by a positivist model of scientific procedure. A more strident nationalism and its extension in European colonialism discouraged the growth of ideas on universal humanism, and in the age of railway expansion and industrialization the notion of man as master of nature inhibited the development of ecosystem concepts, conservation movements and teachings on the harmony of nature, while social Darwinism, based on the doctrine of the struggle for survival, offered a justification for the continued dominance not only of white Europeans over the coloured races but also of the privileged classes over the poor within each country. Humboldt had stood firm against many of the new trends: regarded with some awe to the end of his life, it seems that for his many opponents his death brought considerable relief from embarrassment.

In the following years a noticeable undermining of his influence occurred: even the excessive adulation by admirers proved detrimental to his reputation as a serious scientist and practical reformer, while outside Germany further translation of his work into English virtually ceased with his death. In the case of *Kosmos* no effort was made to translate the last sections and the index in volume V published in 1862, while the four earlier volumes in the standard five-volume version by E. C. Otté of 1849–58 remained, even in the new edition of 1882–4, rather badly translated, often in pompous and inaccurate prose, with its theory of science filtered through a rather crude form of materialist sense-empiricism.[1] Moreover, Humboldt's first biographers in Germany contributed to the decline of his prestige in scientific circles and as Anne Macpherson has indicated, many later misconceptions about his work can be traced not only to changes in science but also to some of the early commentaries: in her view 'most of the damage to Humboldt's reputation can be charged to Dove'. One of the contributors to the 'scientific biography' edited by Karl Bruhns in 1873, Dove regarded Humboldt's science as out of date; his hostility to his subject may have derived as much from personal animosity as from changes in the intellectual milieu, for

Dove's pronounced anti-semitic and nationalistic views were obviously opposed to Humboldt's universalism.[2]

Of the common misconceptions about Humboldt's work, the assumption that his theory of science is outdated bears reconsideration in the present context of paradigm change, while the charge of vague romanticism can now be balanced against his pioneer contributions to ecosystem studies and the development of an alternative to positivist empiricism. Similarly the widely-repeated notion of Humboldt as primarily a physical geographer – in the sense of one concerned with the 'world-minus-man'[3] – must be rejected in the light of his extensive contributions to human geography: the physical geography to which he devoted much of his life included man as part of nature. Furthermore, the kind of science that Humboldt advocated was closely involved with social issues, frequently using meticulous procedures of data collection as a basis for recommending social action or political and economic reform. From the present standpoint his social radicalism can be seen as not merely a kind of aberration in his scientific career but an integral part of his new theory of science.

After Humboldt, geography in all countries reverted to a conservative position; the division between physical and human geography became pronounced towards the end of the century, reflecting the loss of a unitary concept of nature and simultaneous pressure for sciences to conform to positivist methods. This seemed to mark the end for Humboldt's model of the radical humanist geographer advocating synthesis, holistic views and social reform as part of a vigorous scientific empiricism.[4] A notable exception in that respect was the exiled Russian prince, Peter Kropotkin (1842–1921), a leading anarchist and geographer who worked in England from 1886 until his return to Russia after the October Revolution of 1917. As Bob Galois noted, Kropotkin has been omitted from histories of geographic thought, except for a footnote by Preston James in *All Possible Worlds*.[5] His work is now being reassessed with a revival of interest in radical geography, although up to now the extent of his debt to Humboldt does not seem to have been recognized.

In his political views Kropotkin was committed to the ideals of communism, although he broke with Marx on the issue of centralized state socialism and adopted Proudhon's term of *anarchism* (literally 'no government') to indicate his opposition to state control. Where his debt to Humboldt emerges is in aspects of his concept of nature and his view of geography. Like Humboldt he regarded nature as a dynamic whole, subject to constant change and including mankind, social institutions and human reason. In his humanism also he defended the right of all men to equality, freedom and dignity, and his interest in primitive societies and in the general progress of mankind recalls Humboldt's early *Political Essays*. Kropotkin too was aware that the social context affects individual development and he believed that morality is to be sought in nature itself, not from some supernatural source. Here, taking up Darwin's ideas of evolution, he made his own distinctive contribution by rejecting a narrow interpretation of Darwin's struggle for existence and arguing that the forces of co-operation are as essential in the survival of life as the forces of competition. Darwin himself had recognized the existence of co-operation in biological communities, and a similar idea has already been noted in the work of Herder and Humboldt; Kropotkin developed this in his studies on ethics to present the principle of *mutual aid* as the basis for natural morality.[6]

Consonant with this view of nature, Kropotkin saw the role of geography as one of overcoming the division between the 'human sciences – history, economy, politics, morals – and natural sciences', and he noted also the incorporation of aesthetic aspects and a concern for poetry in the approach to nature of Ritter and Humboldt.[7] His references to geography in these respects are clearly dependent on Humboldt's ideas: he saw it as the duty of geography not only to cover the fields of the specialized sciences but also 'to combine in one vivid picture all separate elements of this knowledge; to represent it as a harmonious whole'.[8]

In his views on scientific method Kropotkin was in agreement with Humboldt in his preference for active observation and experiment, his recognition that science involves a constant process of inquiry and change leading to improved knowledge, and his awareness that the so-called laws of nature are not fixed, indeed that the findings of the 'exact' sciences represent a series of approximations. He diverged however from Humboldt in maintaining, with the Marxists, a mechanist and materialist world view which extended to society, and an antipathy to metaphysics which he misinterpreted, in the manner of the earlier positivists, as the area of abstract speculation 'outside physics' or beyond the domain of physical laws or human experience. Like them also he was an admirer of Bacon as 'the great founder of inductive science' who revived 'the sound philosophy of Nature' and provided the basis from which modern science has developed 'the elements of a philosophy of the universe, free of supernatural hypotheses and the metaphysical "methodology of ideas"' – one capable of explaining the universe as a never-ending series of transformations of energy in which the evolution of living matter and even the mechanisms of human thought and feeling, as well as social institutions, can be understood.[9] He frequently praised 'the natural-scientific inductive method' by which both modern natural science and materialist philosophy had been developed, although he apparently had only a superficial comprehension of its implications.[10] Moreover, he seems to have been unaware of the contradiction involved in his own position, accepting Humboldt's theory of geography while maintaining in large part a positivist theory of scientific method and a materialist view of the world. It must be noted how often political radicals prove to be conservative in their support of established scientific procedure.

A new scientific revolution?

For geographers, the general trend in recent scientific thought, away from a narrow positivism and towards the kind of empirical method that Humboldt outlined over a century ago, offers an opportunity to identify with the scientific movement as a whole without a loss of coherence in their own study. After a long interval since the appearance of *Kosmos* there are signs that the philosophical dialogue is being resumed more vigorously in geography. Gunnar Olsson in 1972 noted the growing awareness among geographers that the swing from quantitative and spatial analysis towards 'a search for explanatory statements of relevance to people . . . could not be pursued effectively until some basic epistemological issues had been clarified'; in his opinion 'philosophical and methodological reevaluations of drastic proportions are in the making' and he suggested that similar ideas in other disciplines were an indication of 'an impending social scientific revolution'.[11] More recently Derek Gregory has drawn attention to 'the inadequacies and consequences of a traditional geography' and in the period following 'the positivist revolution' has called for a critical attempt to provide a more satisfactory theory of science as examined discourse of relevance to practical life.[12]

Whether at present the general changes taking place in the current scientific paradigm are sufficiently extensive to justify designating this as a new scientific revolution is a question that probably cannot be resolved at this point, and the term itself requires closer examination. According to Kuhn the scientists of any period form a community of scholars sharing a common paradigm, comprising their accepted models for scientific achievement, along with 'the entire constellation of beliefs, values, techniques' that guide their research. In his view, a drastic change in such a paradigm takes on the character of a revolution, in the course of which a new model is adopted by a process which he regarded as analogous to a gestalt switch in perception, and a new tradition of normal science is established. He associated the names of Copernicus, Newton, Lavoisier and Einstein with such revolutions in the past.[13]

The use of the term scientific revolution to suggest the complete overthrow of an earlier paradigm, however, must be made with some caution, for such a notion, by suggesting the achievement of radical alterations in an entire system of beliefs and methods, can tend to obscure the gradual increments by which many changes are made and the tenacity with which some ideas or customs are retained, even when they are inconsistent with others. The term scientific revolution, well-established for more than a century to describe the transformation that occurred in Western thought and technology after the sixteenth century, has been retained in this discussion, although an organic analogy might well be more suitable in dealing with the changing patterns of human thought. Some changes are clearly dramatic, with the introduction of new theories, concepts or discoveries. However, the identification of a major scientific revolution is difficult and it is doubtful whether this term can be applied, for instance, to the work of Newton or Lavoisier except in relation to the specific field of physics or chemistry.

A study of empirical science as a whole from the time of Copernicus suggests rather that while the term scientific revolution may be usefully applied to the total changes that have occurred, the demarcation of a conclusion to the cycle must be in large part arbitrary. In the course of this period there appears to have been not so much a series of dramatic revolutions in scientific thought as a gradual modification of ideas over three centuries in the context of historical events, with the maintenance of many obsolete or inconsistent notions throughout that time. In many respects the empiricism that provided the chief model for scientific methodology in the seventeenth century can be seen to bear an astonishingly close resemblance to that of the mid twentieth.

The empirical tradition in modern science, as it emerged in the seventeenth century with the work of Galileo and Bacon, was marked by a struggle for freedom from religious authority and it commenced, as this study has shown, with the assertion that all human knowledge concerning the world is derived not from divine inspiration but from experience, or more precisely, from the evidence of the senses. It was argued therefore that scientific knowledge must be based on information obtained through the senses, assisted by the use of experiments and measurement, but with particular care being taken to avoid unsecured opinions or theories received from others. Direct individual contact with the objects of nature came to be regarded as the proper basis for the scientific method, in a period when individualism formed a strong current in Western thought. By the end of the eighteenth century the belief gained wide assent that the primary task of the scientist is to deal with *facts* by observing and classifying objects, and that these facts can as it were speak for themselves if the rules of induction are correctly applied. Philosophers of the time, including Hume and Kant, who expressed scepticism about various aspects of that theory of cognition, evidently made little impact on the popular idea of science as a body of value-free, objective knowledge. Instead there was a growing tendency to claim for science complete independence, not only from religion but from philosophy also.

With the rise of positivism a clear policy was established of dissociating the scientific enterprise from all speculation, emotion, imagination, even normative considerations. Science was justified by its objectivity and utility, becoming almost synonymous with progress as the means for enlarging man's control over the material world, and indeed throughout the nineteenth century that idea was reinforced by the outstanding technological achievements associated with science in the course of the industrial revolution. Significantly it is only with the comparatively recent breakdown of confidence in the outcome of industrialism on a world scale that a concerted effort is being made' to subject the scientific movement itself to closer scrutiny.

In general, with regard to the question of scientific revolution, this study lends support to what Stephen Toulmin called 'an "evolutionary" theory of scientific change (as contrasted with Kuhn's "catastrophism")', an approach which treats scientific change 'as a special case of a more general phenomenon of "conceptual evolution"'.[14] This view, as already indicated, is itself in accordance with Humboldt's view of science. Whether one supports a revolutionary or an evolutionary view, however, there appear to be strong indications that Western science at least is entering a period of transition marked by a reappraisal of goals, methodology and concepts. Increasingly, scientists are exhibiting a greater concern with the concept of the world-wide ecosystem and a concurrent movement away from a simplistic dualism of man versus nature. At the same time stronger demands for social responsibility in science, for practices to be modified in accordance with moral and aesthetic considerations, are undermining the belief that all science contributes to human progress and therefore must be intrinsically beneficial. Along with this there is evidence of fundamental changes in methodology and a growing general reaction against the strict positivist view. Many scientists appear to be well on the way to discarding the materialism and the doctrine of facts inherited from the eighteenth century; the search for exact and universal laws as formulated in the context of seventeenth-century mechanism is no longer considered the prime goal of scientific inquiry, while the ideals of certainty and objectivity that were a legacy of sixteenth-century thought are giving way to a more relativist theory of knowledge. There are signs of the emergence of a new or at least an alternative epistemology, a theory of knowledge in which the development of concepts plays a central part, and with this a theory of the mind that calls into question the faculty model, inherited from Aristotle, which tended to set the rational intellect of man in opposition to the 'animal' faculties of imagination and feelings. While the ideal of scientific universality and impartiality can still be pursued to useful effect, the myth of the detached intellect is being dispelled as far as the scientist is concerned and a stronger justification is now provided for research directed towards understanding social problems.

Social empiricism

Altogether it can be suggested that a number of what at first seem disparate trends are moving in the same direction, one that represents a considerable departure from the notion of science and scientific method formerly given wide assent. Claims for the absolute autonomy of science are receiving sustained attack from two directions: on the one hand, the effects on the environment of uncontrolled individualism can no longer be tolerated; and on the other, the assumptions on which the initial view of empiricism was based are being exposed to a more rigorous questioning. Social considerations in both cases play an important part: not only is there increasing pressure for the activities of science in general to be moderated in accordance with the long-term welfare of society throughout the world, but at the same time the notion of individual autonomy in perception is being replaced by a recognition of the close relationship existing between the so-called senses and the conceptual apparatus that the individual shares with a social group. In combination these are likely to have a lasting impact on science, for just as the new theory of knowledge points out that our perception is conditioned by the problem situation in which we operate, so a recognition of the problem of social responsibility in science will in turn affect the kind of science we pursue.

For this reason it is proposed here that the way out of the empirical crisis for geography and for modern science as a whole lies in developing what may be called a tradition of social empiricism, based on a less narrow concept of experience than that fostered by the Baconian tradition with its emphasis on direct sense-experience and its twin outcomes of positivist objectivism and pronounced specialization in science. The new trend is towards a kind of empiricism that gives assent to the complex process by which scientific research is linked with a paradigm of shared concepts and theories in a social and historical context. This view gives greater recognition to the educative aspect of empirical science, for while its basis in experience is not to be denied, this is no longer interpreted in terms of the individual researcher in some kind of direct contact with his objects of study, but rather the operation of the mind in perception is affirmed, and with it the important role of vicarious or symbolic experience in science.

A key element in the theory of social empiricism is a recognition of the social aspect of knowledge and an attempt to turn this to advantage in formulating a theory of scientific method. In particular, a concern for the sociology of knowledge, previously associated largely with Marxist thought, is now extended to all scientific inquiry. At the same time, in proposing this kind of approach it is important not to overlook the significance of the individual contribution at each stage of the scientific process: ideas from the pool of socially shared knowledge must be internalized and developed by the individual.[15] What is emphasized here is that such individual creativity operates within a social context; far from denying that context, the individual should make more effective use of it, taking full account of the importance of concepts in scientific investigation. All coherent thinking involves the use of concepts, theories, idea-patterns which are themselves built up in a cultural context: the special task of science is to use these intellectual tools to the best advantage, increasing their precision, testing as far as possible their validity and attempting to ensure their consistency. To this end the history of ideas performs an important function, tracing the development of concepts or theories and indicating the complex of possibly conflicting ideas from which these have been formed.

In this area many questions remain to be answered, for comparatively little research has been carried out, the structure and role of concepts being of secondary concern in the Baconian tradition of inquiry. The belief that inquiry commences with sense-impressions can be seen persisting in the teachings of phenomenology as well as in the strict positivism of the materialist–mathematical school. Philosophers also, especially those influenced by logical positivism in the present century, have tended to pay little attention to concepts, Karl Popper being a case in point, while Israel Scheffler also showed little concern with this question.[16] In his attempt to clarify the nature of concepts Lennart Nørreklit commented that these had received only 'a somewhat casual and Cinderella treatment in analytical philosophy'.[17] Although he discussed various usages of the term and considered the importance of systems of inter-connected concepts, Nørreklit himself approached the discussion from the viewpoint of logical or linguistic analysis, and since concepts are rarely articulated fully it is clear that much further inquiry is needed.[18] A useful contribution in that direction was Peter Achinstein's proposal for *a broader concept of theory* in science, to include not only propositions but also concepts, as well as the methods of solving problems and analysing or even structuring situations. He pointed out that 'meaning-dependence' as a philosophical position, supported by studies in the history of science, rejects in effect a fundamental assumption of positivism by claiming that all terms used in science are dependent for their meaning on the scientific theory in which they are used.[19]

Popper himself always insisted that every scientific statement, like all knowledge, is tentative and provisional. Dismissing Bacon's claims to reach certainty through induction – 'his myth of a scientific method that starts from observation and experiment and then proceeds to theories' – Popper noted that this legendary method 'still inspires some of the newer sciences which try to practice it because of the prevalent belief that it is the method of experimental physics'.[20] He emphasized the role of theory in science: 'The empirical sciences are systems of theories . . . observation is always *observation in the light of theories*' [59]. However in *The Logic of Scientific Discovery* his basic concern was a logical analysis of scientific methods, and he set out to make a clear distinction between the 'creative' phase of hypothesis or idea-formation and the logical procedures of scientific examination; accordingly he gave little attention to what he termed 'the initial stage, the act of conceiving or inventing a theory' – a stage that in his view involved 'an irrational element' of intuition or inspiration – for he regarded this as belonging to the sphere of metaphysics [30–2]. While condemning earlier attempts to eliminate metaphysical discussion – he argued that 'positivists, in their anxiety to annihilate metaphysics, annihilate natural science along with it' [36] – nevertheless he too was determined to maintain a clear demarcation between science and metaphysics. He claimed to have found this in the test of *falsifiability*: 'metaphysical concepts and ideas may have helped', he said, 'to bring order into man's picture of the world . . . Yet an idea of this kind acquires scientific status only when it is presented in falsifiable form' [278].

Although scientific theories are never fully verifiable, they are nonetheless testable, according to Popper, and their *objective logical relations* can be examined; the objectivity of scientific statements depends on their testability and their response to what he called inter-subjective criticism. He compared his own use of the term *objective* to that of Kant, who associated objectivity not with the individual's direct sense-impressions of particular objects in nature, as the Baconians had done, but rather with ideas or concepts which receive justification from a number of people in society and therefore, being no longer dependent on a single individual, have as it were an independent existence and are communicated through some form of symbolic representation. The Kantian form of empiricism made it possible to view concepts as objectified experience and in effect emphasized the social aspect of science. In his *Critique of Pure Reason* Kant had argued that scientific knowledge should be justifiable, its validity being established by others independently of the knowing subject. Popper was particularly interested in Kant's recognition that the process of objective justification is most effective when a statement is expressed in the form of a testable theory and when the events to which it refers themselves exhibit some regularity: 'Kant was perhaps the first to realize that the objectivity of scientific statements is closely connected with the construction of theories' [44–5]. Generally, however, the process of concept formation and theory construction received little discussion in Popper's analysis, his emphasis being more on the refutation of hypotheses than on the ordered extension of theory in a science. Indeed, by his own criterion that a scientific statement must in principle be falsifiable he evidently was limited in his ability to integrate into his own work the impressive general concept outlined in the preface to *The Logic of Scientific Discovery*, where he stated what he conceived to be the central concern of both science and philosophy:

It is the problem of cosmology: *the problem of understanding the world – including ourselves, and our knowledge, as part of the world* [15].

Subsequently an evolutionary approach to knowledge was advanced by Popper in his *Objective Knowledge* of 1972, where he postulated a world of objective thoughts in the form of theories, critical arguments and problems which exist independently of individual minds and provide a significant part of the human environment as well as a source for the development of science.[21] It is a view reminiscent of Chardin's concept of the *noosphere* – a term coined by Chardin in 1925 to denote the 'thinking layer', evolved from the biosphere on the earth's surface.[22]

A century earlier the importance of such a concept had already been advanced by Alexander von Humboldt, who argued that the *world of ideas* performs an essential role in the progress of knowledge. He regarded the sphere of human thought as an extension of progressive change in nature and therefore an important element in the total world complex. In particular, as a recent commentator noted, 'Humboldt was the first to define clearly scientific knowledge, gradually acquired, as a distinct part of the whole domain of Nature.'[23] He saw the sphere of ideas as being sustained through the course of history, communicated through education and gradually enlarged, especially through scientific research. In a time of increasing specialization, however, he became aware that a loss of coherence was inevitable without some general concept of the world to which all sciences could relate; he therefore became committed to developing a science, one concerned with a view of the world as a whole, in which that concept might be explored. He saw this as a function of geography, although as already noted he extended his own study beyond the earth to the universe and in *Kosmos* widened his discussion also to consider the history of man's attempts to develop a concept of nature: the historian, in his view, has an important responsibility with regard to the sphere of ideas in enlarging knowledge of not only the sequence of past events but also the way present concepts have been formed.

Geography and the ecosystem concept

Humboldt's world concept was not sustained by geographers after him, although the study of ecology, dealing with the relationship of organisms to their habitat, was developed by Darwin and Haeckel, and maintained chiefly by biologists. By the end of the nineteenth century, as science was directed towards the study and control of the material world and the discovery of physical laws, the emphasis on man in conflict with nature became if anything more pronounced at the very time that evolutionists were demonstrating man's close biological links with the rest of the organic world; any attempt to revive a unitary concept of nature tended to be rejected as unscientific, as evidence of some kind of vague idealism or romanticism. The word *nature* itself fell into disuse in scientific journals, where it was retained mainly as a synonym for the 'world-minus-man' or more specifically for the wilderness untouched by man. In its stead the term *environment*, originating in the seventeenth century from the French *environ*, surroundings, came into wider use, first in biology, then in the social sciences and geography.[24] By the mid twentieth century the term *nature* rarely appeared in a geographical index, reflecting probably what might be called a process of elision – the suppression of a term during a period when its implications are under question.

In geography the gradual introduction of *environment* as a substitute for *nature* occurred at a time when there were strong arguments for physical geography to exclude man in order to progress as an exact science, while Ratzel (1844–1904) with his *Anthropogeographie* of 1882–9 was promoting the idea of a separate study of human societies in relation to the physical earth, and his followers were becoming involved in a series of often rather fruitless debates on the extent to which the natural environment determines human responses.[25] Such problems were difficult to avoid while *environment*, with its implicit dualism of man and habitat, was retained as the chief integrating concept in geography. Even when the notion was expanded in the course of the twentieth century to include man himself and aspects of the cultural landscape in 'the geographic environment' the difficulty remained. In 1923 the idea of geography as human ecology – studying man's relationships to the environment – was proposed by Harlan Barrows.[26] This was opposed, however, by many geographers, either because they associated it with environmental determinism or because like Richard Chorley as late as 1973 they believed such an approach must be too ambitious for analysis and further must underestimate both the complexity of human systems and the extension of man's 'environmental dominance'. In Chorley's view any proposal to 'reconcile the dichotomy between human and physical geography' by this means must be merely a romantic one, 'an attempt to confront our current environmental problems with the visions of Wordsworth and Emerson'. The industrial revolution had made the ecological model out of date, he claimed: 'Man's relation to nature is increasingly one of dominance and control.' Instead he proposed a 'geographical man–machine analogy', emphasizing man's manipulation of the 'machine', the physical and biological environment.[27]

With that concept Chorley made no effort to consider the problem of man's control of man, or the increasing importance of social and intellectual aspects of the human environment in a world where control of urban tension and maintenance of political stability were taking on greater urgency in the nuclear age, and the extreme anthropocentrism which previously had given a useful impetus to science now threatened, as an aspect of an outmoded Western arrogance, to become a hazard to survival. In that context the realization of man's power to destroy his own habitat and the growing concern for action to avoid an *ecocatastrophe* pointed to the need rather for geography to sustain a world concept showing mankind as an element in a dynamic system, currently dominant but probably expendable. Meanwhile the positivist position, as presented by Chorley, was to see mankind as somehow outside the system it is supposed to dominate, in effect reviving a kind of Cartesian dualism and mechanistic outlook at a time when science after three centuries was moving towards a more coherent philosophy.

Already Edward Ackerman in 1963 had noted that in Barrows' time the ecological concept was incompletely formed, emphasizing as it did the adjustment of an organism to environment, whereas 'It has now been replaced by the much more powerful monistic concept of an ecosystem, in which organism and environment are one interacting entity.'[28] Stimulated by the work of A. G. Tansley in 1935, ecosystem studies had been actively developed by biologists, and among them during the 1960s a lead was taken by Marston Bates in promoting a concept of man *in* nature.[29] Writing in 1970 Wilbur Zelinsky applauded the idea of the ecosystem as epochal in the history of modern science, and expressed regret that he found 'scarcely a gesture to adopt the ecosystem concept into either demography or geography'.[30]

The importance to a science of a central organizing concept, a theme already developed by Whewell, had been raised again in 1965 by the economist Kenneth Boulding, who went on to credit geography with responsibility for the wider concept of the whole earth system: 'Of all the disciplines geography is the one that has caught the vision of the study of the earth as a total system, and it has strong claims to be the queen of the human sciences.'[31] Whatever the case for geography's central role, it can be argued that within each science it is possible to distinguish a series or a hierarchy of concepts that give rise to particular research, while themselves being informed by a meaningful general concept – with the whole system subject to a process of constant modification. Unity or coherence is given to a science, then, more by intensifying its central concepts and ensuring their consistency than by restricting the perimeter of the study. In this way, for instance, despite an inevitable increase in specialization, the various aspects of medical science can be related to the concept of the healthy organism, whereas neglect of this has led to criticism of a medical profession which in its preoccupation with analytical scientific methods and an allopathic approach seemed to pay too little attention to the functioning of the mind–body system as a whole. So in geography the growing diversity of specialized research and competing theories need not lead to a disintegration of the subject while they are linked by a general concept which is sufficiently comprehensive to encompass all these and is expressed in such a way as to lead to productive inquiry. Now to build up and articulate a concept of this degree of complexity, and designate it with a suitable term, involves a long and difficult process in human thought, and in the case of a world concept which is significant to the philosophy of the time and to numerous other sciences, this is obviously not a task for geographers alone. The main function of geography as a science is to help make the concept operative, to derive problems and develop techniques of study that will lead to greater understanding and more effective response. In this respect it is clear that the modern ecosystem concept provides a more effective focus for research than earlier concepts of nature and its increasing application in recent geography texts reflects a recognition of this. From that viewpoint, then, geography can be considered to operate as an integrating science not simply because it shares its subject matter with other sciences, or somehow weaves together their research, but because it maintains responsibility for a concept of the earth system that gives coherence to other areas of inquiry.

In conclusion, one further application of the ecological approach remains to be explored, namely its relevance to problems of scientific method and epistemology in general. The sociology of knowledge, to which this study has given support, is based on the concept of society: the question now is to see whether new insights can come from widening this to the concept of the global ecosystem, in the way that seems to have been intimated in Humboldt's final efforts to outline an alternative theory of science.

The ecology of knowledge

The term *ecology of knowledge* has been coined here to emphasize the view that all knowledge occurs not only in society but in the dynamic space–time context of the earth ecosystem; the observer – and the scientist is no exception – thus must be considered always as part of the system being observed, the ideas and actions that issue from such observation are themselves incorporated in the dynamic system, to become part of the future environment. This is in distinct contrast to the pre-Einsteinian positivist view of knowledge, itself adopted from Bacon's sense-empiricism, which considered the scientific observer as an individual detached, intellectually at least, from the objects of observation, with knowledge occurring in a kind of linear sequence: object, sensations, reflection, generalization, leading to knowledge or theory and so to further observation. According to that view the human mind is separate from the material world, so that thought cannot be regarded as part of nature. Similarly for positivists it seemed logical to regard thought as separate from action, theory from practice.

With the development of ecosystem theory it is possible to move away from that linear and rather mechanistic model to a theory of knowledge which views knower and known as part of a single functioning system, in which thoughts actually in some measure change that total system and therefore must themselves be considered as actions. All knowledge thus occurs in a time-sequence of historical continuity: ideas or experiences from the past provide the conceptual apparatus with which new observations and actions are made, and these again create the conditions, both internal and external, for future knowledge. All observation involves a process of interaction with the environment: even introspection is a process that links the individual with his past, with his society and, if only in its impact on the thinker himself, with the future. The ecology of knowledge then looks to a model which is more four-dimensional – even five-dimensional if we dared overstep the positivist limitations on scientific recognition of extra-sensory phenomena; the possibility of a psychic dimension in the earth ecosystem is an intriguing one, although it can only be touched on in the present discussion.

What is emphasized here is that the process of knowledge cannot be divorced from the social–environmental context in which it occurs. Through the use of symbol systems, of course, ideas can be transmitted from the past and the apparent constancy of many such ideas in different societies, over time and over space, has tended to reinforce belief in the existence of 'timeless verities'; nevertheless, all such ideas must be reactivated in the mind of a given individual in a specific context, and furthermore the effort to comprehend such ideas involves an attempt to understand the environment in which they were formulated. That environment moreover is to be seen not simply as the intellectual or social context but as an ecosystem – whether viewed as the total earth ecosystem or one of its smaller systems – in which all elements, organic and inorganic, form a functioning community and in which any change to a part will in some degree affect the entire system. Obviously some aspects of that environment in each case will be more significant than others; nor is this to propose some extreme form of environmentalism such as the climatic determinism of the early Greeks who explained the barbarism of north Europeans by the severity of their winters. Nevertheless, an awareness of the general context in which knowledge occurs is clearly useful: a more relevant example is the current concern to relate science both to politics and to technology, that is, the pattern of tools, materials, artefacts, practices and skills which sustain production in any society.

A further useful application of the ecosystem approach to knowledge is the procedure of considering any problem as an aspect of a complex functioning system where apparently rational solutions may have unforeseen impacts on other parts of the system: the retraining of engineers in eco-awareness, the instruction of medical practitioners in more sympathetic attention to psychosomatic care and community health, and similar efforts to produce lawyers, architects, chemists and physicists who are less destructive of our ecosystems, are all hopeful developments of the post-positivist era. In education, humane teachers for centuries have attempted to teach the whole child: Rousseau and Pestalozzi almost secured the vision of the child as part of the harmony of nature; John Dewey and the later progressives saw learning as problem-solving and action in a social context, although they too did not completely eliminate the classical faculty theory of the mind which justified traditional practices of attempting to introduce a logically-ordered subject matter into the child's intellect while the rest of its animal nature was suppressed, with judicious use of the cane on appropriate parts. A survival of that same model of the mind very likely provided the basis for the rather unsuccessful efforts in the mid twentieth century to transform education through extensive use of programmed teaching machines and audio-visual systems. The implications of the ecology of knowledge for education, however, have yet to be explored.

In the sciences strict positivists claimed that their theory of knowledge was based solely on a method of logical induction from observation and they denied any reference to metaphysics – that is, in their use of the term to any matters or questions outside the range of sense-experience. Clearly, however, they held a theory of man and nature, and this was a dualist one, of mind versus matter, with man seen in opposition to a mechanistic nature which he was intended – whether by divine ordinance or by evolution – to dominate. In contrast, modern ecosystem studies, recalling Humboldt's pioneer work in the early nineteenth century, regard mankind as part of the total earth ecosystem, in which any notion of dominance by one species must be contingent on maintenance of dynamic equilibrium in the whole system. The concept of the global energy balance, for example, shows mankind as one element in a complex system of energy flows and changes which link the organic with the inorganic world: the act of reading these words can be considered then as part of the carbon cycle. Humboldt's concept of nature incorporated human thought and culture; continuing that tradition, the ecology of knowledge affirms that all knowledge occurs within the functioning ecosystem and itself forms an integral part of that system, at least while mankind survives.

A widening concern with these issues and a recurrent effort to clarify new concepts and break away from established patterns of thought on theories of nature, science, and knowledge have become increasingly evident during the last decade. An early contribution to the new epistemology, although one not brought to my attention until after the present volume was completed, can be found in the work of Sir Geoffrey Vickers and Gregory Bateson. Vickers' 1963 paper 'Ecology, planning, and the American dream' emphasized the ecologist's viewpoint as it applied to human society, and introduced the term 'ecology of ideas' to refer to what he called man's *inner world*: 'an ill-defined but inescapable "mental" field' which has its own ecology.[32] Developing the concept proposed by Vickers, Gregory Bateson later used the ideas of communication and information theory to formulate what he termed a new *cybernetic epistemology*, outlining this from 1969 in a series of lectures, collected in his *Steps to an Ecology of Mind* (1972). Here, taking the holist approach, he argued that the individual mind is essentially linked with the surrounding ecosystem and is not limited to the body:

It is immanent also in pathways and messages outside the body; and there is a larger Mind of which the individual mind is only a sub-system. This larger Mind is . . . immanent in the total interconnected social system and planetary ecology.

As he went on to note,

Freudian psychology expanded the concept of mind inwards to include the whole communication system within the body – the autonomic, the habitual and the vast range of unconscious process. What I am saying expands mind outwards.[33]

In a manner reminiscent of Bruno, he stressed the significance of this concept of universal 'Mind', which he saw as not only 'immanent in the large system – the ecosystem', but also 'immanent in the total evolutionary structure'. The basic survival unit of evolution, he argued, is not the organism or species, as Darwin suggested, but the 'flexible organism-in-its-environment' [426]. In the same way, the individual mind is related to the universal mind. And this, he warned, is not the omniscient Mind of an infallible Creator, but rather a mind shared by all mankind and therefore subject to error and prone to disaster, even world-destruction, especially in a society with an advanced technology and a view of man as being distinct from the environment, and entitled to exploit it: 'The most important task today is, perhaps, to learn to think in the new way' [437].

Given the current patterns of change in scientific thought, it can be argued that the age of positivism is over.[34] In this period of transition, then, one of the major challenges is to develop an alternative theory of science that will enable ordered inquiry to continue while avoiding both social determinism and ineffectual relativism. A wholesale rejection of former methods, of course, is neither practical nor desirable, since the positivist philosophy was itself merely one interpretation of the broader tradition of empiricism, and the established methodology still provides useful guidelines for research with its requirements for orderly procedures of data collection and analysis, its concern for validity and its ideal of science as a disinterested search for truth – even if this has to be regarded as ultimately unattainable. An adequate theory, however, must distinguish between research procedure and the entire scientific process which involves wider philosophical, historical and social considerations. In particular, the importance of the moral context is receiving greater recognition in Western science, just as the ecoscience movement and radical revolutions around the world have brought a strong reminder that scientists cannot operate in isolation from the most pressing social and environmental problems of the time. Whether or not the search for Locke's universal moral law, Kant's moral imperative, or Kropotkin's *principle of mutual aid* is to find an answer in ecosystem theory, it is clear that the intensifying of resource conflicts, the alarming growth of inequality, and the danger of ecocide, are giving urgency to environmental issues and promoting greater concern with the impact of private and corporate greed, and the denial of rights to ethnic minorities, to women, and to the poor and disadvantaged in the world's communities. Scientists are centrally involved, and their choice of problems to investigate destroys any myth of value-free research. For the new programme the approach of Humboldt offers a beginning, and the task for geography, as for all science, is to make the most effective use of past traditions in responding to the issues of today.

APPENDIX 1

Translation: Bernhard Varenius, *Geographia generalis* (Amsterdam, 1664)

DEDICATION, INTRODUCTION TO GEOGRAPHY (pp. 1–10)

General Geography

DEDICATION

Most Noble, Distinguished, Honourable and Wise Consuls,

All this universe, which we are accustomed to call the world, has for a long time, as if by a general consent of men, been divided into earth and heaven. This division seems not quite in accordance with logical rules nor with nature itself, partly because it supposes an essential difference between earth and heaven, partly because it compares an extremely small earth with the whole heaven, that is, the centre with the sphere; nevertheless it has been received by practically everybody and approved as if legitimate, for various reasons. Many think the universe should be so divided because the earth is of a quite different nature from the celestial bodies, for it is subject to various changes and processes of decay while these, they believe, remain incorruptible without any loss. Then, they say, the earth is so small if compared with the heaven that it is like a point or centre: but a centre has a quite different nature from the sphere itself, and so earth and heaven are rightly opposed as different and contrary. But these arguments are not in accordance with truth. For neither are the heavenly bodies excepted from change and decay, as the observations of mathematicians of this and the previous century have demonstrated, nor has it so far been shown by firm arguments that the earth does occupy the centre of heaven. But if on account of smallness it should be opposed to heaven, then much more rightly would the moon and certain other stars, which are smaller than the earth, be opposed to it. Therefore, since these arguments suffer through errors, others must be adduced, in accordance with which our ancestors have declared that of this whole universe the heaven is one part and the earth another. Those arguments seem to be the following: first, because the earth is the home of the human race, and therefore should not be considered of less value by us than the whole heaven. Second, because the earth is not only our home, but also the source of our first origin, and it provides us with supports both for our conservation and propagation. Third, because from our earth we contemplate the movements of the heavenly bodies. Fourth, because the heavenly bodies appear to exert their influences on the earth. Fifth, because the constitution of the earth is better known to us than that of the heavenly bodies.

But however those arguments may be, it certainly is conceded by all wise men that knowledge of the earth is not only most worthy of men, but is also most necessary in the republic of letters and for practical life. Without discussing its usefulness for the study of history, whose beacons are not unjustly considered to be chronology and geography, nor its usefulness for theology, natural science, politics and other disciplines, let us consider only those uses and services whereby trade and naval affairs, which both supply the treasury of your republic, that is, the basis for the conduct of affairs, are aided by geography as taught and testified in the daily experience of this city. Surely, if trading is to be successful and profitable, and I add, pleasant, it helps a great deal to know about the regions to which goods are to be sent, or from which they are to be conveyed to this or other places: the location, distance away, the intervening seas, the route, neighbouring places with hostile or friendly inhabitants, and other facts which can be obtained from geographical maps, not without worthwhile intellectual enjoyment. Sailors indeed recognize the very great, even heaven-sent value of geography, when they undertake to plough remote seas and the raging ocean, relying on the accuracy of geographical maps and other rules which geography has provided for them in the direction of the ship. And although today in this your city these are constructed and taught not so much by mathematicians as by technicians for the sake of gain, and by informants of sailors, even by the sailors themselves, yet nobody ought to think on that account that these have originated from such people and not from mathematicians. For the invention of those things, and their proof, on which certainty depends, were much more difficult and cost more effort, than the application of them. The latter, mathematicians made so easy that even those who were not mathematicians could take it up, but the former, I mean the proof, since it derives for the most part from the innermost depths of learning, that is, theoretical geometry, depends on mathematicians alone, since it demands a greater effort of mind than men are willing or able to support, who, intent only on gain, seek merely the application and practical use of knowledge.

Nor does this apply only in geography, for the same thing can be seen in many other mathematical sciences, especially in arithmetic. What is more common than its use and practice today, especially in this your city? What is easier? Which of the merchants, even of the merchants' apprentices, does not know how to calculate readily according to the rules of arithmetic, and even to teach others this art? But indeed, before the rules, in accordance with which practice is established, were extracted by mathematicians from the inner depths of arithmetic and proved, and an efficient method of applying them pointed out, there was absolutely nobody who could do these things. Nor, if you ask one of those who are most ready in practice, why one must conduct a calculation in this or that way, will you get any reason or principle, since proofs are constructed and preserved only by mathematicians. Likewise, if some new rule is to be found for a new practice, none but mathematicians can be consulted. In the same way, in the case of perspective, static mechanics, music, the art of fortification, geodesy and other sciences, it is very easy to show how many benefits have originated from learning for the practical utility of human life, and especially for your city.

But to return to geography, although this is concerned only with the earth, which is like a point if compared with the heavens, nevertheless it forms a no less valuable part of learning than astronomy, and that for the reasons given above. Geography itself falls into two parts: one general, the other special. The former considers the earth in general, explaining its various parts and general affections.* The latter, that is, special geography, observing general rules, considers, in the case of individual regions, their site, divisions, boundaries and other matters worth knowing. But those who have so far written on geography have discussed at length special geography alone, practically without exception, and have explained very little relating to

---

* *Affectiones*: the term *affection* for Varenius can indicate a property, a state or condition, a relationship or influence, or a change of condition. It is therefore often difficult to translate, and has been retained throughout where it occurs. [Tr.]

general geography, with much that is necessary being neglected and omitted, so that young men, learning special geography, are for the most part ignorant of the bases of this discipline, and geography itself scarcely preserves the title of a science. When I noticed this, in order to provide a remedy for this evil, I began to turn my thoughts to making good this defect, that is, to the writing of a general geography, nor did I rest from the labour undertaken until the work was finished, so that up to the poor limit of my strength, granted to me by God and cultivated by years of diligence in the mathematical disciplines, I might serve the republic of letters and the studies of youth, or at least prove my will to serve.

I have been induced by many considerations to inscribe and dedicate my work to you, most distinguished men. First, because in the whole world there is no state which needs and uses a knowledge of geography more than this of yours, on account of your splendid voyages to all corners of the earth. Second, through the voyages of your republic, the study of geography has grown in no small degree, and as a result of this many doubts and uncertainties have been removed. Third, I know that you are patrons and promoters of all learning, so that I do not doubt at all that you set the greatest value on geography also. Fourth, and this is the principal reason, because this work of mine has been both completed and published in your city. It should not, therefore, see the light under the favour of any patronage other than yours. For after my fatherland, having first been afflicted by various kinds of damage in war, collapsed at last entirely into sparks and ashes, it happened that by some good fate, or rather through the providence of God, best and greatest, I took myself to your state, where I have found a convenient opportunity for writing about many matters which I could scarcely have treated properly in any other place. For not a few matters, either omitted or wrongly transmitted by other writers, concerning the condition of parts of the earth and ocean, their location, winds, and the customs of the people, had to be found out from sailors or others who had lived in those regions. Many passages in this book bear witness to that: it is equally evident in the notes which I published last year on the kingdom and religion of Japan, and its difference from the other religions of the world. It will show still more clearly, if by your favour and liberality you aid my studies so that it may be possible for me to finish them, along with other researches of mine on natural observations in various parts of the earth, on the food and drink of various peoples, on the essential nature and material substance of food and drink, and also on the various medicines of different peoples, including which medicaments are easy to prepare, as well as other matters on which I have begun to comment. It is to your honour, leading citizens, that you willingly encourage those who are industrious and seek by their labour to serve the republic, demonstrating this by their works themselves, that you foster them, nourish them and seek to promote their studies: grant to me also, whose efforts an adverse fortune has done its best by wars and fire to prevent, that I shall be able to commend your liberality and ascribe to you the promotion of those works which I wish to publish for the public good, so that the whole republic of letters may realize how much is owed to your wisdom. Should you do this I shall endeavour to ensure that you may see you have not conferred your beneficence on a man who is idle and makes improper use of your liberality and favour. May God preserve you and your republic. Farewell. From my study on the first day of August, 1650.

To your magnificence, greatness, and wisdom,
Most devoted
Bernhard Varenius,
Doctor of Medicine.

[1] GENERAL GEOGRAPHY: ABSOLUTE PART
FIRST SECTION
PRECOGNITA: INTRODUCTION TO GEOGRAPHY

Chapter 1
Geography: its Definition, Division, Method,
and other preliminary Remarks

There has been a long-standing custom, that those who treat some science or discipline as a whole should make certain prefatory remarks about the conditions, method and properties of this doctrine. This I think is not done without reason, provided only that it is done without sophistry, since through such information the reader's mind conceives some idea of the whole discipline, or at least its arguments, and gets to know how one conducts oneself in this discipline. We therefore in this first chapter will make a few prefatory remarks about the constitution of geography and its nature.

Definition
Geography, called one of the mixed mathematical sciences, teaches those affections of the earth and its parts which depend on quantity, namely shape, location, size, motion, celestial phenomena and other related properties.

By certain people it is less strictly taken as merely the description of the regions of
[2] the earth and their distribution. By others on the contrary it is too widely extended, when they add a political description of individual regions. These however are easily excused since they do this to retain and arouse the interest of their readers, who are generally bored with a bare enumeration and description of regions without an explanation of the customs of the people.

Divisions
We subdivide geography into general and special, or universal and particular. (Golnitzius suggests a twofold division of geography into exterior and interior, but this classification is improper and abusive to language, and is adopted without reason since the terms general and special are more apt.) General or universal geography is that which considers the earth in general and explains its affections without regard to particular regions. Special or particular geography is that which teaches the constitution of individual regions of the earth: it is twofold, consisting of chorography and topography. Chorography is concerned with description of a region that is at least of medium size. Topography describes some small tract of the earth or a place.

In this book we shall set out general geography, which we believe should be subdivided into three parts, namely the *absolute part*, the *respective part*, and the *comparative part*. In the absolute part we shall consider the actual body of the earth and its parts, along with its affections and properties such as shape, size, motion, land masses, seas, rivers, etc. In the respective part we shall consider those affections and occurrences which happen to the earth through celestial causes. Finally the comparative part will contain an explanation of those properties which emerge from a comparison of the various places of the earth.

[3]   *Object*
The object of geography, or its subject matter, is the earth, especially its surface and parts.

Affections

There are, it appears three kinds of things, namely *terrestrial, celestial,* and *human,* which in individual regions deserve consideration, and indeed in special geography make explanation of individual regions possible, with profit to students and readers. I call celestial affections those which depend on the apparent motion of the sun and stars, and in my reckoning they seem to be eight. First, *polar elevation, the distance of a place from the equator and from the pole.* Second, *the obliquity of the diurnal movement of the stars above the horizon of that place.* Third, *the length of the longest and shortest day.* Fourth, *the climate†and zone.* Fifth, *heat, cold, and seasons of the year: also rain, snow, winds and other atmospheric events.* For although these can be referred to the properties of the earth, yet since they have a strong connection with the four seasons of the year and the sun's motion, we have for that reason referred them to the class of celestial events. Sixth, *the rising of the stars, their appearance and movement above the horizon.* Seventh, *the stars that pass through the zenith of the place.* Eighth, *the amount or speed of motion* of any place, which according to the Copernican hypothesis occurs through rotation in a single hour. According to astrologers a *ninth affection* could be added, since they set over each region one of the twelve signs of the zodiac with a planet of this sign. I have always thought this doctrine worthless, nor do I see any foundation for this doctrine: nevertheless at the end of the special geography we shall review this distribution of theirs.

So much for the class of celestial affections. I call *terrestrial* those aspects that are considered within the region itself, and of these I count ten. First, *borders and boundaries.* Second, *shape.* Third, *size.* Fourth, *mountains.* Fifth, *waters, that is, rivers, springs, bays of the sea.* Sixth, *woods and deserts.* Seventh, *fertility and sterility, along with kinds of produce.* Eighth, *minerals or products of mining.* Ninth, *animals.* Tenth, *the longitude of a place,* which can be added to the first terrestrial property, namely to its boundary.

The third affection, which deserves to be considered in individual regions, I make the *human,* which depends on men, that is, the inhabitants of regions. Approximately ten of these also can be established. First, *the size of the inhabitants, their appearance, colour, length of life, origin, food and drink.* Second, *the sources of income and occupations of the inhabitants, their trading and the merchandise which that region sends to others.* Third, *their virtues, vices, learning, intelligence, schools,* etc. Fourth, *their customs concerning childbirth, marriages, funerals.* Fifth, *the speech or language which the inhabitants use.* Sixth, *their political organization.* Seventh, *religion and state of ecclesiastical affairs.* Eighth, *cities and more famous places.* Ninth, *notable historical events.* Tenth, *distinguished men, craftsmen, and inventions of the inhabitants of individual regions.*

These are the three kinds of affections to be explained in special geography, and though those which make up the third class are less correctly referred to geography, something however must be conceded to custom and the convenience of students. Apart from these we shall add to the special geography many sections about the practical application of geography.

In the general geography, however, which we shall set out in this book, we consider first the absolute affections of the earth and the construction of its parts.

---

† *Clima* from the Greek *klinō*, slope, had the meaning for Varenius of a belt or zone of the earth between two parallels of latitude. Later the term *climate* came to refer also to the weather associated with the frigid, temperate or torrid zone, and subsequently was restricted to atmospheric conditions alone. [Tr.]

Next we shall consider in general the celestial properties, which are then to be applied in the special geography to individual regions. Finally in the comparative part will be set out those matters which emerge from the comparison of one place with another.

[5]   *Principles*
The principles which geography uses to confirm the truth of its propositions are threefold. First, the propositions of geometry, arithmetic and trigonometry. Second, the precepts and theorems of astronomy, even though it seems incredible that to get to know the nature of the earth in which we live, we must make use of the heavenly bodies which are so many tens of thousands of miles away from us. Third, experience, for the greatest part of geography, particularly special geography, rests solely on the experience and observation of men who have described individual regions.

Order
The order which I think it is convenient to observe in this geographical discipline has been discussed in the sections on Divisions and Affections. However, there arises a certain difficulty in observing the order in the explanation of these affections, namely whether the affections should be assigned to individual regions, or whether the affections having been explained in general terms the regions are to be assigned to them. Aristotle in Book I of his *History of Animals*, as in Book I concerning the *Parts of Animals*, raises a similar doubt and argues at length: whether the properties of animals are to be listed in relation to individual species of animals, or whether these properties are to be explained in general terms and the animals on which they are founded added by way of appendix. A similar difficulty occurs in other branches of philosophy. In our general geography we have explained in general terms certain affections which in the special geography we shall relate to the explanation of individual regions.

[6]   *Method*
Concerning method, that is, the manner of proving the truth of geographical teachings, it should be explained that in general geography very many are confirmed by demonstrations properly so-called,‡ above all those concerning celestial affections. In special geography, however, practically everything is explained without demonstration (excepting celestial affections, which can be demonstrated) since experience and observation, that is, the evidence of the senses, confirm most things; indeed, they cannot be proved in any other way. For science is taken in three ways. Firstly, for any kind of knowledge even if it derives only from what is probable. Secondly, for certain knowledge, whether this certitude depends on the strength of demonstrations or on the testimony of the senses. Thirdly, for knowledge solely by demonstration: which use is most strict and is appropriate to geometry, arithmetic, and other mathematical sciences, except chronology, astrology and geography, to which the word science in the second sense is applicable.

Many propositions are also proved by the artificial terrestrial globe and through geographical maps, and of these propositions which are so shown, some can be confirmed by legitimate demonstrations (which however are omitted on account of the reader's comprehension), others cannot be proved in that way but are accepted

‡ Demonstration: proof by logical argument, here in effect a mathematical proof. [Tr.]

because we suppose that everything located on the globe and maps is arranged as on the earth itself. In these, however, we are rather following descriptions composed by geographical authors: the globe and maps serve the purpose of illustration and easier comprehension.

[7]

Origin of Geography

The origin of geography is not new, nor was it brought to light at a single birth, nor does it originate from a single man: but its beginnings were laid down many centuries ago, although ancient geographers were occupied solely with describing regions, that is, with chorography and topography. The Romans, after conquering and subjugating some province, used to illustrate its chorography on a tablet, illuminated with clear drawings, and these were displayed in a triumph to spectators. There were moreover at Rome in the portico of Lucullus many geographical tablets displayed for the contemplation of all. The Roman senate, about a century before the birth of Christ, sent surveyors and geographers to different parts of the world to survey the whole earth, but they scarcely completed the survey of a twentieth part. Neco, King of the Egyptians, many centuries before the birth of Christ, ordered the exploration by the Phoenicians of the whole outer coast of Africa in a three year period. Darius ordered an examination of the mouths of the Indus and the East Ethiopian Sea. Alexander the Great in his Asian expedition took with him, according to Pliny, two surveyors of routes, Diognetus and Beto, and from their notes and itineraries geographers of the following centuries have taken much. For when the pursuit of practically all other studies is damaged by wars, geography alone, I might say, grows as a result of them, along with the art of fortification as it is called.

But the geography of the ancients was very faulty, imperfect and full of many errors, because they were ignorant of what is not an insignificant but a most important part of knowledge, namely, the parts of the Earth (or at least they did not [8] have reliable experience of them): 1. All of America. 2. Northern lands. 3. Terra Australis and Magellanica. 4. That the earth can be circumnavigated and that ocean surrounds the lands (I do not deny that some of the ancients were of this opinion, but I do deny that they had a certain knowledge of it). 5. That the Torrid Zone is habitable and is inhabited by countless people. 6. The true size of the earth, although much has been written on this matter. 7. The southern part of Africa, and the fact that Africa can be circumnavigated. 8. Concerning remote regions, both the Greeks and the Romans lacked true descriptions, and left innumerable false and fictitious accounts about peoples at the end of Asia and in northern regions. 9. They were ignorant of the motion of the sea, the different directions of its flow, and its general flux. 10. Indeed the Greeks and Aristotle himself had no knowledge of the ebb and flow of the sea. 11. With regard to the winds their insight was limited to a few varieties, and a general view was altogether unknown to them. 12. The remarkable property of the magnet which shows north and south was concealed from them, although they had got to know about its other property of attracting iron. Anaximander, however, who lived in about the 400th year before Christ, is recorded as being the first to have tried to measure the earth.

The Importance of Geography

The study of geography is commended by, 1. Its value, for it is in the highest degree suitable for man as an inhabitant of the earth and endowed with reason beyond other animals. 2. It is also pleasant and indeed a worthy recreation to contemplate the regions of the earth and its properties. 3. Its remarkable utility and necessity, since neither theologians, nor medical men, nor lawyers, nor historians, nor other educated persons can do without knowledge of geography if they wish to advance in their studies without hindrance. These points have been sufficiently shown by others and can be illustrated by many examples.

[9]   I attach here two tables, of which the first sets out before the eyes the contents of this book, namely the general geography: the other the order to be followed in the special geography for the explanation of individual regions.

Chapter II
Certain items from Geometry and Trigonometry
which students of Geography ought to know

Plato wisely called geometry and arithmetic the wings by which men's minds flew up to heaven, that is, investigated the motions and affections of sun and stars. In geography those disciplines are equally necessary if anyone wants to learn it with judgment and free of impediment. Geography, however, consists of fewer elements than astronomy. And since many are attracted by the study of geography who are not skilled in those disciplines, we shall here introduce a few points from them which we think necessary to the student of geography so that he may engage more successfully and readily in this discipline. We do not approve at all that bad custom whereby youths who have not yet studied geometry and arithmetic turn their attention to other branches of philosophy: but the cause of that rests with teachers and professors, of whom the majority are themselves ignorant of these sciences and for that reason do not warn the young men about this erroneous habit. From arithmetic we assume in the reader a knowledge of the four mathematical operations, namely addition, subtraction, multiplication, division, and moreover the golden rule of three. We shall not talk about these for the present since most youths are instructed in that knowledge and if they are ignorant of it would rather learn it from the living
[10]   voice of a teacher than from a book. Instead we shall discuss *geometry.**

* In the remainder of the chapter, Varenius deals with geometry and trigonometry, following this with a comparison of various measures of distance used in Europe. [Tr.]

APPENDIX 2

Translation: Alexander von Humboldt, *Kosmos.*
Entwurf einer Physischen Weltbeschreibung,
volume V (Stuttgart, 1862)
INTRODUCTION (pp. 3–9, 19–22)

Kosmos, volume V

[3] CONTINUATION OF THE SPECIAL RESULTS OF
OBSERVATION IN THE FIELD OF
TERRESTRIAL PHENOMENA

*Introduction**

The fifth and last volume of *Kosmos*, for which I intend this *introduction*, concludes the portrayal of *terrestrial* phenomena in their purest objectivity. Together with the fourth volume, of which it is meant to be the continuation, according to the original plan of my work, it forms as it were a complete whole: what is usually called *physical geography*. It was for a long time my wish to publish this fifth volume as a *second section of the fourth*, appearing together with the *first section* as a counterpart to the single third, *uranological* volume; but the unpleasant delay in publication that this would have entailed, was an obstacle to the fulfilment of that wish.

In the case of the *astronomical* volume there are only the reciprocally disturbing and compensatory movements of heavenly bodies to be described, and (apart from
[4] the contact of meteor-asteroids circling in our solar system) our observation is concerned only with the activity of *homogeneous* matter; whereas the *earthly* part of the cosmos reveals, alongside the dynamic operations of moving forces, the influence of powerful and wonderfully complex *specific material diversity* (*Kosmos*, vol. III, pp. 4 and 594). In these differences in the complexity and relative abundance of the material to be treated lies part of the reason (I dare not say the justification) for the extremely great interval in time between the appearance of the individual volumes. The main reason for the long delay lies however in the decreasing vitality of an almost ninety year old man who, despite the same nightly industry, cannot but advance less and with decreased cheerful confidence. Thus since the time that in the foreword to the first volume of *Kosmos* I called, 'the late evening of an eventful life', more than twelve years have already passed by.

When Descartes worked on his Cosmos, *Le Traité du Monde*, which was supposed to include the 'whole world of phenomena (the heavenly spheres together with everything he knew about living and non-living nature)' he repeatedly, in letters to his friend, Father Mersenne, that were published by Baillet in 1691, made bitter complaints about the slow progress of his work and the great difficulty of arranging so many objects in order (*Oeuvres de Descartes*, publiées par Victor Cousin, 1824, vol.

* In this translation the phrasing and use of italics, as well as all citations, follow the original as closely as possible. In accordance with Humboldt's practice, his own notes are placed after the text. [Tr.]

284

I, p. 101; *Kosmos*, vol. III, p. 20). How much more bitter would have been the complaints of this versatile philosopher and anatomist if he had been able to ex-
[5] perience in the middle of the nineteenth century the almost disheartening spectacle of the expanding, richly filled spheres of heavens and earth! Ten years ago, as my *Kosmos* at the end of the second volume (p. 398) shows, I was still living in the mistaken hope of being able to combine in one single last volume the main results of special observations, which now will fill three volumes. It is easier, if one wants to keep some charm of form, to design a general *world picture* within predetermined limits, than to attempt to classify and examine the individual elements, on which, up to a certain period of time in our scientific knowledge, it was believed that results are based.

With the completion of a work carried through at least with lasting diligence, it may be permissible for the author to touch once more on the question, whether his book of the cosmos has remained true to the originally prescribed plan, that is to say, the *limits* that seemed advisable in his own view, according to his acquaintance with the condition of knowledge achieved at the present time. I have striven in the book for: a thoughtful consideration of *empirical*[1] phenomena, the combination into a *natural whole* of elements that have the capacity for evolution. The generalization of views concerning the transitions one into the other of real, continuous, active *natural processes* (one of the most magnificent outcomes of our age!) leads to the exploration of *laws* wherever they are to be recognized or at least anticipated. Clarity and liveliness of language in the objective representation of phenomena, as in the reflex
[6] of external nature on the *intellectual life in the cosmos*, on the world of thought and feeling, belong to the essential conditions of such a composition which, I dare say, has never yet been accomplished (*Kosmos*, vol. II, pp. 3–8, 50–2; vol. III, pp. 6–8). The enumeration of my endeavours leads inevitably, by their very nature, to the suggestion of connections between my attempts and the hazardous ventures of a *metaphysical natural science* that stands for what profound thinkers call the *philosophy of nature* as opposed to the *philosophy of the mind*. I have already stated, frankly and in contradiction to several of my highly respected friends in this country, that despite my great preference for generalizations, to me the establishment of a *rational science of nature* (a philosophy of nature developed, in accordance with its promise, to such an extent as to be a reasoned understanding of the phenomena of the universe) seemed to be up to now an impossible undertaking. How much of what has been recognized by sense-perception remains foreign to any mathematical development of thought! The succession in size, density, position of axes and eccentricity of orbits of the planets and satellites has apparently been deprived of all lawfulness; in the configuration of the continents coastal forms and elevation of the land are probably the results of very recent cosmic *events*, as was the case with the *occurrence* in our time (December 1845) of the permanent division of Biela's comet (*Kosmos*, vol. III, pp. 24 and 568–70). Moreover we are far from knowing all the substances and forces (actions) of nature. The boundless spheres of observation, which through newly discovered means (instruments) daily become wider, and indeed the *incompleteness* of perception for every single moment of speculation,
[7] make the task of a *theoretical philosophy of nature* to some extent an *uncertain* one.

Description of nature at present leads only in particular groups of phenomena to an *explanation of nature*.[2] The most active endeavour of research (I repeat this here) must be centred on the conditions under which occur the actual processes in the great and complex community that we call nature and world, and on the laws that can be recognized with assurance in particular groups. It is not always possible, however, to ascend from the laws to the *causes* themselves. The investigation of a partial *causal relationship* and the gradual increase of *generalizations* in our physical knowledge are for the present the highest objects of cosmical work.

Already in the Hellenic world of ideas the penetrating minds of the mighty Heraclitus of Ephesos,[3] of Empedocles[4] and of the Clazomenean [Anaxagoras][5] were presented with insoluble problems in the specific *diversity of matter* and *metabolism* (the change of one element into another). So we in our time are faced with the problem of the diversity of matter in the numerous so-called *simple bodies* of chemistry and the *allotropisms* of coal (with diamond and graphite), of phosphorus, and of sulphur. Although I have described rather vividly the uncertainty and difficulty of the task of a theoretical philosophy of nature, I am still far from dissuading anyone from the attempt to achieve success in this noble and important part of the world of thought. The *Metaphysical Foundations of Natural Science* of the immortal philosopher of Königsberg [Kant] certainly belongs to the remarkable products of [8] this great genius. He seemed himself to want to *restrict* his plan, as he stated in a foreword 'that *metaphysical natural science* extends no further than where mathematics can be connected with metaphysical propositions'.* Jacob Friedrich Fries, a thinker enthusiastically devoted to the Kantian view, and long a friend of mine, feels obliged to explain, at the close of his *History of Philosophy*, 'that of the wonderful advances made by natural science up to 1840, everything has depended upon observation and the art of geometry, the art of mathematical analysis; the philosophy of nature has contributed nothing at all towards these discoveries'. May the evidence of *former* unfruitfulness not destroy all hope for the future! For it does not become the free spirit of our time to reject as groundless hypothesis every philosophical attempt to enter deeper into the linkage of natural phenomena, when such attempts are also based on induction and analogy. Considering these noble talents which nature has granted to man, it does not become him to condemn either reason meditating on causal relationships or the lively power of imagination, indispensable for stimulating all discovery and creation.[6]

I for my part believe I have done what I could set myself to undertake according to the nature of my inclinations and the extent of my powers. I wished to produce a work after the great example of the *Exposition du Système du Monde* of Laplace, in whose inspiring company, in Arcueil and in the Bureau des Longitudes of the Paris observatory, I had the good fortune to live, along with Gay-Lussac and Arago, for more than twenty years. If in the *mechanism of the heavens*, despite the simplicity of the functioning forces, we cannot in many conditions of *being* of the heavenly bodies [9] recognize also their *becoming*; if even in the numerical relations of the planets in their distances from one another, in their succession in mass and size, in the inclination of their axes, as also in the shape of the star clusters and nebulae, if in these nearly everything escapes the mathematical development of thought (perhaps because, as I mentioned previously, these relationships are results of extremely diverse, special *heavenly events*):[7] so in the terrestrial zone, where the *diversity of matter* actively enters and complicates the problems there could be little hope that

---

* The quotation as given by Humboldt – 'dass *metaphysische Naturwissenschaft* nicht weiter lange, als wo Mathematik mit metaphysischen Sätzen verbunden werden könne' – does not appear in the foreword to Kant's *Metaphysische Anfangsgründe der Naturwissenschaft* (Riga, 1786). Kant however does argue here that all science of nature, including 'special metaphysical natural science (physics and psychology)' – namely, that part of the metaphysics of nature where principles established in the transcendental part are applied to the objects of the inner and outer sense – 'is possible only by means of mathematics'. [Tr.]

the *description of the world* would be at the same time an *explanation of the world*. Even Plato's intellectual generalizing power would be there insufficient:[8] when in any point of time, with each higher level of knowledge, the attempt at a solution still lacks the conviction that it is possible to understand all conditions under which phenomena appear, all the substances whose active forces exhibit themselves so mysteriously. I have not wanted to neglect touching freely myself on the most important of all the reproaches that have been directed against the scientific and literary composition of my Kosmos. Such a renewed justification was required of me by my obligation to the public which now for more than half a century has given my work such stimulating attention.

(Written in July 1858)

Notes

[1] (p. 5) 'Aristotle', says Brandis in his *Geschichte der Griechisch-Römischen Philosophie* (vol. II, part 2, p. 45), 'is the most decided advocate of the claims of experience; he is both Lord Bacon's predecessor and at the same time his superior in depth and compass of intellect. Proceeding from the empirical was a necessity for him, because he was convinced that the human mind can apprehend the world of actuality not *from* the concept but *only by means of* the concept; and indeed, only to the extent that the latter is developed in its correlation with the data of experience.'† Hegel, also, calls the Stagirite as a natural philosopher a *complete* but at the same time a *thinking empiricist* (*Vorlesungen über die Geschichte der Philosophie*, edited by Michelet, vol. II, 1833, p. 340).‡ On the long struggle between *realism* and *idealism*, the historical phases of empirical philosophy, and on the stages in the development of empiricism in general, see the gifted Kuno Fischer in his *Franz Baco von Verulam und das Zeitalter der Realphilosophie* (1856), pp. 383–8, especially pp. 468–72.

[2] (p. 7) In the stronger sense of the word and in greater generalization of ideas, '*Description of the world* is the history of nature and humanity. *Explanation of the world* is *science*, which interprets the information given by history' (*Franz Baco von Verulam*, p. 165).*

[3] (p. 7) In the Heraclitean natural system the process of becoming consists of a constant alternation of strict contraries; 'the death of fire is the birth of air': for destruction is only the transformation into their opposites of the things destroyed. As in organic bodies, so in the whole universe a constant process of transformation prevails. Living and dying were to the Ephesian identical natural processes, *living* indeed *a process of everlasting dying*: an expression that reminds me of Dante's in the *Purgatorio* (XXXIII, 54):

     Del viver, ch'è un correre alla morte.

The physical life-process of the individual consists in the transition from being to non-being; in a movement like that of a river, a *flowing*. Even the sun is always new,

---

† See Christian August Brandis, *Handbuch der Geschichte der Griechisch-Römischen Philosophie*, 3 vols in 4 (Berlin, 1835–66), vol. II, part 2.1 (1853), p. 45. [Tr.]

‡ See *Hegel's Lectures on the History of Philosophy*, translated by Haldane and Simpson (London, 1955), II, p. 133. [Tr.]

* For an alternative translation see *Francis Bacon of Verulam. Realistic Philosophy and its Age*, translated by John Oxenford (London, 1857), p. 183. [Tr.]

included in the continual process of being extinguished and rekindled. Every flame, like the *flame of the sun*, has its being in its becoming. See *Die Philosophie Herakleitos des Dunklen von Ephesos* presented by Ferd. Lassalle, 1858, vol. I, pp. 157–63, vol. II, pp. 104–10.† In this book the author also shows the remarkable influence of *Heraclitus the Obscure* on Hippocrates *De diaeta*; see Lassalle, vol. I, pp. 165–71. Hegel (*Geschichte der Philosophie*, ed. Michelet, vol. I, 1833, p. 333) says: 'it is a great conception of Heraclitus, to turn from *being* to *becoming*'. Aristotle, too, recognizes that all becoming and passing away, all change, develops *antithetically* by means of so-called *deprivation* ('Aristotle and his academic contemporaries', by Aug. Brandis in the *Geschichte der Philosophie*, vol. II, part 2, 1857, pp. 704 and 716). Even according to the ancient saying (*Gâthâs*) of the Bactrian Zarathustra (translated by Martin Haug, I, p. 101) 'the entire content of earthly life is the antithesis of *being* and *non-being*'.

⁴ (p. 7) Empedocles was shown by Aristotle, according to a passage in the 1st Book of the *Metaphysics* (I. 4. 985a 32; also I. 3. 984a 8), to be the actual originator of the fourfold division of the elements (*roots of matter*): such a limitation to the number four was foreign to the Milesian Anaximander and Anaximenes (Brandis, vol. I, 1835, p. 196).

⁵ (p. 7) In order to explain, in *becoming*, qualitative changes or transformations of condition, Anaxagoras, who was censured by Aristotle, accepted, instead of the four elements, 'an infinite variety of simple, qualitatively defined elements (*seeds of matter*), different from one another: so that contraries might be able to develop from contraries'. According to a statement of Simplicius, the Clazomenean censured the Greeks on account of the common view of *becoming* and *passing away*; for nothing becomes and passes away, but *existing* matter becomes combined and separated, and one could rightly call becoming a combined-becoming, and passing away a separated-becoming. The totality of matter remains the same (Brandis, vol. I, p. 240; *Kosmos*, vol. IV, p. 12). The *all in all* (πάντα ἐν πᾶσιν, or ἐν παντὶ παντὸς μοῖρα ἔνεστι) of Anaxagoras is related to the phenomena of *combustion*. According to Sextus Empir. (*Pyrrhoniarum hypotyposeon*, Bk. I, 13, 33) when Anaxagoras stated that the water from which snow forms is black, the conclusion should be drawn that snow is black; Cicero (*Lucull.*, 31) on the other hand concludes only that snow is not white; and Galen too (*De simpl. medicam.*, II, 1) only attributes to it the latter assertion: thus it remains very doubtful whether the Clazomenean himself so decidedly called snow black, as was later accepted. (See on this Jul. Ideler, *Meteorol. Graec. et Rom.*, 1832, p. 147, and his edition of the *Meteorologica of Aristotle*, vol. II, 1836, p. 481.) Anaxagoras probably taught only that everything existing contains in itself parts of others (or of the whole). – Compare Schelling, distinguished for thoughtfulness and language (*sämmtl. Werke*, part I, vol. 2, 1857, pp. 267–73; I, 3, 1858, pp. 24–6).

⁶ (p. 8) The philosopher, who thought he had demonstrated the possibility of a *nature-philosophy* or *speculative physics* (*Schelling's sämmtliche Werke*, part I, vol. 3, p. 274), himself admits (p. 105): 'that the force which governs the whole of nature, and through which nature is maintained in its identity, has *hitherto not been discovered* (deduced). We find ourselves driven to the same conclusion; yet this force remains always only a hypothesis, capable of an endless number of modifications which can be as different as the conditions under which it operates.' Matter, endowed with immutable forces (indestructible in terms of our present means), is now

† See Ferdinand Johann Gottlieb Lassalle, *Die Philosophie Herakleitos des Dunklen von Ephesos*, 2 vols (Berlin, 1858). [Tr.]

referred to in our scientific language as chemical elements (Helmholtz, *Über Erhaltung der Kraft*, 1847, p. 4).‡

[7] (p. 9) Laplace, *Expos. du Syst. du Monde* (5th edn, 1824) pp. 389–95 and 414.

22] [8] (p. 9) 'It has been repeatedly urged by continental critics', says a writer personally unknown to me, but a very well-disposed reviewer of *Kosmos* (*Atlas*, 9 Jan. 1858), 'that Bn Humboldt has not entirely solved his cosmographical axiom; still, *Kosmos* is a gorgeous accumulation of facts, the results of immense experience, study, and research, combined with some equally grand *aperçus*, *points de vue*, and *theories*. It is an improved *Pliny* of the present time, just such a work as a *savant* and traveller of his rank could produce. Whether such acquirements could be combined with the high generalising genius of Plato, and the still older Greek sages, we have no means of judging, as no such constellation has yet appeared amongst the ranks of man.'*

‡ See Hermann Ludwig Ferdinand von Helmholtz, *Über die Erhaltung der Kraft, eine physikalische Abhandlung, vorgetragen in der Sitzung der physikalischen Gesellschaft zu Berlin am 23sten Juli 1847* (Berlin, 1847). [Tr.]

* Quoted by Humboldt in the original English. [Tr.]

NOTES

Introduction

1 Wilbur Zelinsky, 'The demigod's dilemma', *Annals AAG* 65, no. 2 (1975), 123–43 (p. 139).
2 David Harvey, *Explanation in Geography* (London, 1969), p. 481.
3 Sir George Clark, *The Seventeenth Century*, 2nd edn (Oxford, 1963), p. 242.
4 A narrower use of *paradigm* as the approved scientific model was proposed by Thomas Kuhn, *The Structure of Scientific Revolutions* (Chicago, 1962; 2nd edn, 1970).
5 For a review of recent developments in psychology, sociology and education see the *International Review of Education* 25, no. 3 (1979), ed. James Bowen.
6 Fred Schaefer, 'Exceptionalism in geography: a methodological examination', *Annals AAG* 43, no. 3 (1953), 226–49.
7 John Kirtland Wright, '*Terrae incognitae*: the place of the imagination in geography', *Annals AAG* 37 (1947), 1–15; reprinted in his *Human Nature in Geography* (Cambridge, Mass., 1966), pp. 68–88.
8 David Lowenthal, 'Geography, experience, and imagination: towards a geographical epistemology', *Annals AAG* 51, no. 3 (1961), 241–60.
9 Michael Polanyi, *Personal Knowledge* (Chicago, 1958).
10 William Kirk, 'Historical geography and the concept of the behavioural environment', *Indian Geographical Journal* (1952), 152–60, and 'Problems of geography', *Geography* 48 (1963), 357–71.
11 Harold Brookfield, 'On the environment as perceived', in C. Board, R. J. Chorley, D. Stoddart, and P. Haggett (eds), *Progress in Geography* (London, 1969).
12 Preston James, 'On the origin and persistence of error in geography', *Annals AAG* 57 (1967), 1–24.
13 Preston James, *All Possible Worlds: A History of Geographical Ideas* (New York, 1972).
14 Wayne Davies (ed.), *The Conceptual Revolution in Geography* (London, 1972).
15 Alexander von Humboldt, *Kosmos. Entwurf einer physischen Weltbeschreibung*, 5 vols (Stuttgart, 1845–62). Hereafter called *Kosmos*.
16 See e.g. Kenneth Boulding, J. D. Roslansky (ed.), *The Control of Environment* (Amsterdam, 1967), p. 52.
17 A useful monograph is Joseph May, *Kant's Concept of Geography and its Relation to Recent Geographical Thought* (Toronto, 1970); while C. J. Glacken, *Traces on the Rhodian Shore: Nature and Culture in Western Thought from Ancient Times to the End of the Eighteenth Century* (Berkeley, California, 1967),

and James, *All Possible Worlds* are significant general works; however, neither treats this period intensively.

18 A. Downes, 'The bibliographic dinosaurs of Georgian geography (1714–1830)', *Geographical Journal* 137, part 3 (1971), 379.

19 Schaefer, 'Exceptionalism in geography', 234–8.

20 David Stoddart, 'Organism and ecosystem as geographical models', in R. Chorley and P. Haggett (eds), *Models in Geography* (London, 1967), p. 523

21 David Hooson, review of André Meynier, *Histoire de la Pensée Géographique en France (1872–1969)* (Paris, 1969), in *Geog. Rev.* 61, no. 3 (1971), 460.

22 M. J. Bowen, 'Scientific method – after positivism', *Australian Geographical Studies* 17, no. 2 (1979), 210–16.

1 Foundations of modern empiricism

1 Giordano Bruno, *De l'Infinito Universo et Mondi* (London, 1584), translated by Dorothea W. Singer as *On the Infinite Universe and Worlds* in her *Giordano Bruno: His Life and Thought* (New York, 1950), pp. 226–378.

2 William Gilbert, *De magnete, magneticisque corporibus, et de magno magnete Tellure; physiologia nova, plurimus et argumentis et experimentis demonstrata* (London, 1600). First English translation, New York, 1893. See *On the Magnet*, facsimile of the London, 1900, translation by Silvanus P. Thompson et al. (New York, 1958).

3 Johannes Kepler, *Astronomia nova ΑΙΤΙΟΛΟΓΗΤΟΣ, sev physica coelestis, tradita commentariis de motibus stellae Martis. Ex observationibus G. V. Tychonis Brahe*; vol. III in Johannes Kepler, *Gesammelte Werke*, ed. Max Caspar (Munich, 1937).

4 Arthur Koestler, *The Sleepwalkers: A History of Man's Changing Vision of the Universe* (Harmondsworth, Middlesex, 1968), pp. 54 and 87.

5 William C. D. Dampier-Whetham, *A History of Science and its Relations with Philosophy and Religion* (Cambridge, 1929), pp. 28–31. Dampier-Whetham's impressive argument against materialism and in favour of a more philosophic science was not strengthened by his antipathy to Plato.

6 Plato, *The Republic*, translated by Paul Shorey, Loeb Classical Library, 2 vols (London, 1953–6), II, pp. 186–90. Professor Shorey expressed his defence of Plato as early as 1927 in an article 'Platonism and the history of science', *American Philosophical Soc. Proceedings*, 66, 171ff.

7 Plato, *Republic*, VII, 530 B–C. Translated by Shorey (II, pp. 186–90).

8 Aristotle, *De Partibus Animalium*, I, 1:641b 20, translated by William Ogle, vol. V in *The Works of Aristotle*, ed. W. D. Ross (Oxford, 1947). Quotation from *De Caelo* translated by J. L. Stock, vol. II in the Oxford series.

9 Aristotle, *The Physics*, II, 2: 194a. Translated by P. H. Wicksteed and F. M. Cornford, Loeb Classical Library, 2 vols (London, 1960–3), I, pp. 120–1. A more satisfactory translation of the *Physica*, by R. P. Hardie and R. K. Gaye, is in the Oxford *Works of Aristotle*, vol. II. Modifications of both are used here.

10 Aristotle, *Physics*, II, 2: 194a–b. For a further statement of Aristotle's architecture analogy and his view that the science of nature is concerned not with the material elements but with their composition, see also *De Partibus Animalium*, I, 5: 645a.

11 The customary translation of this as *On the Soul* is somewhat misleading, the term *soul* being now more restricted in connotation than the Greek *psychē* which

referred to such signs of life in living things as breath, heart, activity and sensations, as well as mind or spirit.

12 Aristotle, *On the Soul*, I, 4: 408b. Greek text, with translation by W. S. Hett, in Aristotle, *On the Soul, Parva Naturalia, On Breath*, Loeb Classical Library (London, 1957), pp. 48–9. Some translations modified.

13 For Aristotle's hierarchy of organisms, see *De Anima*, 414a 29–415a 13. In his important study, *The Great Chain of Being* (New York, 1960), pp. 55–9, Arthur Lovejoy argued that the concept of the universe as a single chain of being is derived from Aristotle, who thus introduced into Western science two opposing doctrines: the discrete classification of nature into fixed species, and the continuity of natural forms (a belief which led Aristotle to admit the limitations of classification).

14 Aristotle, *Physics*, I, 1: 184a23. The Oxford translation adds 'masses' after 'confused' (*Physica*, vol. II); while the Loeb translation (I, p. 11) reads: 'Now the things most obvious and immediately cognizable by us are *concrete and particular, rather than abstract and general*; whereas elements and principles are only accessible to us afterwards, as derived from the *concrete data* when we have analyzed them' (my italics). A materialist interpretation is not justified here.

15 *The Geography of Strabo*, translated by H. L. Jones and J. R. S. Sterrett, 1917, Loeb Classical Library (Cambridge, Mass., and London, 1969). Translations modified where necessary.

16 *Geography of Strabo* 1.1.20–1, Loeb, I, pp. 40–1: Jones translates *koinē ennoia* as 'our intuition' or 'intuitive knowledge'; however, the term here seems to be used rather in the sense of a commonly accepted concept or notion.

17 *Claudii Ptolemaei Geographia*, ed. C. F. A. Nobbe (Hildesheim, 1966), I. 2.2. Selections translated in M. Cohen and I. E. Drabkin (eds.), *A Source Book in Greek Science* (New York, 1948), pp. 163–81. The first complete English translation, by E. L. Stevenson, was the *Geography of Claudius Ptolemy* (New York, 1932).

18 *Geography of Strabo*, 1.1.16.

19 Cosmas Indicopleustes (Cosmas of Alexandria), *Topographia Christiana*, A.D. 535–47.

20 See A. C. Crombie, 'The significance of medieval discussions of scientific method for the Scientific Revolution', in Marshall Clagett (ed.), *Critical Problems in the History of Science* (Madison, Wisconsin, 1962), pp. 83–4.

21 J. Bowen, *A History of Western Education*, II (London, 1975), pp. 207–31.

22 L. Bagrow and R. A. Skelton, *History of Cartography* (London, 1964), p. 77.

23 R. V. Tooley, *Maps and Map-Makers* (New York, 1961), p. 6.

24 Joseph Fischer argued for the authenticity of the regional maps of the early manuscripts, while admitting that the world map copied as Ptolemy's should be attributed to a later Alexandrian geographer: see his introduction to *Geography of Claudius Ptolemy*, translated by E. L. Stevenson (New York, 1932), pp. 3–6.

25 See Waldseemüller [?], *Cosmographiae introductio* (St Dié, 1507); facsimile reproduction, with a translation by Joseph Fischer and Franz von Wieser: *Introduction to Cosmography, with certain necessary Principles of Geometry and Astronomy. To which are added The Four Voyages of Amerigo Vespucci* (Ann Arbor, Michigan, 1966), p. 70.

26 Humboldt, *Examen critique de l'histoire de la géographie de Nouveau Continent et des progrès de l'astronomie nautique aux XVe et XVIe siècles* (Paris, 1814–34), facsimile (Amsterdam and New York, 1971). Not yet translated into English.

27 Humboldt, *Kosmos*, II, p. 490. His researches had aroused much interest on their publication in Germany in 1836: see Humboldt's *Briefwechsel mit Heinrich Berghaus* (Leipzig, 1863), II, pp. 142–8.

28 Waldseemüller is cited without question as the author of the *Cosmographiae introductio* in the 1966 Ann Arbor facsimile and also, for example, in Bagrow and Skelton, *Cartography*, p. 126, and Tooley, *Maps and Map-Makers*, p. 26.

29 F. Laubenberger, 'Ringmann oder Waldseemüller? Eine kritische Untersuchung über den Urheber des Namens America', *Erdkunde* 13, Heft 3 (1959), 163–79. In his introduction to a facsimile of the Strasburg Ptolemy of 1513, R. A. Skelton argued that Laubenberger has not disproved collaboration between the two men in the writing of the *Cosmographiae introductio*. See *Claudius Ptolemaeus, Geographia*, vol. IV in *Theatrum Orbis Terrarum*, 2nd series (Amsterdam, 1966), p. viii.

30 See J. N. L. Baker, 'Geography and its history', *Advancement of Science* 12 (1955), 188–98; reprinted in *The History of Geography* (Oxford, 1963), pp. 84–104 (p. 99).

31 See J. K. Wright, 'Map makers are human: comments on the subjective in maps', *Geographical Review* 32 (1942), 527–44; reprinted in his *Human Nature in Geography* (Cambridge, Mass., 1966), pp. 33–52 (p. 33).

32 Humboldt, *Kosmos*, II, p. 324, and III, p. 10.

33 C. T. Onions (ed.), *The Oxford Dictionary of English Etymology* (Oxford, 1966), p. 471.

34 See Aristotle, *Physics* I, 2: 185a14; V, 1: 224b30; V, 5: 229b3.

35 Vives, *Opera*, Bk II, line 605 (Basle, 1555), translated by R. M. Blake in Blake, Ducasse, and Madden, *Theories of Scientific Method* (Seattle, 1960), p. 8.

36 Blake, ibid., p. 8.

37 Bacon, *Novum Organum*, 1620, Bk I, Aphorisms 11–18; translated by R. Ellis and J. Spedding, *c.* 1858 (London, n.d.), pp. 62–3.

38 Bacon, *The Great Instauration*, Proemium (first published 1620, with the *Novum Organum*); translated in *Bacon: Selections*, ed. M. T. McClure (London, 1928), pp. 20–2.

39 Bacon, *The Advancement of Learning*, 1605 (London, 1954), p. 124.

40 Humboldt, *Kosmos*, II, p. 515. See also *Kosmos*, IV, p. 57.

41 Bacon, *The Great Instauration*, p. 24. Bacon repeated this view, *Novum Organum*, I, Aph. 50 (p. 77).

42 Bacon, *Novum Organum*, I, 97. My italics.

43 An examination of Bacon's occasional references to geography is given by Joseph May, in *Kant's Concept of Geography* (Toronto, 1970), pp. 34–8.

44 Bacon, *Advancement of Learning*, pp. 79–80.

45 Bacon, *Novum Organum*, I, 98–9. See also *Advancement of Learning*, pp. 70–3, and *Great Instauration*, pp. 27–30, for similar statements on natural history.

46 Bacon, *The Great Instauration*, p. 29.

47 Bacon, *Advancement of Learning*, pp. 147–8. Bacon was following Aristotle here fairly closely, see *On the Soul*, III, 10: 433b.

48 Bacon, *Sylva Sylvarum; or, A Natural History in Ten Centuries* (London, 1627), Preface. Published with this was the unfinished *New Atlantis*.

2 Science and geography: the seventeenth-century encounter

1 Galileo, *Dialogue Concerning the Two Chief World Systems – Ptolemaic and*

Copernican (Florence, 1632), translated by Stillman Drake, foreword by Albert Einstein (Berkeley, California, 1953), pp. 55–6.

2 Ibid., p. 406.

3 Einstein in ibid., pp. xvii–xix. Authorized translation by Sonja Bargmann.

4 *Descartes: Philosophical Letters*, translated by Anthony Kenny (Oxford, 1970), p. 22.

5 Humboldt, *Kosmos*, III, pp. 19–20, and *Kosmos*, V, pp. 4–5.

6 Descartes, *Principles of Philosophy*, Principles XXXVII–XLVII, in *The Philosophical Works of Descartes*, translated by E. S. Haldane and G. R. T. Ross (Cambridge, 1967), pp. 273–4.

7 Ibid., Part IV, Principle CCIII (p. 299).

8 See *Oeuvres de Descartes*, ed. Charles Adam and Paul Tannery (Paris, 1909), vol. XI, *Le Monde*, pp. i–iv.

9 Descartes, *Philosophical Letters*, p. 28.

10 Descartes, *Discourse on Method*, Part II, in *Philosophical Works*, p. 92.

11 Ibid., Part III, p. 101.

12 Descartes, *Principles of Philosophy*, Part I, Principle VIII, *Philosophical Works*, p. 221.

13 Ibid., Principle XXX (p. 231). This argument was presented earlier in his *Meditations on the First Philosophy*, 1641 (p. 191).

14 Descartes, *Principles of Philosophy*, Part I, Principles IX, XXXIX–XLI (pp. 222–35).

15 Descartes, *Philosophical Letters*, p. 36.

16 Ibid., Letter to Mersenne, April 1630, p. 11. Stated again in *Principles of Philosophy*, Part I, Principle X (p. 222).

17 A. C. Crombie, *Augustine to Galileo* (Harmondsworth, Middlesex, 2nd edn, 1969), II, p. 321.

18 See Josef Schmithüsen, *Geschichte der geographischen Wissenschaft von den ersten Anfängen bis zum Ende des 18. Jahrhunderts* (Mannheim, 1970), p. 113.

19 George Abbot, *A Briefe Description of the whole Worlde, wherein is particularly described, all the Monarchies, Empires, and Kingdomes of the same with their severall titles and scituations thereunto adioyning*, London, 1599 (facsimile: Amsterdam, 1970). Similar in many respects is the German work, M. Quad's *Geographisch Handtbuch*, Cologne, 1600 (facsimile: Amsterdam, 1969).

20 E. G. R. Taylor, *Late Tudor and Early Stuart Geography, 1583–1650* (London, 1934), p. 37.

21 Bartholomew Keckermann, *Systema Compendiosum: totius mathematices, hoc est Geometriae, Opticae, Astronomiae, et Geographiae* (Hanover, 1617).

22 Keckermann, *Systema Geographicum . . . Adiecta . . . Problemata Nautica* (Hanover, 1611).

23 Keckermann, *Systema Compendiosum* (Oxford, 1661), p. 357. Page references in the text are to this edition: the *General Geography* occupies pp. 350–440, the *Special* pp. 445–66, and *On Navigation* pp. 467–510.

24 Manfred Büttner, *Die Geographia Generalis vor Varenius* (Wiesbaden, 1973). No translation or detailed analysis of Keckermann's work exists in English; Baker, *History of Geography*, pp. 6 and 113, drew attention to its significance as a model, although his references contain minor errors.

25 Peter Heylyn, *Microcosmus* (Oxford, 1621), pp. 1–10. For notes on the teaching of geography at Oxford in this period see J. N. L. Baker, 'Academic geography in

the 17th and 18th centuries', *Scottish Geographical Magazine* 51 (1935), 129–43; reprinted in Baker, *History of Geography*, pp. 14–32.

26 *Geography of Strabo*, 1.1.16 (1969, pp. 28–9): 'And to this knowledge of the nature of the country and of the species of animals and plants, we must add a knowledge of all that relates to the sea; for in a sense we are amphibious, and belong no more to the land than to the sea.' See also Strabo, 1.3.4–12.

27 Heylyn, *Microcosmus* (1621), pp. 11–18.

28 Philip Cluverius, *Introductio in universam geographiam, tam Veterem quam Novam* (Amsterdam, 1683).

29 R. E. Dickinson, *The Makers of Modern Geography* (London, 1969), p. 9.

30 Nathaniel Carpenter, *Geography delineated forth in two Bookes* (Oxford, 1625), I, pp. 75–7 and 99. Subsequent page references in the text.

31 Even J. N. L. Baker, in drawing attention to the neglect of Carpenter's work, commented, 'There is much in Carpenter's first book, which is not strictly geographical', and he admitted turning to the second book 'with some relief'. See 'Nathaniel Carpenter and English geography in the seventeenth century', *Geographical Journal* 71 (1928), 261–71; reprinted in Baker, *History of Geography*, pp. 1–11.

32 William Pemble, *A Briefe Introduction to Geography. Containing a Description of the Grounds, and generall Part thereof* (Oxford, 1630), p. 1.

33 Robert Stafford [John Prideaux], *A Geographicall and Anthologicall Description of all the Empires and Kingdomes, both of Continent and Ilands in this terrestriall Globe* (London, 1634). MS note on the Bodleian Library copy, also referred to by Baker, *History of Geography*, pp. 1–11.

34 Johannes Comenius, *Orbis Sensualium Pictus*, 3rd edn, London, 1672 (facsimile, Sydney, 1967), p. 11.

35 Bernhard Varenius, *Geographia generalis, In qua affectiones generales Telluris explicantur*, 1650, 2nd edn (Amsterdam, 1664). Later Elzevier editions from Amsterdam were in 1671 and 1672.

36 J. N. L. Baker, 'The geography of Bernhard Varenius', *Transactions of the Institute of British Geographers* 21 (1955), 51–60; reprinted in Baker, *History of Geography*, pp. 105–18. The translation used by Baker is *A Compleat System of General Geography: explaining The Nature and Properties of the Earth*, translated by Dugdale and Shaw, 3rd edn (London, 1736); the fourth and final edition appeared in 1765. In listing the Latin editions of Varenius, Baker made the point that 'the note in Humboldt's *Cosmos*, referring to the *Geographia generalis*, is misleading'; however, he himself used Otté's faulty translation of *Kosmos*, which exaggerated the error. See *Kosmos*, I, p. 74.

37 See Appendix 1, pp. 276–83. The generous help of Mr Alan Treloar, Reader in Comparative Philology in the University of New England, is gratefully acknowledged in making this translation.

38 Schmithüsen, *Geschichte der geographischen Wissenschaft*, p. 118.

39 Varenius, *Geographia generalis* (1664), p. 1. Subsequent page references in the text are to this edition: see Appendix 1.

40 On this point note that Humboldt has been translated as stating that Varenius 'was the first to distinguish between *general and special geography*': see *Cosmos: A Sketch of a Physical Description of the Universe*, translated by E. C. Otté, 1849–58 (London, 1882–4), I, p. 48. Humboldt himself however made no such claim, stating simply, 'He distinguished very clearly between *general* and *special* geography.' *Kosmos*, I, p. 60.

41 Abraham Gölnitz, *Compendium Geographicum* (Amsterdam, 1643).
42 Baker, 'Geography of Bernhard Varenius', pp. 51–7. Baker traced the opposing interpretation to H. R. Mill in his Presidential Address of 1901 to Section E, British Association, Glasgow (*Report*, 1901, p. 703).
43 *Claudii Ptolemaei Geographia*, ed. C. F. A. Nobbe (Hildesheim, 1966), I.1.2.
44 May, *Kant's Concept of Geography*, pp. 54–6. The passage in question was omitted altogether from Blome's translation of 1682.
45 Varenius, *Geographia generalis*, Dedication. See translation, Appendix 1, pp. 276–8 above.
46 *Geography of Strabo*, 1.1.15.
47 Taylor, *Late Tudor and Early Stuart Geography*, p. 30, p. 141.
48 Varenius, *Geographia generalis*, p. 8. Recent scholarship has established earlier dates for Anaximander c. 611–547 B.C.
49 Noting this, Humboldt added that the entire general description of the earth in the absolute part can be regarded as a comparative study in Ritter's sense: *Kosmos*, I, p. 74.
50 Varenius, *Geographia generalis*, p. 6. Varenius himself included tables and occasional diagrams in his *General Geography*, but no maps.
51 Richard Blome, *Cosmography and Geography* (London, 1682). Note that the date was given as 1693 (actually the date of the third edition) in Baker, 'Geography of Bernhard Varenius', p. 53, an error repeated by Dickinson, *Makers of Modern Geography*, p. 7.
52 Varenius, *Geographia generalis, In qua affectiones generales Telluris explicantur*, Summâ curâ quam plurimus in locis Emendata, et XXXIII Schematibus Novis, Aere incisis . . . Ab Isaaco Newton, 2nd edn, corrected (Cambridge, 1681)
53 Edmund Bohun, *Geographical Dictionary* (London, 1688).

3 Geography in decline: the age of Newton

1 Oskar Peschel, *Geschichte der Erdkunde*, 2nd edn, 1878 (facsimile: Amsterdam, 1961), p. 805.
2 Bartoldi Feind, *Cosmographia*, 5th edn (Hamburg, 1694).
3 British Museum, *General Catalogue of Printed Books* (London, 1961), vol. 103, p. 430. The date of 1649 for the first edition, given by Taylor, *Late Tudor and Early Stuart Geography*, pp. 139 and 294, appears to be an oversight.
4 William Warntz, *Geography Now and Then* (New York, 1964), p. 8.
5 Peter Heylyn, *Cosmography* (London, 1682).
6 Ibid., Preface and p. 1. See also E. W. Gilbert, 'Geographie is better than Divinity', *Geographical Journal* 128, no. 4 (1962), 494–7.
7 Taylor, *Late Tudor and Early Stuart Geography*, p. 140.
8 Heylyn, *Cosmography* (1682), pp. 4–23. Ptolemy actually defined geography as a representation 'of the portion of the earth known to us', stipulating only later that 'geography must first consider the form and size of the whole earth'. Similarly, Ptolemy assigned 'villages' to chorography and 'large cities' to geography: *Claudii Ptolemaei Geographia*, I.1.1–8.
9 John Newton, *Cosmographia, or A View of the Terrestrial and Coelestial Globes, . . . To which is added an Introduction unto Geography* (London, 1679), p. 226.
10 Samuel Clarke, *A Mirrour or Looking-Glass both for Saints and Sinners . . . collected out of the antient Fathers . . . Chronicles . . . etc.* (London, 1646).

11 Samuel Clarke, *A Mirrour or Looking-Glass both for Saints, and Sinners, . . . Whereunto are added a Geographical Description of all the Countries in the known World: As also the Wonders of God in Nature* (London, 1671), title-page, Part I.

12 Ibid., Part II, p. 89.

13 Richard Blome, *Cosmography and Geography. In Two Parts* (London, 1682), Preface.

14 Sanson d'Abbeville, *A Geographical Description of all the World, Taken from the Notes and Works of . . . Monsieur Sanson*, translated by R. Blome (London, 1680), p. 493. Bound as Part II of Blome's *Cosmography and Geography* (1682). With regard to geography in general, Sanson elsewhere suggested three divisions: *Astronomical, Natural* (divisions of land and water), and *Historical* (political regions, and the distribution of religions and principal languages). *Introduction a la Geographie* (Utrecht, 1692).

15 Blome, *Cosmography* (1682), title-page and Dedication.

16 Edmund Bohun, *A Geographical Dictionary. Representing the Present and Ancient Names of all Countries, Provinces, Remarkable Cities, Universities, Ports, Towns, Mountains, Seas, Streights, Fountains, and Rivers of the Whole World: Their Distances, Longitudes and Latitudes. With a short Historical Account of the same; and their present state . . . Very necessary for the right understanding of all Modern Histories, and especially the divers Accounts of the present Transactions of Europe* (London, 1688), Preface.

17 Quoted in Laurence Echard, *A Most Compleat Compendium of Geography, General and Special*, 6th edn (London, 1704), Preface. In the same letter Bohun noted also, 'the second impression of my *Geographical Dictionary* was lately Printed without my knowledge as *Corrected* and Enlarged; when in truth it is neither'.

18 Such works were popular at this time, some combining history, geography and literature, e.g. Charles Estienne's substantial Latin *Dictionarium historicum geographicum poeticum* (Oxford, 1670); and Rhodes, Meredith et al. (eds), *The Great Historical, Geographical and Poetical Dictionary* (London, 1694), which claimed to include 'the last Five Years Historical and Geographical Collections of Edmond Bohun'.

19 For an interesting study of history in its relation to the social sciences and to eighteenth-century conceptions of science, see F. J. Teggart, *Theory of History* (Yale, 1925), reprinted in his *Theory and Processes of History* (Berkeley, California, 1960).

20 R. E. Dickinson, *Makers of Modern Geography*, p. 11. On the idiographic method in history and geography, see also David Harvey, *Explanation in Geography* (London, 1969), pp. 49–54.

21 R. Hartshorne, 'The Region in Theory and Practice', visiting lecture, School of Geography, Oxford University, 19 November 1973. Compare his 'The nature of geography', *Annals AAG* 29, nos 3, 4 (1939), 173–658 (p. 379), and *Perspective on the Nature of Geography* (London, 1959), p. 149.

22 Statutes of the Royal Society, quoted in J. D. Bernal, *Science in History* (Harmondsworth, Middlesex, 1969), II, p. 455.

23 Stephen F. Mason, *A History of the Sciences* (New York, 1962), pp. 264–265.

24 Newton, letter to Coles, London, 1713, in *Newton's Philosophy of Nature*, ed. H. S. Thayer (New York, 1953), pp. 6–7.

25 John Herman Randall Jr, in *Newton's Philosophy of Nature*, pp. ix–xiv.

26 *Philosophical Transactions of the Royal Society*, xxiv, 1919; quoted in Baker, *History of Geography*, pp. 227–36.
27 A. Wolf, *A History of Science, Technology, and Philosophy in the 16th and 17th Centuries*, 2nd edn (London, 1950), p. 608.
28 See especially Humboldt's *Essai Politique sur le Royaume de la Nouvelle-Espagne* (Paris, 1811).
29 Carpenter, *Geography delineated* (1625), p. 3.
30 Humboldt, *Kosmos*, II, p. 391.
31 Thomas Burnet, *The Theory of the Earth: Containing an Account of the Original of the Earth* (London, 1684), Preface.
32 Ibid., Dedication.
33 Burnet, *The Theory of the Earth: the two Last Books concerning the Burning of the World and concerning the New Heavens and New Earth* (London, 1690).
34 Burnet, *Theory of the Earth* (London, 1691), p. 84.
35 John Ray, *The Wisdom of God Manifested in the Works of the Creation*, 1691 (12th edn, London, 1759).
36 Ibid., Preface. Page references in the text are to this edition.
37 Ralph Cudworth, *The true Intellectual System of the Universe . . . wherein All the Reason and Philosophy of Atheism is Confuted*, 1678 (2nd edn, London, 1743).
38 For a review of research in geography on probability see Leslie Curry, 'Chance and landscape' (1966), in P. W. English and R. C. Mayfield (eds), *Man, Space, and Environment* (New York, 1972), pp. 611–21.
39 Pliny, *Natural History*, translated by H. Rackham (Cambridge, Mass., 1942), book VII, chapter 1. See p. 511 for alternative translation.
40 John Ray, letter to Lhwyd, Keeper of the Ashmolean Museum, Oxford, October 1695, *Further Correspondence of John Ray*, ed. R. W. T. Gunther (London, 1928), p. 260. Lhwyd introduced Ray to fossil samples; see letters 1693 to 1697, *Further Correspondence*, pp. 238, 254, 259, 269.
41 Ibid., letters to Lhwyd, 1692, 1693: pp. 236–7, 242. As late as May 1691 Ray claimed not to have read Burnet's *Theory of the Earth*, and in March 1692 denied that his own *Discourses* had been directed against that work. See *Further Correspondence*, pp. 218, 236.
42 Ibid., April 1695, p. 257. See also p. 242.
43 Ibid., April 1695, p. 256. See also p. 277.
44 Ibid., 1699, p. 277. Edmund Halley had suggested a similar comet theory in a paper to the Royal Society in 1694.
45 William Whiston, *A New Theory of the Earth, from its Original to the Consummation of all Things*, 5th edn (London, 1737).
46 William Derham, *Physico-Theology: or, a Demonstration of the Being and Attributes of God, from his Works of Creation*, 3rd edn (London, 1714), Preface.
47 John C. Greene, *The Death of Adam: Evolution and Its Impact on Western Thought*, 1959 (New York, 1961), pp. 23–4.
48 Clarence Glacken, *Traces on the Rhodian Shore* (Berkeley, California, 1967), p. 423. See also ibid., pp. 379 and 535.
49 Arthur O. Lovejoy, *The Great Chain of Being*, 1936 (New York, 1960), p. 52.

4 Eighteenth-century empiricism: Locke, Berkeley and Hume

1 L. G. Crocker, *The Age of Enlightenment* (London, 1969), pp. 1–2.
2 A. Downes, 'The bibliographic dinosaurs of Georgian geography (1714–1830)',

Geographical Journal 137, part 3 (1971), 379–87. Harry Robinson, 'Geography in the Dissenting Academies', *Geography* 36 (1951), 179–86.

3 Isaiah Berlin, *The Age of Enlightenment* (New York, 1956), p. 30. Crocker, *Age of Enlightenment*, p. 14.

4 Thomas Hobbes, *Leviathan*, 1651 (Oxford, 1955), p. 82. In illustration, Hobbes added, 'the savage people in many places of America . . . live this day in that brutish manner' (p. 83).

5 John Locke, *Essays on the Law of Nature*, ed. and translated by W. von Leyden (Oxford, 1954), p. 133.

6 Locke, *Of Human Understanding*, Bk 1, Chapter 1, Section 2, in *The Works of John Locke* (London, 1823), I, pp. 1–2.

7 J. Bronowski and B. Mazlish, *The Western Intellectual Tradition* (Harmondsworth, Middlesex, 1963), p. 235.

8 J. L. Axtell (ed.), *The Educational Writings of John Locke* (Cambridge, 1968), p. 74.

9 Locke referred elsewhere to 'the train of our own ideas' [II, 14, 21]. Earlier, Hobbes had suggested that a 'train of thoughts', when directed towards some end, is the means by which the mind seeks either causes or effects; however, he did not claim that this was a form of induction leading to certainty: *Leviathan* (1955), pp. 14–15.

10 George Berkeley, *A Treatise concerning the Principles of Human Knowledge* (Dublin, 1710), Part I, §8; reprinted (Chicago, 1920), p. 33.

11 Karl Popper, *Conjectures and Refutations: The Growth of Scientific Knowledge* (London, 1963), pp. 169–72. Berkeley's views on relative motion, space, and time were expressed in his *Treatise of Human Knowledge* (1710), I.97–9, I.110–17, and in his *De Motu* (1721): see *Of Motion*, translated by A. A. Luce in *The Works of George Berkeley, Bishop of Cloyne*, eds A. A. Luce and T. E. Jessop, vol. IV, pp. 31–52 (London, 1951).

12 Berlin, *Age of Enlightenment*, pp. 160–1.

13 David Hume, *A Treatise of Human Nature: Being An Attempt to introduce the experimental Method of Reasoning into Moral Subjects* (London, 1739); reprinted (Oxford, 1949), pp. xvii–xviii. Further page references are to this edition.

14 Berlin, *Age of Enlightenment*, pp. 185–90.

15 Curt J. Ducasse made this second point, in Blake, Ducasse, and Madden, *Theories of Scientific Method* (Seattle, 1960), p. 144.

16 H. H. Price, *Hume's Theory of the External World* (Oxford, 1940), pp. 6–8, p. 25.

17 A. Einstein, *Geometrie und Erfahrung*, pp. 3 ff., translated in Popper, *The Logic of Scientific Discovery* (London, 1959), p. 314.

18 Hume, *An Enquiry concerning Human Understanding*, in *Hume: Theory of Knowledge*, ed. D. C. Yalden-Thomson (Edinburgh, 1951), pp. 3–12.

19 Humboldt, *Kosmos*, V, p. 7.

20 See *Kosmos*, I, p. 69.

21 Berlin, *Age of Enlightenment*, pp. 13–26.

22 Thomas Kuhn, *The Structure of Scientific Revolutions*, 2nd edn (Chicago, 1970), pp. 1–9.

5 On the margins of science: eighteenth-century geography texts

1 Schmithüsen, for example, in his *History of Geographical Science*, noted that in 1726 Polycarp Leyser (1690–1728), at the University of Helmstadt, proposed in

his *Commentatio de vera geographiae methodo* the idea of a *pure geography* dealing not only with political boundaries but also with physical features and the character of regions; however, in the following year Eberhard David Hauber (1695–1765), in his *Nützlicher Discours von dem gegenwärtigen Zustand der Geographie, besonders in Teutschland* (Useful discourse on the current position of geography, especially in Germany), criticized the lack of development in general geography and pointed out that the best source remained the work of Varenius, now eighty years old. See Josef Schmithüsen, *Geschichte der geographischen Wissenschaft von den ersten Anfängen bis zum Ende des 18. Jahrhunderts* (Mannheim, 1970), pp. 132–5.

2 Bernhard Varenius, *A Compleat System of General Geography*, translated by Dugdale, revised and corrected by Peter Shaw, M.D., 2 vols, 3rd edn (London, 1736).

3 William Warntz, *Geography Now and Then: Some Notes on the History of Academic Geography in the United States* (New York, 1964), pp. 9 and 67.

4 Laurence Echard, *A Most Compleat Compendium of Geography*, 6th edn (London, 1704), p. 1.

5 Laurence Echard, *Gazetteer's or Newsman's Interpreter: Being a Geographical Index of . . . Europe* (London, 1692). This proved very successful, reaching a 3rd edition by 1695 and continuing to a 17th in 1751.

6 Neither Varenius nor Carpenter, however, was mentioned in Echard's appended list of writers of geography, although it included Münster, Mercator, Cluverius, and Heylyn [220]. Echard, at Cambridge a decade after Newton's edition of the *Geographia generalis* was produced there, must have been familiar with Varenius, but his own concern was chiefly with cartography and special geography.

7 J. K. Wright, in R. Miller and W. Watson (eds), *Geographical Essays in Memory of A. G. Ogilvie* (Edinburgh, 1959), p. 147. Discussed in Downes, 'Bibliographic dinosaurs of Georgian geography', p. 379.

8 Patrick Gordon, *Geography Anatomiz'd: or, the Geographical Grammar. Being a short and Exact Analysis Of the whole Body of Modern Geography*, 17th edn (London, 1741), Preface.

9 Edward Wells, *A Treatise of Antient and Present Geography* (Oxford, 1701). For notes on Wells see Baker, *History of Geography*, p. 27.

10 Locke (1703), in J. L. Axtell (ed.), *The Educational Writings of John Locke* (Cambridge, 1968), pp. 400–1. Locke went on (p. 402) to recommend more strongly books of travel, including Churchill's *Collection of Voyages* (1704), for which Locke himself has been credited with authorship of the Introduction: see *Works of John Locke* (1823), I, Preface by the Editor.

11 Locke, *Some Thoughts concerning Education*, §178–82; in Axtell, *Educational Writings of John Locke*, pp. 289–92.

12 *Works of John Locke* (1823), III, pp. 303–30. Newton's participation in this project was discussed by Axtell, *Educational Writings of John Locke*, p. 76.

13 Herman Moll, *The Compleat Geographer*, 4th edn (London, 1723), '*Advertisement* to this *New Edition*'.

14 Moll, *Atlas Geographus: or, a Compleat System of Geography, Ancient and Modern*, 5 vols (London, 1711–17), Preface.

15 Nicolas de Fer, *A Short and Easy Method to understand Geography . . . From the French of Mr. A. D. Fer, Geographer to the French King* (London, c. 1715), p. 311.

16 Nicolas Lenglet Dufresnoy, *The Geography of Children: or, A Short and Easy*

Method of Teaching or Learning Geography . . . by way of Question and Answer. Translated from the French of Abbot Lenglet Dufresnoy . . . with the addition of a more particular Account of Great Britain and Ireland (London, 1737), Preface.

17 Gawin Drummond, *A Short Treatise of Geography, General and Special. To which is added a brief Introduction to Chronology*, 3rd edn (Edinburgh, 1740), Preface.

18 J. Gregory, *A Manual of Modern Geography . . . To which is added . . . Hydrography*, 4th edn (London, 1760).

19 Harry Robinson, 'Geography in the dissenting academies', *Geography* 36 (1951), 179–86 (p. 185); Warntz, *Geography Now and Then*, pp. 9 and 68.

20 Isaac Watts, *The Knowledge of the Heavens and the Earth made Easy: or, The First Principles of Astronomy and Geography*, 6th edn (London, 1760), pp. 1–3.

21 George Gordon, *An Introduction to Geography, Astronomy, and Dialling. Containing the most Useful Elements of the said Sciences . . . with an Introduction to Chronology* (London, 1726), Preface.

22 Locke, *Some Thoughts concerning Education*, §182; in Axtell, *Educational Writings of John Locke*, p. 292.

23 Locke, *Of Human Understanding*, II, 14.21–2; in *Works* (1823), I, pp. 184–6.

24 H. C. Darby, 'On the relations of geography and history', *Trans. Inst. Br. Geogr.* 30 (1953), 6.

25 David Harvey, *Explanation in Geography* (London, 1969), pp. 410–18.

26 A. O. Lovejoy, *The Great Chain of Being*, 1936 (New York, 1960), p. 259.

27 Thomas Salmon, *A New Geographical and Historical Grammar*, 11th edn (London, 1769), Preface.

28 Oskar Peschel, *Geschichte der Erdkunde* (1878), p. 805.

29 Johann Hübner, *A new and easy Introduction to the Study of Geography, By Way of Question and Answer . . . translated from the High Dutch of the late celebrated Mr. Hubner*, 2nd edn (London, 1742), pp. 1–2.

30 Peschel, *Geschichte der Erdkunde* (1878), pp. xvi and 803.

31 Anton Friedrich Büsching, *Erdbeschreibung*, 8th edn, 10 vols (Hamburg, 1787–92), I, p. 9. A modest edition, with neither maps nor illustrations. In the foreword (1787), Büsching drew attention to improvements made since 1752, and to subsequent translations made into English, Dutch, French and Italian, as well as parts into Russian, Swedish, and Polish.

32 Busching, *A new System of Geography*, 6 vols (London, 1762), I, pp. iii–iv. Subsequent page references are to this edition.

33 K. F. R. Schneider, *Handbuch der Erdbeschreibung und Staatenkunde* (Glogan and Leipzig, 1857).

34 Philippe Buache, *'Essai de géographie physique'*, *Mémoires de l'Académie Royale des Sciences* (Paris, 1752), pp. 399–412.

35 Andrew Brice, *A Universal Geographical Dictionary* (London, 1759), title-page.

36 John Barrow, *A new Geographical Dictionary*, 2nd edn (London, 1762), pp. iii–xii.

37 Demarville (ed.), *The Young Gentleman and Lady's Geography; containing, an accurate Description of the several Parts of the known World . . . To which is prefixed, an introduction to geography* (Dublin, 1766). Subsequent references are to this edition.

38 John Mair, *A brief Survey of the Terraqueous Globe* (Edinburgh, 1775).

39 Richard Turner, *A View of the Earth: Being a short but comprehensive System of Modern Geography* (London, 1762), pp. 9–13.

40 Benjamin Martin, *Physico-Geology: or, A New System of Philosophical Geography* (London, 1769), p. 33.

41 Thomas Malthus, *An Essay on Population* (London, 1798). For an excellent account of the design argument and concepts of nature in the eighteenth century, see part four of Clarence J. Glacken, *Traces on the Rhodian Shore* (Berkeley, 1967).

42 William Guthrie, *A New Geographical, Historical and Commercial Grammar and present State of the several Kingdoms of the World*, 8th edn (London, 1783), p. 6.

43 Downes, 'Bibliographic dinosaurs of Georgian geography', pp. 380–3.

44 Robert Heron *et al.*, *A New and Complete System of Universal Geography*, 2 vols (Edinburgh, 1796), I, p. v.

45 Guthrie, *A New Geographical, Historical and Commercial Grammar* (1783), pp. 7–10.

46 Richard Gadesby, *A new and easy Introduction to Geography, by way of Question and Answer*, 3rd edn (London, 1787).

47 *Geography for Youth, or, a Plain and easy introduction to the Science of Geography, for the use of young Gentlemen and Ladies*, 3rd edn (London, 1787).

48 John Blair, *The History of the Rise and Progress of Geography*, (London, 1784), p. 3.

49 These works are discussed by Downes, 'Bibliographic dinosaurs of Georgian geography', pp. 383–4.

50 William Guthrie, *A New System of Modern Geography: or, A Geographical, Historical, and Commercial Grammar*, 4th edn (London, 1788).

51 Alexander Kincaid (ed.), *A New Geographical, Commercial, and Historical Grammar*, 2 vols (Edinburgh, 1790).

52 Michael Adams, *The New Royal Geographical Magazine; or A Modern, Complete, Authentic, and Copious System of Universal Geography* (London, c. 1794).

53 W. Peacock, *A Compendious Geographical and Historical Grammar: Exhibiting a Brief Survey of the Terraqueous Globe* (London, 1795), p. iii. This book measures barely 6 × 9 cm.

54 Heron *et al.*, *Universal Geography* (1796), pp. iii–iv.

55 C. T. Middleton, *A New and Complete System of Geography. Containing a full, accurate, authentic and interesting Account and Description of Europe, Asia, Africa, and America* (London, 1778), title-page and Preface.

56 William Guthrie, *A New System of Modern Geography: or A Geographical, Historical and Commercial Grammar*, 2 vols, first American edn (Philadelphia, 1794), pp. 3–10.

57 Jedidiah Morse, *The American Geography*, 2nd edn (London, 1792), p. iii.

58 Warntz, *Geography Now and Then*, pp. 27–56.

59 Ibid., pp. 43–145.

60 Arnold Guyot, *The Earth and Man*, 1849 (London, 1875), pp. 2–3.

61 Preston James, *All Possible Worlds* (New York, 1972), p. 193.

6 Science and philosophy: enlightenment conflicts in Europe

1 C. J. Ducasse, in Blake, Ducasse, and Madden, *Theories of Scientific Method* (Seattle, 1960), pp. 188–217.

2 Leszek Kolakowski, *Positivist Philosophy: From Hume to the Vienna Circle* (Harmondsworth, Middlesex, 1972), pp. 42–59.

3 A. N. Whitehead, *The Concept of Nature*, 1920 (Cambridge, 1955).

4 Locke, *Of Human Understanding*, IV, 15, I–IV, 16, 6.

5 Leclerc de Buffon, *Histoire naturelle, générale et particulière*, vol. I, *Théorie de la Terre*, 1749 (Paris, 1799), p. 71.

6 Ibid., I, pp. 60–5. This part of Buffon's introductory *Discourse on the Manner of Studying and Treating Natural History* is translated by L. G. Crocker in *The Age of Enlightenment* (London, 1969), pp. 104–8.

7 Isaiah Berlin, *The Age of Enlightenment* (New York, 1956), pp. 19–20.

8 *Works of John Locke* (London, 1823), Preface by the Editor, pp. xvi–xvii.

9 Helvétius, *On the Mind*, translated in Crocker, *Age of Enlightenment*, p. 148.

10 La Mettrie, *L'Homme machine*, 1747, translated in Crocker, *Age of Enlightenment*, pp. 98–9.

11 Diderot, *De l'Interprétation de la nature*, 1753, translated in Crocker, *Age of Enlightenment*, pp. 289–90.

12 Diderot, *The Encyclopedia*, 1751–66, translated by S. Gendzier (New York, 1967), pp. 92–5.

13 Adam Smith, *An Inquiry into the Nature and Causes of the Wealth of Nations*, 1776 (London, 1893).

14 Wilbur Zelinsky, 'Beyond the exponentials, the role of geography in the Great Transition', *Economic Geography* 46, 3 (1970), 498–535 (p. 524).

15 J. Bronowski and B. Mazlish, *The Western Intellectual Tradition* (Harmondsworth, Middlesex, 1963), pp. 380–401.

16 Ibid., p. 391.

17 Antoine Lavoisier, *Traité élémentaire de chimie*, 1789, in *Oeuvres de Lavoisier* (Paris, 1864), I, p. 208.

18 J. R. Partington, *A History of Chemistry*, Part 1: *Theoretical Background* (London, 1970), I, pp. xv–xvi.

19 Leibniz, 'New system of nature and of the communication of substances, as well as of the union of soul and body', *Journal des Savans* (June, 1695), translated in *Leibniz: Selections*, ed. P. Wiener (New York, 1951), pp. 107–11.

20 Crocker, *Age of Enlightenment*, pp. 11–12.

21 Stephen Mason expressed this view in *A History of the Sciences* (New York, 1962), p. 337. Arthur Lovejoy, on the other hand, argued that Leibniz in many of his writings introduced the idea of universal progress, of individual, biological, and cosmical evolution through time, although this actually conflicted with his own theory of immutable monads and his principle of sufficient reason, and in effect split his system of philosophy completely in two: *The Great Chain of Being* (New York, 1960), pp. 252–62.

22 Wiener, *Leibniz: Selections*, p. xv.

23 Leibniz, letters to Samuel Clarke (1715–16), translated in Wiener, *Leibniz: Selections*, pp. 223–46.

24 Leibniz, notes on 'Advancing the sciences', 1680, and 'Discourse touching the method of certitude, and the art of discovery', translated in Wiener, *Leibniz: Selections*, pp. 40–50.

25 Leibniz, *Reflections on Knowledge, Truth, and Ideas* (1684), translated in Wiener, *Leibniz: Selections*, p. 285.

26 Leibniz, *Essay on Dynamics*, 1695, translated in Wiener, *Leibniz: Selections*,

pp. 132–3. See also *ibid.*, pp. 192 and 278, for additional comments by Leibniz in 1705, and 1715–16.

27 Leibniz, *New Essays concerning Human Understanding*, translated by A. G. Langley, 1896 (La Salle, Illinois, 1949), p. 43.

28 Ibid., pp. 43–4 and 71.

29 Leclerc de Buffon, *Histoire naturelle*, 1749 (Paris, 1799), I, pp. 13–23.

30 Translator's preface, *Buffon's Natural History [Barr's Buffon]* (London, 1797), I, p. vi.

31 Buffon, *Epochs of Nature*, 1778, translated in Crocker, *Age of Enlightenment*, p. 109.

32 Lovejoy, *Great Chain of Being*, pp. 242–55.

33 Discussed in Mason, *History of the Sciences*, pp. 339–41.

34 On this account, charges of determinism in *The Spirit of Laws* are denied by K. M. Kriesel, 'Montesquieu: possibilistic political geographer', *Annals, AAG* 58, no. 3 (1958), 557–74.

35 William Falconer, *Remarks on the Influence of Climate, Situation, Nature of Country, Population, Nature of Food, and Way of Life, on . . . Mankind* (London, 1781), pp. 2–3.

36 *Kant's Cosmogony, as in his Essay on the Retardation of the Rotation of the Earth, and his Natural History and Theory of the Heavens*, translated by W. Hastie (Glasgow, 1900), p. lxxv.

37 Kant, *Observations on the Feeling of the Beautiful and Sublime*, 1764, translated by J. T. Goldthwait (Berkeley, California, 1960), p. 6.

38 Pierre Simon de Laplace, *Exposition du Système du Monde*, 1796, vol. 6 in *Oeuvres complètes*, 6th edn (Paris, 1835), p. 138.

39 Hastie, *Kant's Cosmogony*, p. xviii. See also *Immanuel Kant's Sämmtliche Werke*, ed. Rosenkranz and Schubert (Leipzig, 1839): vol. V, *Allgemeine Naturgeschichte und Theorie des Himmels*.

40 Bronowski and Mazlish, *Western Intellectual Tradition*, pp. 341–5.

41 Ibid., p. 335. *Romantic* was a seventeenth-century term referring to the nature of *romances* – fictitious tales, often extravagant and imaginative, composed from the medieval period in the Romance vernaculars of France, Spain and Italy: cf. C. T. Onions (ed.), *The Oxford Dictionary of English Etymology* (Oxford, 1966), pp. 772–3.

42 Jean-Jacques Rousseau, *Émile ou de l'éducation*, 1762 (Paris, 1964), pp. 104–105.

43 Kant, *Prolegomena to any future Metaphysics that will be able to present itself as a Science*, 1783, translated by P. G. Lucas (Manchester, 1953), pp. 5–9.

44 Joseph May, *Kant's Concept of Geography* (Toronto, 1970), p. 68.

45 Kant, *Beobachtungen über das Gefühl des Schönen und Erhabenen* (Königsberg, 1764), translated in Goldthwait, *Observations*, pp. 115–16.

46 Kant, *Critique of Practical Reason*, 1790, in *Kant's Critique of Practical Reason and other Works on the Theory of Ethics*, translated by T. K. Abbott, 1873 (London, 1963), p. xcvii.

47 Kant, *Dissertation on the Form and Principles of the Sensible and Intelligible World*, 1770, in *Kant's Inaugural Dissertation and Early Writings on Space*, translated by John Handyside (Chicago, 1929), pp. 35–46.

48 Kant, letter to Marcus Herz, 7 June 1771, in *Selected Pre-Critical Writings and Correspondence with Beck*, translated by G. B. Kerferd and D. E. Walford (Manchester, 1968), pp. 108–9.

49 Kant, *Kritik der reinen Vernunft*, 1st edn (A) 1781, 2nd edn (B) 1787, translated by Norman Kemp Smith: *Critique of Pure Reason* (London, 1952), A51, B75. See also A258, B314.

50 Kemp Smith, ibid., p. 580.

51 May, *Kant's Concept of Geography*, pp. 108–13.

52 P. F. Strawson, *The Bounds of Sense* (London, 1966), pp. 245–72.

53 Kant, *Prolegomena to any future Metaphysics*, 1783 (1953), p. 15.

54 Kant, *Metaphysical Foundations of Natural Science*, translated by James Ellington (Indianapolis and New York, 1970), pp. 4–5.

55 May, *Kant's Concept of Geography*, p. 4.

56 Kant, *Entwurf und Ankundigung eines Collegii der physischen Geographie*, 1757, translated by May, *Kant's Concept of Geography*, pp. 64–5. Although he did not discuss the origins of Kant's ideas, May noted (p. 15) that Büsching's work was included in Kant's own list of sources, along with Varenius, Buffon, Linnaeus, Leibniz, Woodward, Whiston and others.

57 May, *Kant's Concept of Geography*, pp. 72–5.

58 Kant, *Physische Geographie*, Introduction, in *Gesammelte Schriften* (Berlin, 1902–66), IX, pp. 156–65; translated by May, *Kant's Concept of Geography*, pp. 255–64 (p. 256). Page references in the text are to this translation.

59 Kant, *Critique of Pure Reason*, translated by N. Kemp Smith, p. 254.

60 May, *Kant's Concept of Geography*, pp. 114–15.

61 Roger Minshull, *The Changing Nature of Geography* (London, 1970), p. 110. A similar comment, introduced as an editorial footnote, in M. J. Bowen, 'Mind and nature: the physical geography of Alexander von Humboldt', *Scottish Geographical Magazine*, 86, No. 3 (1970), 222–33 (p. 233 n. 12), must also be rejected as an attempt to limit Kant to a materialist definition of geography.

62 May, *Kant's Concept of Geography*, p. 9, p. 102. The tripartite classification, suggested by Alfred Hettner, *Die Geographie, Ihre Geschichte, ihr Wesen, ihre Methoden* (Breslau, 1927), was adopted by Richard Hartshorne, 'The nature of geography', *Annals AAG* 29, nos 3, 4 (1939), 173–658. See also Hartshorne, 'The concept of geography as a science of space, from Kant and Humboldt to Hettner', *Annals AAG* 48, no. 2 (1958), 97–108 (p. 99).

63 Kant, 'Nachricht von der Einrichtung seiner Vorlesungen in dem Winterhalbenjahre von 1765–1766', *Gesammelte Schriften*, II, pp. 303–13; translated in May, *Kant's Concept of Geography*, p. 68.

64 Kant, *Kritik der Urtheilskraft* (Berlin, 1790), translated by J. C. Meredith: *The Critique of Judgement* (Oxford, 1952), pp. 92–100.

65 Humboldt's debt to Kant in that respect is discussed at length by Anne Macpherson, 'The Human Geography of Alexander von Humboldt' (unpublished Ph.D. dissertation, University of California, Berkeley, 1972), pp. 34–152.

7 Geography revived: the age of Humboldt

1 Humboldt was called 'monarch of the sciences' and 'the new Aristotle' by nineteenth-century admirers: Hanno Beck, *Alexander von Humboldt*, 2 vols (Wiesbaden, 1959–61), I, p. 231. This, the most detailed modern biography of Humboldt, has not yet been translated into English.

2 George Forster, *A Voyage Round the World in His Britannic Majesty's Sloop, Resolution* (London, 1777), pp. 13–14.

3 George Forster, *Ansichten vom Niederrhein, von Brabant, Flandern, Holland,*

England und Frankreich in April, Mai und Junius 1790, Berlin, 1791 (Berlin, 1958), p. 19.

4 Humboldt, *Mineralogische Beobachtungen über einige Basalte am Rhein* (Braunschweig, 1790).

5 Humboldt, *Kosmos*, II, p. 72.

6 Humboldt, 'Mes confessions', 1806; in A. Leitzmann, *Georg und Therese Forster und die Brüder Humboldt* (Bonn, 1936), p. 201. See also Beck, *Humboldt*, I, pp. 32–3.

7 Humboldt, *Florae Fribergensis* (Berlin, 1793), pp. ix–x. This statement was repeated, in the original Latin, as a footnote in *Kosmos*, I, pp. 486–7. A translation by R. Hartshorne, *Annals AAG* 48, no. 2 (1958), 100, has been modified here.

8 Humboldt, *Kosmos*, I, p. 487.

9 Humboldt, *Essai géognostique sur le gisement des roches dans les deux hémisphères* (Paris, 1823), translated as *A Geognostical Essay on the Superposition of Rocks* (London, 1823).

10 Humboldt, *Aphorismi ex doctrina physiologiae chemicae plantarum*, in *Florae Fribergensis*, 1793, pp. 133–51.

11 Wilhelm von Humboldt, letter to Brinkmann, 18 March 1793; in Beck, *Humboldt*, I, pp. 45–6.

12 *Letters from Goethe*, translated by M. von Herzfeld and C. Melvil Sym (Edinburgh, 1957), p. 252.

13 R. Gray, *Goethe: A Critical Introduction* (Cambridge, 1967), pp. 122–4.

14 Humboldt, *Kosmos*, I, p. 19.

15 Humboldt, *Die Lebenskraft oder der Rhodische Genius*, 1957. Reprinted in Humboldt's *Ansichten der Natur*, 2nd edn (Tübingen, 1826), and 3rd edn (Stuttgart and Tübingen, 1849).

16 Humboldt, *Kosmos*, II, p. 75.

17 Johann Gottfried von Herder, *Ideen zur Philosophie der Geschichte der Menschheit*, 1784; translated by T. Churchill: *Outlines of a Philosophy of the History of Man*, 1800 (facsimile, New York, n.d.), pp. vi–ix.

18 Eberhard August Wilhelm von Zimmermann, *Geographische Geschichte des Menschen und der allgemeinen-verbreiteten vierfüssigen Thiere*, 3 vols (Leipzig, 1778–83). Humboldt, in his comments on plant and animal geography in *Florae Fibergensis* seems to have followed Herder closely.

19 Humboldt, letter to J. F. Pfaff, November 1794; in Beck, *Humboldt*, I, 256.

20 Humboldt, *Correspondence scientifique et littéraire*, ed. M. de la Rocquette (Paris, 1865), p. 4; in Anne Macpherson, 'The Human Geography of Alexander von Humboldt' (unpublished Ph.D. dissertation, University of California, Berkeley, 1972), p. 120.

21 Macpherson, 'Human Geography of Alexander von Humboldt', pp. 132–52.

22 Humboldt, *Versuche über die gereizte Muskel- und Nervenfaser*, 2 vols (Posen and Berlin, 1797). This work aroused much interest, a French translation, by J. F. N. Jadelot, appearing within two years: *Expériences sur le Galvanisme* (Paris, 1799).

23 Humboldt, letter to Freiesleben, 4 June 1799; *Lettres américaines d'Alexandre de Humboldt 1798–1807*, ed. E.-T. Hamy (Paris, 1905), p. 18.

24 Humboldt, letter to von Moll, 5 June 1799; *Lettres américaines*, p. 18.

25 Humboldt, letter to W. von Humboldt, 17 October 1800; *Lettres américaines*, pp. 86–7.

26 Humboldt, *Voyage aux régions équinoxiales du Nouveau Continent*, Paris, 1805–34 (facsimile: Amsterdam and New York, 1971–3).

27 Humboldt, *Essai sur la Géographie des plantes, accompagné d'un Tableau physique des régions équinoxiales, et servant d'introduction à l'Ouvrage*, Paris, 1807 (facsimile: Amsterdam and New York, 1973), pp. v–vi.

28 Humboldt, *Aus Schelling's Leben. In Briefen* (Leipzig, 1870), II, p. 50; in Karl Bruhns (ed.), *Life of Alexander von Humboldt*, 1872, translated by Lassell (London, 1873), I, p. 203.

29 Humboldt, *Ansichten der Natur mit wissenschaftlichen Erläuterungen*, 1808, 3rd edn (Stuttgart and Tübingen, 1849); translated by Mrs Sabine: *Aspects of Nature* (London, 1850), pp. vii–xiii.

30 A. Gode- von Aesch, *Natural Science in German Romanticism*, 1941 (New York, 1966), p. 27.

31 A. N. de Condorcet, *Sketch for a Historical Picture of the Progress of the Human Mind*, 1795, translated by Barraclough (New York, 1955).

32 Humboldt, *Vues des Cordillères* (Paris, 1810); translated by H. M. Williams: *Researches Concerning the Institutions and Monuments of the Ancient Inhabitants of America, with Descriptions of Some of the Most Striking Scenes in the Cordilleras* (London, 1814).

33 A perceptive analysis of this work is given in Macpherson, 'Human Geography of Alexander von Humboldt', pp. 395–428.

34 Humboldt, *Essai Politique sur le Royaume de la Nouvelle-Espagne; avec un Atlas géographique et physique* (Paris, 1811); translated by J. Black: *Political Essay on the Kingdom of New Spain*, 4 vols, 1811, reprinted (New York, 1966), I, pp. 140–57.

35 Humboldt, *Relation historique du Voyage aux Régions équinoxiales du Nouveau Continent*, 3 vols, Paris, 1814–25; vol. I, 1814 (facsimile: Amsterdam and New York, 1973), pp. 2–3.

36 Humboldt, *Personal Narrative of Travels to the Equinoctial Regions of America during the years 1799–1804*, translated by T. Ross (London, 1851), I. p. x.

37 Humboldt, *Examen critique de l'histoire de la géographie du Nouveau Continent et des progrès de l'astronomie nautique aux XVe et XVIe siècles*, Paris, 1814–34 (facsimile: Amsterdam and New York, 1971).

38 Humboldt, *Essai géognostique sur le gisement des roches dans les deux hémisphères* (Paris, 1823), p. vi. See also *A Geognostical Essay on the Superposition of Rocks in both Hemispheres*, pp. vi–vii. Humboldt repeated this passage in the last pages of *Kosmos*: V, p. 92.

39 For a more detailed list of geography texts published in Great Britain and Ireland during this period, see R. A. Peddie and Q. Waddington (eds.), *The English Catalogue of Books . . . 1801–1836* (London, 1914; reprinted New York, 1963).

40 John Pinkerton, *Modern Geography* (London, 1802), I, iii–v.

41 See A. A. Wilcock, ' "The English Strabo" the geographical publications of John Pinkerton', *Trans. Inst. Br. Geogr.* 61 (1974), 35–45.

42 A similar use of the terms Astronomical, Natural, and Historical geography by Sanson in the late seventeenth century has already been noted. See above, p. 297 n. 14.

43 Humboldt, *Political Essay on the Kingdom of New Spain* (New York, 1966), I, pp. 263–4.

44 Conrad Malte-Brun, *Précis de la Géographie Universelle*, 2nd edn (Paris, 1812–29), I, pp. 1–2.

45 W. R. Mead, 'Luminaries of the North. A reappraisal of the achievements and influence of six Scandinavian geographers', *Trans. Inst. Br. Geogr.* 57 (1972), 2–4.

46 August Zeune, *Gea. Versuch die Erdrinde* (1808), 3rd edn (Berlin, 1830), p. vi.

47 Eberhard A. W. von Zimmermann, *Die Erde und ihre Bewohner nach den neuesten Entdeckungen. Ein Lesebuch für Geographie, Völkerkunde, Produktenlehre und den Handel* (The earth and its inhabitants after the newest discoveries. A textbook for geography, ethnology, product-studies and commerce) Leipzig, 1810–11, 3rd [?] edn, completed by F. Rühs and H. Lichtenstein, 7 vols (Stuttgart, 1816–20), I, pp. 138–72.

48 Carl Ritter, *Comparative Geography*, translated by W. L. Gage (Edinburgh and London, 1865), p. 24. Page references in the text are to this translation of Ritter's *Einleitung zur allgemeinen vergleichenden Geographie* (Berlin: Reimer, 1852).

49 Carl Ritter, *Die Erdkunde im Verhältniss zur Natur und zur Geschichte des Menschen; oder, allgemeine vergleichende Geographie*, 2nd edn (Berlin, 1822–59). On Ritter's *Erdkunde* and his debt to Pestalozzi see Hanno Beck, *Carl Ritter: Genius der Geographie* (Berlin, 1979).

50 *Briefwechsel Alexander von Humboldt's mit Heinrich Berghaus aus den Jahren 1825 bis 1858*, 3 vols (Leipzig, 1863), I, pp. 1–5.

51 Humboldt, *Relation historique*, Paris, 1825 (facsimile: Amsterdam and New York, 1973), III, pp. 387–9.

52 Wilbur Zelinsky, 'The demigod's dilemma', *Annals AAG* 65, No. 2 (1975), 123–43, p. 135.

53 Humboldt, *Briefwechsel . . . mit Heinrich Berghaus* (1863), I, pp. 117–19.

54 *Letters of Alexander von Humboldt, written between the years 1827 and 1858, to Varnhagen von Ense*, authorized translation [no tr.] (London, 1860), p. 3. Note that an alternative translation of 'Physische Weltbeschreibung' is 'description of the natural world'.

55 Humboldt, *Asie Centrale. Recherches sur les chaînes de montagnes et la climatologie comparée*, 3 vols (Paris, 1843), I, pp. xi–xii.

56 *Hegel's Lectures on the History of Philosophy*, translated by E. S. Haldane (London, 1955), II, pp. 133–55.

57 Humboldt, *Letters to Varnhagen von Ense*, p. 2.

58 Ibid., pp. 33–4. Humboldt's library, after his death, included the 1837 and 1840 editions of Hegel's *Philosophy of History*, as well as his *Lectures on the History of Philosophy* (1833) and on *Aesthetics* (1835–8). See Henry Stevens, *The Humboldt Library*, London, 1863 (facsimile: Leipzig, 1967), pp. 290–1.

59 William Whewell, *The Philosophy of the Inductive Sciences founded upon their History*, 2 vols (London, 1840), I, p. iv.

60 William Whewell, *Novum Organon Renovatum*, 3rd edn (London, 1858), pp. 46–53.

61 Johann Friedrich Herbart, *Outlines of Educational Doctrine* (1835), translated by A. F. Lange (New York, 1901), p. 15. Discussed in James Bowen, *A History of Western Education*, III (London, 1981), chapter 8.

62 Humboldt, *Letters to Varnhagen von Ense*, pp. 15–18.

63 Ibid. Humboldt had already discussed this new project with Arago as early as 1829: see *Correspondance d'Alexandre de Humboldt avec François Arago (1809–1853)*, ed. E.-T. Hamy (Paris, 1907), pp. 44, 94, and 134–5.

64 Eduard Buschmann (ed.), *Kosmos*, V, p. 102.

65 Humboldt, *Letters to Varnhagen von Ense*, p. 275.

66 Humboldt, *Kosmos,* V, pp. 3–4. See Appendix 2, pp. 284–9 above.

67 Mary Somerville, *Physical Geography* (London, 1848); Arnold Guyot, *The Earth and Man,* 1849 (London, 1875).

68 Humboldt, *Kosmos,* I, p. xii.

69 See, for example, Carl Troll, 'Selenographie und Geographie: ein Ruckblick auf das Jahr 1969', *Erdkunde* 23 (1969), 326–8.

70 A. N. Strahler, *The Earth Sciences* (New York, 1963).

71 Humboldt, *Briefwechsel . . . mit Heinrich Berghaus* (1863), III, pp. 39–61.

72 Humboldt, *Examen critique* (Paris, 1834), p. v.

73 Karl Bruhns (ed.), *Alexander von Humboldt. Eine wissenschaftliche Biographie,* 3 vols (Leipzig, 1872), II, p. 513.

74 Macpherson, 'Human Geography of Alexander von Humboldt', pp. 466–8.

75 Beck, *Humboldt,* II, pp. 69–70, 269–70. The *Tagebuch* remains unpublished but from his study of the manuscript Hanno Beck reports that the journal does not contain the kind of material, especially the social and political commentary, which might have been expected to form part of the final volume of the *Relation historique.* (Personal communication, May 1979.)

76 Francis Darwin (ed.), *Life and Letters of Charles Darwin,* 3 vols (London, 1887), quoted in L. Kellner, *Alexander von Humboldt* (London, 1963), p. 34.

77 Kellner, *Alexander von Humboldt,* pp. 222–3.

78 Humboldt, *Letters to Varnhagen von Ense,* pp. 85 (letter of 1842) and 133 (letter of 1845).

79 Hanno Beck (ed.), *Kosmos für die Gegenwart* (Stuttgart, 1978), p. x.

80 Humboldt, *Letters to Varnhagen von Ense,* p. 153 (letter of 1846).

81 Robert Payne, *Marx* (London, 1968), pp. 116–17.

82 Karl Marx and Friedrich Engels, *Briefwechsel* (Berlin, 1949), II, pp. 128–9; translated by D. Torr in Karl Marx and Friedrich Engels, *Correspondence 1846–95: A Selection,* 2nd edn (London, 1936), pp. 77–8.

83 Beck, *Humboldt,* II, p. 308.

84 Alfred Schmidt, *The Concept of Nature in Marx* (London, 1971); translated by Ben Fowkes from *Der Begriff der Natur in der Lehre von Marx* (1965).

85 Marx and Engels, *The Holy Family,* in *Collected Works* (London, 1975), IV, pp. 124–34.

86 *John Stuart Mill's Philosophy of Scientific Method,* ed. Ernest Nagel (New York, 1950), pp. 177–81.

87 A. Comte, *A General View of Positivism,* 1848, translated by J. H. Bridges (Stanford, n.d.).

88 Humboldt, *Briefwechsel . . . mit Heinrich Berghaus* (1863), III, p. 1. Apart from this statement, Humboldt seems to have avoided discussing philosophical issues with the geographer.

89 Humboldt, *Letters to Varnhagen von Ense,* pp. 208–9.

90 The preceding discussion of Humboldt's thought in large part follows M. J. Bowen, 'Mind and nature: the physical geography of Alexander von Humboldt', *Scottish Geographical Magazine* 86, no. 3 (1970), 222–33.

91 Humboldt, *Kosmos,* V, p. 19. Humboldt's reference is to C. A. Brandis, *Handbuch der Geschichte der Griechisch-Romischen Philosophie,* 3 vols (Berlin 1835–66), II, p. 45.

92 Brandis, *Handbuch,* II, p. 45.

93 Humboldt, *Kosmos,* V, p. 19. See Appendix 2. Humboldt was quoting from Hegel's *Lectures on the History of Philosophy,* delivered between 1822 and 1831,

and published posthumously: *Vorlesungen über die Geschichte der Philosophie* (1833) II, p. 340, translated by Haldane and Simpson as *Hegel's Lectures on the History of Philosophy* (London, 1955), II, p. 133.

8 Epilogue: the way ahead

1 Humboldt, *Cosmos: A Sketch of a Physical Description of the Universe*, translated by E. C. Otté (1849–58) new edn, with tr. by B. H. Paul (vol. IV) and W. S. Dallas (vol. V), 5 vols (London, 1882–4).

2 Anne Macpherson, 'The Human Geography of Alexander von Humboldt' (unpublished Ph.D. dissertation, University of California, Berkeley, 1972), pp. 3–15.

3 A view maintained, e.g. in Peter Haggett, *Geography: A Modern Synthesis* (New York, 1972) and 3rd edn, 1979, p. 599.

4 The belief that Wilhelm von Humboldt but not Alexander is to be recognized as a humanist persists: see e.g. Yi-Fu Tuan, 'Humanistic geography', *Annals AAG* 66, no. 2 (1976), 276.

5 Bob Galois, 'Ideology and the idea of nature: the case of Peter Kropotkin', *Antipode* 8, no. 3 (1976), 1–16; reprinted in Richard Peet (ed.), *Radical Geography* (London, 1977), 66–93. See also Preston James, *All Possible Worlds: A History of Geographical Ideas* (New York, 1972), p. 286.

6 Peter Kropotkin, *Mutual Aid: A Factor of Evolution*, 1902 (London, 1972).

7 Kropotkin, 'On the teaching of Physiography', *Geographic Journal* 2 (1893), 350–9.

8 Kropotkin, 'What Geography ought to be', *Nineteenth Century* (December 1885), p. 949; compare with *Kosmos*, V, 3–9.

9 Kropotkin, *Ethics: Origin and Development*, 1924 (New York, 1968), pp. 2–6.

10 Kropotkin, 'The Place of Anarchism in Modern Science' in *Kropotkin's Revolutionary Pamphlets*, 1927 (New York, 1970), pp. 146–93.

11 Gunnar Olsson, Preface, special issue: 'Geography, Epistemology, and Social Engineering', *Antipode* 4, no. 1 (1972).

12 Derek Gregory, *Ideology, Science and Human Geography* (London, 1978), pp. 170–1.

13 Thomas Kuhn, *The Structure of Scientific Revolutions*, 2nd edn (Chicago, 1970).

14 Stephen Toulmin, 'Does the distinction between normal and revolutionary science hold water?', in I. Lakatos and A. Musgrave (eds), *Criticism and the Growth of Knowledge* (Cambridge, 1972), pp. 39–47.

15 A view advanced by Michael Polanyi, *Personal Knowledge* (Chicago, 1958).

16 See Karl Popper, *Logik der Forschung* (1934), translated as *The Logic of Scientific Discovery* (London, 1959); and *Conjectures and Refutations: The Growth of Scientific Knowledge* (London, 1963). See also Scheffler, *Science and Subjectivity* (New York, 1967).

17 Lennart Nørreklit, *Concepts: Their Nature and Significance for Metaphysics and Epistemology* (Odense, 1973), pp. 11–13.

18 See M. Polanyi, *The Tacit Dimension* (London, 1967).

19 Peter Achinstein, *Concepts of Science* (Baltimore, 1968), pp. 154 and 91.

20 Popper, *Logic of Scientific Discovery*, pp. 278–81.

21 K. Popper, *Objective Knowledge: An Evolutionary Approach* (Oxford, 1972).

22 Pierre Teilhard de Chardin, *The Phenomenon of Man*, translated by Bernard

Wall (London, 1967), p. 202. Halford Mackinder in 1943 suggested the term *Psychosphere*: see his 'Global geography', *Geography* 28 (1943), 69–71.

23 Robert Van Dusen, *The Literary Ambitions and Achievements of Alexander von Humboldt* (Berne, 1971), p. 55.

24 Poole's *Index to Periodical Literature, 1802–96* (New York, 1938), for example, shows no article on environment before 1887; after that date the use of this term increased while articles on nature declined in number and no longer dealt with a wide range of scientific or philosophic problems.

25 For a discussion of environmentalist viewpoints see G. R. Lewthwaite, 'Environmentalism and determinism: a search for clarification', *Annals AAG* 56 (1966), 1–23.

26 Harlan H. Barrows, 'Geography as human ecology', *Annals AAG* 13 (1923), 1–14.

27 Richard Chorley, 'Geography as human ecology', in R. J. Chorley (ed.), *Directions in Geography* (London, 1973), pp. 155–69.

28 Edward Ackerman, 'Where is a research frontier?', *Annals AAG* 53, no. 4 (1963), 429–40; reprinted in Wayne K. D. Davies (ed.), *The Conceptual Revolution in Geography* (London, 1972), pp. 266–79.

29 Marston Bates, *Man in Nature* (Englewood Cliffs, New Jersey, 1962).

30 Wilbur Zelinsky, 'Beyond the exponentials, the role of geography in the Great Transition', *Economic Geography* 46, no. 3 (1970), 509.

31 Kenneth Boulding, *The Meaning of the Twentieth Century: the Great Transition* (London, 1965), pp. 27 and 108.

32 Geoffrey Vickers, 'Ecology, planning, and the American dream', in Leonard J. Duhl (ed.), *The Urban Condition* (New York, 1963); reprinted in Geoffrey Vickers, *Value Systems and Social Process* (New York, 1968), pp. 28–51.

33 Gregory Bateson, 'Form, substance and difference', *General Semantics* Bulletin, no. 37 (1970); reprinted in his *Steps to an Ecology of Mind*, 1972 (London, 1978), pp. 423–42 (p. 436).

34 M. J. Bowen, 'Scientific method – after positivism', *Australian Geographical Studies* 17, no. 2 (1979), 210.

BIBLIOGRAPHY

The following select bibliography is divided into two sections:

1 *Comprehensive texts in geography, 1599 to 1859.* This includes those works located in the course of the present study: the formation of a more exhaustive bibliography is a task for future research.

2 *General bibliography.* Here the remaining primary sources and references have been collected; Humboldt's works are listed in this section.

Note on citations
Selections from extended sub-titles in early works are provided to supplement footnotes in the text; the date of a first edition, where appropriate, is given in parentheses after the title. The following abbreviations are used:

Annals AAG Annals of the Association of American Geographers.
Trans. Inst. Br. Geogr. Transactions, Institute of British Geographers.

1 Geography texts 1599–1859

Abbot, George. *A Briefe Description of the whole Worlde, wherein is particularly described, all the Monarchies, Empires, and Kingdomes of the same* (London, 1599), facsimile, Amsterdam: Theatrum Orbis Terrarum, 1970
Adam, Alexander. *A Summary of Geography and History, both Ancient and Modern . . . To which is prefixed, an historical account of the progress and improvements of astronomy and geography . . . to the time of Sir Isaac Newton . . . Designed chiefly to connect the study of classical learning with that of general knowledge*, Edinburgh: A. Strahan and T. Cadell, 1794. (Bound with this is Adam's *Geographical Index*, Edinburgh: 1795)
Adams, Michael. *The New Royal Geographical Magazine; or A Modern, Complete, Authentic, and Copious System of Universal Geography: containing A Complete . . . History and Description of the several Parts of the Whole World; as divided into empires, kingdoms, states, provinces, . . . continents, islands, oceans . . . etc. Together with new accounts of their Soil, Situation, Extent, and Bounds, in Europe, Asia, Africa, and America: With a very particular Account of their Subdivisions and Dependencies; Their Cities . . . Commerce . . . Customs, etc. . . . And whatever is found curious, useful, and entertaining, at Home and Abroad. To which will be added, A New and Easy Guide to Geography and Astronomy . . . with an Account of the Rise and Progress of Navigation . . . Together with Chronological Tables of the Sovereigns of the Whole World. Including every Interesting Discovery and Circumstance in the Narratives of Captain Cook's Voyages Round the World,*

Together with all the recent Discoveries made in the Pelew Islands, New Holland, New South Wales . . . etc. . . . . compiled from the Late Journals . . . of Captain Phillips, King, Ball, Hunter . . . etc. London: Alex. Hogg [n.d.: *c.* 1794]

Barrow, John. *A new Geographical Dictionary containing a full . . . account of the several Parts of the known World, as it is Divided into Continents, Islands, Oceans, Seas, Rivers, Lakes, etc. . . . . To which is prefixed An Introductory Dissertation, explaining the Figure and Motion of the Earth, the Use of the Globes, and Doctrine of the Sphere, in order to render the Science of Geography easy and intelligible to the meanest Capacity,* 2nd edn, London: Coote, 1762

Blome, Richard. *Cosmography and Geography. In Two Parts: The First, Containing the General and Absolute Part of Cosmography and Geography, being a Translation from that Eminent and much Esteemed Geographer Varenius, Wherein are at large handled All such Arts as are necessary to be understood for the true knowledge thereof. To which is added the much wanted Schemes omitted by the Author. The second part Being a Geographical Description of all the World, Taken from the Notes and Works of the Famous Monsieur Sanson, Late Geographer to the French King: To which are added About an Hundred Cosmographical, Geographical and Hydrographical Tables of several Kingdoms and Isles in the World, with their Chief Cities, Seaports, Bays, etc. drawn from Maps of the said Sanson,* London: Printed by S. Roycroft for Richard Blome, 1682

Bohun, Edmund. *A Geographical Dictionary,* London: Charles Brome, 1688

Brice, Andrew. *A Universal Geographical Dictionary; or Grand Gazetteer; of General, Special, Antient and Modern Geography: Including a comprehensive View of various Countries of Europe, Asia, Africa and America; More especially of the British Dominions and Settlements throughout the World,* 2 vols, London: Robinson and Johnston, etc., 1759

Buache, Philippe. 'Essai de géographie physique', *Mémoires de l'Académie Royale des Sciences,* Paris: 1752, pp. 399–412

Bucher, August Leopold. *Betrachtungen über die Geographie und über ihr Verhältniss zur Geschichte und Statistik,* Leipzig: Gerhard Fleischer, 1812

Büsching, Anton Friedrich. *Erdbeschreibung* (1754–92), 8th edn, 10 vols, Hamburg: Bohn, 1787–92

(Büsching). *A new System of Geography: in which is given, A General Account of the Situation and Limits, the Manners, History and Constitution, of the several Kingdoms and States in the known World; And a very particular Description of their subdivisions and Dependencies; their Cities and Towns, Forts, Sea-ports, Produce, Manufactures and Commerce,* translated [by P. Murdoch] from the last German edn, 6 vols, London: Millar, 1762

Carpenter, Nathaniel. *Geography delineated forth in two Bookes. Containing the Sphaericall and Topicall Parts thereof,* Oxford: Printed by John Lichfield and William Turner for Henry Capps, 1625

Clarke, Samuel. *A Mirrour or Looking-Glass both for Saints and Sinners . . . Examples . . . collected out of the antient Fathers . . . Chronicles . . . etc.,* London: Bellamy, 1646

A Mirrour or Looking-Glass both for Saints, and Sinners, Held forth in some Thousands of Examples; . . . Whereunto are added a Geographical Description of all the Countries in the known World; As also the Wonders of God in Nature; and the Rare, Stupendious, and Costly Works made by the Art, and Industry of Man. As the most famous Cities, Temples, Structures, Statues, Cabinets of Rarities, etc. which have been, or are now in the World, 4th edn, enlarged,

London: Printed by Tho. Milbourn for Clavel, Passinger, Cadman, Whitwood, Sawbridge and Birch, 1671

Cluverius (Clüver), Philip. *Introductio in universam geographiam* (1624), ed. Johannis Bunonis, Amsterdam: Gulielmum Gulielmi, 1683

Demarville (ed.). *The Young Gentleman and Lady's Geography; containing, an accurate Description of the several Parts of the known World; Their Situation, Bounds, Chief Towns, Air, Soil, Manners, Customs, and Curiosities. Compiled from the Writings of the most eminent Authors, with particular Attention to the Modern State of every Nation. To which is prefixed, an Introduction to Geography; wherein the terms made use of in that Science, and the Method of speedily acquiring a thorough Knowledge of Maps, are explained in so concise a Manner, as to render the Whole perfectly easy to be attained Without the Assistance of a Teacher. Also, An Astronomical Account of the Motion and Figure of the Earth, the Vicissitudes of Night and Day, and the Four Seasons of the Year*, Dublin: J. Hoey, Sen., S. Cotter, W. Steater, J. Hoey, jun., and J. Williams, 1766

Drummond, Gawin. *A Short Treatise of Geography, General and Special. To which is added a brief Introduction to Chronology with Tables of the principal Coins in Europe and Asia . . . Collected from the best Authors . . . for the use of schools* (1708), 3rd edn, Edinburgh: Gawin Drummond, 1740

Echard (Eachard), Laurence. *A Most Compleat Compendium of Geography; General and Special; Describing all the Empires, Kingdoms, and Dominions, in the Whole World . . . Together with an Appendix of General Rules for making a large Geography, together with the great Uses of that Science* (1691), 6th edn, London: Nicholson and Ballard, 1704

Estienne, Charles. *Dictionarium historicum geographicum poeticum*, new edn, Oxford: Downing, Williams, 1670

Feind, Bartoldi. *Cosmographia, Das ist: Gründliche Anleitung zur Betrachtung des gantzen Welt-Kreyses. In zwey Theile . . . : I. Eine Betrachtung der Stern-Kunst; II. Eine Beschreibung des Erd-Kreises . . . in deutscher Sprache*, 5th edn, Hamburg: Gottfried Liebernickel, 1694

Fer, Nicolas de. *A Short and Easy Method to understand Geography . . . Made English by a Gentleman of Cambridge. From the French of Mr. A. D. Fer, Geographer to the French King*, London: H. Banks and T. Woodward [n.d.: c. 1715]

Gadesby, Richard. *A new and easy Introduction to Geography, by way of Question and Answer, Divided into Lessons. Principally designed for the Use of Schools*, 3rd edn, enlarged, London: R. Gadesby, 1787

Geography for Youth, or, a Plain and easy introduction to the Science of Geography, for the use of young Gentlemen and Ladies: containing An Accurate Description of the Longitude and Latitude of the most remarkable Places on the Terraqueous Globe. Illustrated by twelve Maps, on which are delineated The New Discoveries made by Commodore Byron, And the Captains Wallis, Carteret, and Cook, 3rd edn, London: W. Lowndes, 1787

Gölnitz (Golnitzius), Abraham. *Compendium Geographicum succincta methodo adornatum*, Amsterdam: Elzevir, 1643

Gordon, George. *An Introduction to Geography, Astronomy, and Dialling. Containing the most Useful Elements of the said Sciences, Adapted to the Meanest Capacity, by the Description and Uses of the Terrestrial and Celestial Globes with an Introduction to Chronology*, London: Senex, Strahan, Innys . . . Gordon, 1726

Gordon, Patrick. *Geography Anatomiz'd: or, the Geographical Grammar. Being a short and Exact Analysis Of the whole Body of Modern Geography, After a New and Curious Method. Comprehending I. A general view of the Terraqueous Globe, Being a Compendious System of the true Fundamentals of Geography; Digested . . . II. A particular view of the Terraqueous Globe, . . . of all remarkable Countries . . . their Situation, Extent, Division, Sub-division, Cities, Chief Towns, Name, Air, Soil, Commodities, Rarities, Archbishopricks, Bishopricks, Universities, Manners, Languages, Government, Arms, Religion.* Collected from the best Authors (1693), 17th edn, corrected, enlarged and new maps by Mr Senex, London: Midwinter, Ward and others, 1741

Gregory, J. *A Manual of Modern Geography, containing a Short . . . Account of all the known World . . . 4th edition Corrected etc. To which is added . . . Hydrography; or an Account of Water, the Ocean, Seas, Gulphs, Lakes, Streights, and the most considerable Rivers in all Parts of the World. By J. Gregory. Whereunto is added a table of latitudes and longitudes of some of the principal cities and towns in the world . . . by Emanuel Bowen, Geographer to His Majesty,* 4th edn, London: John Ward and J. Roe, 1760

Guthrie, William. *A New Geographical, Historical and Commercial Grammar and present State of the several Kingdoms of the World. The Astronomical Part by James Ferguson* (1770), 8th edn, London: Dilly and Robinson, 1783

A New System of Modern Geography: or, A Geographical, Historical, and Commercial Grammar, 4th edn, London: Dilly and Robinson, 1788

A New System of Modern Geography: or A Geographical, Historical and Commercial Grammar . . . The Astronomical Parts corrected by Dr. Rittenhouse, 2 vols, 1st American edn, Philadelphia: Printed by Mathew Carey, 1794

Guy, Joseph. *Guy's School Geography, on a new and easy Plan: comprising not only a complete general Description, but much Topographical Information, in a well digested Order,* 25th edn, London: Cradock, Whittaker, and Simpkin, Marshall and Co., 1858

Guyot, Arnold. *The Earth and Man; or Comparative Physical Geography In its Relation to the History of Mankind* (1849), new edn, London: Chatto and Windus, 1875

Heron, Robert, et al. *A New and Complete System of Universal Geography: containing A Full Survey of the Natural and Civil State of the Terraqueous Globe; Exhibiting all the latest and most authentic Information concerning Europe, Asia, Africa, and America, . . . etc. As also an accurate explanation of those Principles of Geography which depend upon the Discovery of Astronomy: and a Philosophical View of Universal History; the last Article written by Robert Heron,* 2 vols, Edinburgh: Morison and Son, 1796

Heylyn, Peter. *Microcosmus or A little Description of the Great World. A Treatise Historicall, Geographicall, Politicall, Theologicall,* Oxford: Lichfield and Short, 1621

[Dr.]. *Cosmography in Four Books. Containing the Chorography and History of the whole World: and all the Principal Kingdoms, Provinces, Seas, and the Isles there of* (1652), London: Printed for Passenger, Tooke, and Sawbridge, 1682

Hübner, Johann. *A new and easy Introduction to the Study of Geography, By Way of Questions and Answer. Principally designed for the Use of Schools: In Two Parts. Containing I. An Explication of the Sphere; or of all such Terms as are any ways requisite for the right understanding of the Terraqueous Globe. II. A*

General Description of all the most remarkable Countries throughout the World, translated by J. Cowley, 2nd edn, London: Cox and Hodges, 1742

Keckermann, Bartholomew. *Systema Geographicum Duobus libris adornatum et publicè olim praelectum, Adiecta sunt in fine aliquot Problemata Nautica eiusdem Authoris*, Hanover: Haeredes Guilielmi Antonii, 1611

Systema Compendiosum: totius mathematices, hoc est Geometriae, Opticae, Astronomiae, et Geographiae. Publicio Praelectionibus Anno 1605 in Celeberrimo Gymnasio Dantiscano propositum. In fine accessit brevis commentatio nautico, ab eodem autore ibidem proposita Anno 1603, Hanover: Apud Pertrum Antonium, 1617

Systema Compendiosum Totius Mathematicae, hoc est Geometriae, Opticae, Astronomiae, et Geographiae . . . Item Methodus facilio Arithmeticae Practicae Per Gemmam Frisium, Oxford: Francis Oxclad, 1661

Kincaid, Alexander (ed.). *A New Geographical, Commercial, and Historical Grammar; and Present State of the several Empires and Kingdoms of the World . . . executed on a plan similar to that of W. Guthrie, Esq. By a Society in Edinburgh. The Astronomical part collected from the works of James Ferguson, F.R.S.*, 2 vols, Edinburgh: Alexander Kincaid, 1790

Lenglet Dufresnoy, Nicolas. *The Geography of Children: or, A Short and Easy Method of Teaching or Learning Geography . . . by way of Question and Answer. Translated from the French of Abbot Lenglet Dufresnoy, just published in Paris; with the addition of a more particular Account of Great Britain and Ireland*, London: Edward Littleton and John Hawkins, 1737

Mair, John. *A brief Survey of the Terraqueous Globe . . . republished, with great Additions . . . and Improvements by an Able Hand and Illustrated with Maps of the Ancient and Modern Worlds, Engraved by T. Kitchen* (1762), Edinburgh: John Bell and William Creech, 1775

Malte-Brun, Conrad (Malthe Conrad Bruun). *Précis de la Géographie Universelle, ou Description de toutes les parties du monde, sur un plan nouveau, d'après les grandes divisions naturelles du globe; Précédée de l'Histoire de la Géographie chez les Peuples anciens et modernes, et d'une Théorie générale de la Géographie Mathematique, Physique et Politique*, 2nd edn, 8 vols, Paris: Buisson, 1812–29

Universal Geography, or a Description of All the Parts of the World, on a new Plan, according to the great natural Divisions of the Globe; . . . with Analytical, Synoptical, and Elementary Tables, 9 vols, Edinburgh: Adam Black; London: Longman, Hurst, Rees, 1822–33

Martin, Benjamin. *Physico-Geology: or, A New System of Philosophical Geography. Containing: A new and general Description of the Terraqueous Globe . . . etc. In this Treatise the Countries are now first illustrated with a new and accurate Set of Maps . . . engraved by E. Bowen, Geographer to His late Majesty*, London: W. Owen, 1769

Middleton, Charles Theodore. *A New and Complete System of Geography. Containing a full, accurate, authentic and interesting Account and Description of Europe, Asia, Africa, and America; as consisting of Continents, Islands, Oceans . . . and divided into Empires, Kingdoms, States, and Republics. With Their Limits, Boundaries, Climate, Soil, . . . Productions, Religion, Laws, Government, Revenues, Forces . . . etc. . . . Provinces, Cities . . . Mountains, Mines, . . . Universities, etc. . . . And all that is interesting relative to the Customs, Manners, Genius, Tempers, Habits, Amusements, Ceremonies, Commerce, Arts, Sciences, Manufactures, and Language of the Inhabitants . . . Kinds of Birds,*

Beasts . . . etc. . . . Voyages and Travels . . . Also a Compendious History of every Empire, Kingdom, State, etc. A New and Easy Guide to Geography, the Use of the Globes, etc. with an Account of . . . Navigation, 2 vols, London: J. Cooke, 1778

Mitchell, Samuel Augustus. *Mitchell's Geographical Question Book; comprising Geographical Definitions, and containing Questions on all the Maps of the Mitchell's School Atlas. To which is added an Appendix, embracing valuable Tables in Mathematical and Physical Geography*, Philadelphia: Thomas, Cowperthwait and Co., 1852

Moll, Herman. *Atlas Geographus: or, a Compleat System of Geography, Ancient and Modern. Containing What is of most Use in Bleau, Varenius, Cellarius, Cluverius, Baudrand, Brietius, Sanson, etc. with the Discoveries and Improvements of the best Modern Authors to this Time. Illustrated with about 100 New Maps, done from the latest Observations by Herman Moll, Geographer*, 5 vols, London: John Nutt, 1711–17

The Compleat Geographer: or, the Chorography and Topography of all the known Parts of the Earth. To which is premis'd an Introduction to Geography, and a Natural History of the Earth and the Elements, 4th edn, London: Knapton, Knaplock, 1723

Moore, Henry. *A New and Comprehensive System of Universal Geography*, London: Whellier [n.d.: *c.* 1810]

Morse, Jedidiah. *The American Geography: or A View of the Present Situation of the United States of America . . . with a particular Description of Kentucky, The Western Territory, and Vermont* (1789), 2nd edn, London: Stockdale, 1792

Newton, John, D. D. *Cosmographia, or A View of the Terrestrial and Coelestial Globes, . . . To which is added an Introduction unto Geography*, London: Passinger, 1679

Peacock, W. *A Compendious Geographical and Historical Grammar: Exhibiting a Brief Survey of the Terraqueous Globe*, London: Printed for W. Peacock, 1795

Pemble, William. *A Briefe Introduction to Geography. Containing a Description of the Grounds, and generall Part thereof*, Oxford: Printed by John Lichfield Printer to the Famous University for Edward Forrest, 1630

Pinkerton, John. *Modern Geography. A Description of Empires, States, and Colonies, with the Ocean, Seas, and Isles; in all Parts of the World: including the most recent discoveries, and political alterations. Digested on a new Plan. The Astronomical Introduction by the Rev. S. Vince*, 2 vols, London: Cadell, Davies, Longman and Rees, 1802

Quad, M. *Geographisch Handtbuch* (Cologne: Bey Johan Buxemacher, 1600), facsimile, Amsterdam: Theatrum Orbis Terrarum, 1969

Rhodes, Henry, with Luke Meredith, John Harris, and Thomas Newborough (eds). *The Great Historical, Geographical and Poetical Dictionary; Being A Curious Miscellany of Sacred and Profane History . . . Collected from the best Historians, Chronologers, and Lexicographers*, London: Rhodes, Meredith, Harris, and Newborough, 1694

Salmon, Thomas. *A New Geographical and Historical Grammar: wherein the Geographical Part is truly Modern; and the Present State of the several Kingdoms of the World Is so Interspersed As to render the Study of Geography both Entertaining and Instructive. Containing I. A Description of the Figure and Motion of the Earth. II. Geographical Definitions and Problems, being a necessary Introduction to this Study. III. A general Definition of the Globe into Land and Water. IV.*

The Situation and Extent of the several Countries contained in each Quarter of the World; their Cities, Chief Towns, History, Present State, respective Forms of Government, Forces, Revenues, Taxes, Revolutions and Memorable Events (1749), 11th edn, London: Johnston, Strahan, etc., 1769

Sanson, d'Abbeville. *Introduction a la geographie, où font la geographie astronomique, Qui explique La correspondence du Globe Terrestre avec la Sphere. La geographie Naturelle, Qui donne Les Divisions de toutes les Parties de la Terre et de l'Eau, . . . La Geographie Historique, Qui considere la Terre. Par les Estats Souverains, Par l'Estenduë des Religions, Et par l'Estenduë des principales Langues*, Utrecht: Francois Halma, Imprimeur ordinaire de l'Université, 1692

Schneider, Karl Friedrich Robert. *Handbuch der Erdbeschreibung und Staatenkunde in ihrer Verbindung mit Natur- und Menschenkunde*, 4 vols in 3, Glogan and Leipzig, 1857

Somerville, Mary. *Physical Geography*, 2 vols, London: John Murray, 1848

Stafford, Robert [John Prideaux]. *A Geographicall and Anthologicall Description of all the Empires and Kingdomes, both of Continent and Ilands in this terrestriall Globe. Relating their Scituations, Manners, Customes, Provinces, and Governments* (1618?), London: Waterson, 1634

Turner, Richard. *A View of the Earth: Being a short but comprehensive System of Modern Geography . . . To which is added, a description of the Terrestrial Globe . . .*, London: S. Crowder & Co.; Worcester: S. Gamidge, 1762

U.K. Commissioners of National Education in Ireland. *Compendium of Geography; being an abridgment of the larger work, entitled an Epitome of Geographical Knowledge, Ancient and Modern, compiled for the Use of the Teachers and advanced classes of the National Schools in Ireland*, new edn, revised, Dublin: Her Majesty's Stationery Office, 1854

Varenius, Bernhard. *Geographia generalis, In qua affectiones generales Telluris explicantur* (1650), 2nd edn, Amsterdam: Elzevir, 1664

Bernhardi Vareni Geographia generalis, In qua affectiones generales Telluris explicantur, ed. Isaac Newton, 2nd edn, Cambridge: John Hayes, 1681

General Geography, translated by Richard Blome, in Blome, *Cosmography and Geography*, London: Blome, 1682

Geographia generalis, In qua affectiones generales Telluris explicantur. Adjecta est Appendix, praecipua Recentiorium inventa ad Geographiam spectantia continens, ed. Jacob Jurin, Cambridge: Cornelius Crownfield, at the University Press, 1712

A Compleat System of General Geography: explaining The Nature and Properties of the Earth; viz. It's Figure, Magnitude, Motions . . . The Uses and Making of Maps, Globes, and Sea-Charts. The Foundations of Dialling; the Art of Measuring Heights and Distances; the Art of Ship-Building, Navigation, and the Ways of Finding the Longitude at Sea, Translated by Dugdale, revised by Peter Shaw, 2 vols, 3rd edn, London: Stephen Austen, 1736

Watts, Isaac. *The Knowledge of the Heavens and the Earth made Easy: or, The First Principles of Astronomy and Geography Explain'd by the Use of Globes and Maps . . . Written several Years since for the Use of Learners* (1726), 6th edn, corrected, London: Longman, Buckland, Fenner and others, 1760

Zeune, August. *Gea. Versuch die Erdrinde sowohl im Land- als Seeboden mit Bezug auf Natur- und Völkerleben zu schildern* (1808), 3rd edn, Berlin: Nauck, 1830

Zimmermann, Eberhard von. *Die Erde und ihre Bewohner nach den neuesten Entdeckungen. Eine Lesebuch für Geographie, Völkerkunde, Produktenlehre*

und den Handel (Leipzig, 1810–11), 3rd [?] edn, completed by F. Rühs and
H. Lichtenstein, 7 vols, stuttgart: August Friedrich Macklot, 1816–20

2 *General bibliography*

Achinstein, Peter. *Concepts of Science. A Philosophical Analysis*, Baltimore: Johns
Hopkins Press, 1968
Ackerman, Edward A. 'Where is a research frontier?', *Annals AAG* 53, no. 4
(1963), 429–440
Agassiz, Louis. *His Life and Correspondence*, ed. Elizabeth Cary Agassiz, 2 vols,
Boston: Houghton, Mifflin, 1887
Amaldi, Ginestra. *The Nature of Matter: Physical Theory from Thales to Fermi*,
translated by Peter Astbury, Chicago: University of Chicago Press, 1966
Amaral, Daniel J. and Wisner, Ben. 'Participant-observation, phenomenology, and
the rules for judging sciences: a comment', *Antipode* 2, no. 1, (1970) 42–51
Arago, François. *Meteorological Essays. With Introduction by Baron Alexander von
Humboldt*, translated by Edward Sabine, London: Longman, 1855
Aristotle. *On the Soul, Parva Naturalia, On Breath*, translated by W. S. Hett, Loeb
Classical Library, London: William Heinemann, 1957
The Physics, translated by P. H. Wicksteed and F. M. Cornford (1929–34), Loeb
Classical Library, 2 vols, London: William Heinemann, 1960–3.
The Works of Aristotle, ed. W. D. Ross, vol. II: *Physica*, translated by R. P.
Hardie and R. K. Gaye, *De Caelo*, trans. J. L. Stocks, *De Generatione et
Corruptione*, trans. H. H. Joachim (1930); vol. V: *De Partibus Animalium*,
trans. W. Ogle (1938); reprinted, Oxford: Clarendon Press, 1947
Ayer, A. J. and Winch, Raymond. *British Empirical Philosophers: Locke, Berkeley,
Hume, Reid and J. S. Mill*, London: Routledge and Kegan Paul, 1952
Bacon, Francis. *The Advancement of Learning* (1605), ed. G. W. Kitchin, London:
J. M. Dent and Sons, 1954
Novum Organum (1620), translated by R. Ellis and J. Spedding, London:
Routledge [n.d.: *c.* 1858]
Bacon: Selections, ed. Matthew Thompson McClure, London: Charles Scribner's
Sons, 1928
Sylva sylvarum; or, A Natural History in Ten Centuries, with *New Atlantis*, ed.
W. Rawley, London: W. Lee, 1627
Bagrow, Leo, and Skelton, R. A. *History of Cartography*, translated from Leo
Bagrow's *Geschichte der Kartographie* (Berlin, 1951) by D. L. Paisley and
revised by R. A. Skelton, London: C. A. Watts, 1964
Baker, John Norman Leonard. 'The geography of Bernhard Varenius', *Transactions
of the Institute of British Geographers* 21 (1955), 51–60; reprinted in Baker, *The
History of Geography*, pp. 105–18
The History of Geography: The Collected Papers of J. N. L. Baker, Oxford: Basil
Blackwell, 1963
'Mary Somerville and geography in England', *Geographical Journal* 3 (1948),
207–22
Balmer, Heinz. *Beitrage zur Geschichte der Erkenntnis des Erdmagnetismus*, Aarau:
Sauerländer: Schweizer Ges. für Gesch. der Med. und der Naturwiss., 1956
Barnes, Barry (ed.). *Sociology of Science. Selected Readings*, Harmondsworth,
Middlesex: Penguin Books, 1972
Barrows, Harlan H. 'Geography as human ecology', *Annals AAG* 13 (1923), 1–14

Barzun, Jacques. *Science: The Glorious Entertainment*, London: Secker and Warburg, 1964

Bates, Marston. *Man in Nature*, Englewood Cliffs, New Jersey: Prentice-Hall, 1962

Bateson, Gregory. *Steps to an Ecology of Mind: Collected Essays in Anthropology, Psychiatry, Evolution and Epistemology*, London: Granada, 1973, reprinted 1978

Beck, Hanno. *Alexander von Humboldt*, 2 vols: vol. I (1959), *Von der Bildungsreise zur Forschungsreise 1769–1804*; vol. II (1961), *Vom Reisewerk zum 'Kosmos' 1804–1859*, Wiesbaden: Franz Steiner, 1959–61

(ed.). *Alexander von Humboldt, Kosmos für die Gegenwart*, Stuttgart: Brockhaus, 1978

Carl Ritter: Genius der Geographie. Zu seinem Leben und Werk, Berlin: Dietrich Reimer, 1979

Geographie: Europäische Entwicklung in Texten und Erläuterungen, Freiburg/Munich: Karl Alber, 1973

Gespräche Alexander von Humboldts, Berlin: Akademie-Verlag, 1959

Grosse Reisende: Entdecker und Erforscher unserer Welt, Munich: Georg D. W. Callwey, 1971

Berkeley, George. *A Treatise concerning the Principles of Human Knowledge* (1710), reprint of first edn, with modifications from 2nd edn (1734), Chicago: Open Court Publishing Co., 1920

De Motu (1721), translated as *Of Motion* by A. A. Luce in *The Works of George Berkeley, Bishop of Cloyne*, ed. A. A. Luce and T. E. Jessop, vol. IV, London: Nelson, 1951

Berlin, Isaiah. *The Age of Enlightenment. The 18th Century Philosophers selected, with Introduction and Interpretive Commentary*, New York: New American Library, 1956

Berry, Brian J. L. 'Approaches to regional analysis: a synthesis', *Annals AAG* 54, no. 2 (1964), 2–11

Blair, John. *The History of the Rise and Progress of Geography*, London: Cadell and Ginger, 1784

Blake, Ralph M., Ducasse, Curt J. and Madden, Edward H. *Theories of Scientific Method: The Renaissance through the Nineteenth Century*, ed. E. H. Madden, Seattle: University of Washington Press, 1960

Board, C., Chorley, R. J., Stoddart, D. and Haggett, P. (eds). *Progress in Geography: International Reviews of Current Research*, vol. I, London: Edward Arnold, 1969

Botting, Douglas. *Humboldt and the Cosmos*, London: Sphere Books, 1973

Boulding, Kenneth Ewart. *The Meaning of the Twentieth Century: The Great Transition*, London: Allen and Unwin, 1965

Bowen, James. *A History of Western Education*, 3 vols, London: Methuen, 1972–81

(ed.). *International Review of Education* 25, no. 3 (1979)

Bowen, Margarita J. 'Mind and nature: the physical geography of Alexander von Humboldt', *Scottish Geographical Magazine* 86, no. 3 (1970), 222–33

'Scientific method – after positivism', *Australian Geographical Studies* 17, no. 2 (1979), 210–16

Bowra, C. M. *The Romantic Imagination*, Oxford: Oxford University Press, 1949

Brandis, Christian August. *Handbuch der Geschichte der Griechisch-Römischen Philosophie*, 3 vols, Berlin: Reimer, 1835–66

Bronowski, J. and Mazlish, Bruce. *The Western Intellectual Tradition: from Leonardo to Hegel* (1960), Harmondsworth, Middlesex: Penguin Books, 1963

Bruhns, Karl C. (ed.). *Alexander von Humboldt. Eine wissenschaftliche Biographie*, 3 vols, Leipzig: Brockhaus, 1872

(ed.). *Life of Alexander von Humboldt*, translated by Jane and Caroline Lassell, 2 vols, London: Longmans, Green and Co., 1873

Bruno, Giordano. *On the Infinite Universe and Worlds* (1584), in D. W. Singer, *Giordano Bruno: His Life and Thought: With Annotated Translation of His Work: On the Infinite Universe and Worlds*, pp. 226–378, New York: Henry Schumann, 1950

Buchanan, Keith. 'The white north and the population explosion', *Antipode* 5, no. 3 (1973), 7–15

Buffon, George Louis Leclerc de. *Histoire naturelle, générale et particulière* (1749–1804), ed. C. S. Sonnini, 127 vols, Paris: Dufart, 1799–1808

Buffon's Natural History, Containing a theory of the earth, a general history of man, of the brute creation, and of vegetables, minerals etc. [Barr's Buffon], English translation, ed. J. S. Barr, 16 vols, London: Barr, 1797–1808

Bunge, William. 'Ethics and logic in geography', in R. J. Chorley (ed.), *Directions in Geography*, London: Methuen, 1973

'Theoretical Geography', *Lund Studies in Geography* Series C, no. 1, Lund: Gleerup, 1962

Burckhardt, Jacob. *The Civilization of the Renaissance in Italy* (1860), translated from the 15th German edn by S. G. C. Middlemore, 2 vols, New York: Harper, 1958

Burnet, Thomas. *The Theory of the Earth: Containing an Account of the Original of the Earth and of all the General Changes which it hath already undergone, or is to undergo Till the consummation of all Things. The First 2 Books, Concerning the Deluge and Concerning Paradise*, London: Kettilby, 1684

The Theory of the Earth, 2nd edn, London: Kettilby, 1691

A Review of the Theory of the Earth and of its Proofs; especially in reference to scripture, London: Kettilby, 1690

The Theory of the Earth: the two Last Books concerning the Burning of the World and concerning the New Heavens and New Earth, London: Kettilby, 1690

Burton, Ian. 'The quantitative revolution and theoretical geography', *The Canadian Geographer* 7, no. 4 (1963), 151–62

Büsching, Anton Friedrich. *Magazin für die neue Historie und Geographie*, 4 vols in 2: vol. I, Hamburg: F. C. Ritter, 1767; vol. II, Hamburg: J. N. C. Buchenröders, 1769

Buttimer, Sister Annette. *Values in Geography*, Association of American Geographers Resource Paper no. 24, Washington, D.C.: Commission on College Geography, 1974

Büttner, Manfred. *Die Geographia Generalis vor Varenius. Geographisches Weltbild und Providentialehre*, Wiesbaden: Franz Steiner, 1973

Chalmers, A. F. *What is this Thing called Science?*, St Lucia: University of Queensland Press, 1976

Chardin, Pierre Teilhard de. *The Phenomenon of Man* (1955), introduction by Julian Huxley, translated by Bernard Wall, London: Collins, 1967

Chorley, Richard J. (ed.). *Directions in Geography*, London: Methuen, 1973

Chorley, Richard J. and Haggett, Peter (eds.). *Frontiers in Geographical Teaching* (1965), 2nd edn, London: Methuen, 1970
Models in Geography, London: Methuen, 1967
Clark, George N. *The Seventeenth Century* (1929), Oxford: Clarendon Press, 1963
Clark, Robert T., Jr. *Herder: His Life and Thought*, Berkeley: University of California, 1955
Claval, Paul. *La Pensée géographique. Introduction à son histoire*, Paris: Société d'édition d'enseignement supérieur, 1972
Cohen, Morris R. *Reason and Nature: An Essay on the Meaning of Scientific Method* (1931), 2nd edn, Glencoe, Illinois: Free Press, 1953
Cohen, Morris R. and Drabkin, I. E. (eds.). *A Source Book in Greek Science*, New York: McGraw-Hill, 1948
Comenius, Johannes Amos. *Orbis Sensualium Pictus* (1659), facsimile, 3rd London edn (1672), ed. James Bowen, Sydney: Sydney University Press, 1967
Comte, Auguste. *A General View of Positivism* (Paris, 1848), translated by J. H. Bridges, Stanford: Academic Reprints [n.d.]
Conant, James B. *Science and Common Sense*, Oxford: Oxford University Press, 1951
Condorcet, A. N. de. *Sketch for a Historical Picture of the Progress of the Human Mind* (1795), translated by Barraclough, New York: Noonday Press, 1955
Copernicus, Nicolas. *De revolutionibus orbium coelestium, libri VI*, Nuremberg: apud Ioh. Petreium, 1543
Crocker, Lester G. (ed.). *The Age of Enlightenment*, London: Macmillan, 1969
Crombie, A. C. *Augustine to Galileo* (1952), 2 vols, revised edn, Harmondsworth, Middlesex: Penguin Books, 1969
'The significance of medieval discussions of scientific method for the Scientific Revolution', in Marshall Claggett (ed.), *Critical Problems in the History of Science*, Madison: University of Wisconsin Press, 1962
Crone, G. R. *Background to Geography*, London: Museum Press, 1964
Modern Geographers. An Outline of Progress in Geography since 1800 A.D., London: Royal Geographical Society, 1951
Crosland, Maurice. *The Society of Arcueil. A View of French Science at the Time of Napoleon I*, London: Heinemann, 1967
Cudworth, Ralph. *The true Intellectual System of the Universe: the first part; wherein All the Reason and Philosophy of Atheism is Confuted, and Its Impossibility Demonstrated* (1678), 2nd edn, London: J. Walthoe, and others, 1743
Cuvier, Georges Leopold de. *Histoire des progrès des sciences naturelles, depuis 1789 jusqu'a ce jour*, 5 vols, Paris: Librairie encyclopédique de Roret, 1826–36
Dampier-Whetham, William Cecil Dampier. *A History of Science and its Relations with Philosophy and Religion*, Cambridge: Cambridge University Press, 1929
Darby, H. C. 'The problem of geographical description', *Trans. Inst. Br. Geogr.* 30 (1962), 1–11
Darwin, Charles. *The Origin of Species by Means of Natural Selection or The Preservation of favoured Races in the Struggle for Life* (1859), 6th edn (1872), reprinted, New York: Mentor, 1958
Darwin, Francis (ed.). *Life and Letters of Charles Darwin, including an Autobiographical Chapter* (London, 1887), 2 vols, New York: Appleton, 1898
Davies, Wayne K. D. (ed.). *The Conceptual Revolution in Geography*, London: University of London Press, 1972
De Jong, G. *Chorological Differentiation as the Fundamental Principle of Geo-*

graphy: An Inquiry into the Chorological Conception of Geography, Groningen: Wolters, 1962

Delvaille, Jules. *Essai sur l'histoire de l'idée de progrès jusqu'à la fin de XVIIIe siècle* (1910), reprinted, Geneva: Slatkine Reprints, 1969

de Martonne, Emmanuel. *Traité de géographie physique* (1909), vol. I: *Notions générales, Climat, Hydrographie*, Paris: Librairie Armand Colin, 1957

Derham, William. *Physico-Theology: or, a Demonstration of the Being and Attributes of God, from his Works of Creation* (1713), 3rd edn, corrected, London: W. Innys, 1714

Descartes, René. *Oeuvres de Descartes*, ed. Charles Adam and Paul Tannery, Paris: Léopold Cerf, 1909

Descartes: Philosophical Letters, translated and edited by Anthony Kenny, Oxford: Clarendon Press, 1970

The Philosophical Works of Descartes, translated by E. S. Haldane and G. R. T. Ross (1911), 2 vols, Cambridge: Cambridge University Press, 1967

Philosophical Writings, a selection translated and edited by E. Anscombe and P. T. Geach, Introduction by Alexandre Koyré (1954), London: Nelson, 1966

Dickinson, Robert E. *The Makers of Modern Geography*, London: Routledge and Kegan Paul, 1969

Dolan, E. F. *Green Universe: The Story of Alexander von Humboldt*, New York: Dodd, Mead, 1959

Döring, Lothar. *Wesen und Aufgaben der Geographie bei Alexander von Humboldt* (Frankfurter Geographische Hefte, 1931. Heft 1), Frankfurt-on-Main: Möll, 1931

Downes, Alan. 'The bibliographic dinosaurs of Georgian geography (1714–1830)', *Geographical Journal* 137, part 3 (1971), 379–87

East, W. G. 'An eighteenth century geographer, William Guthrie', *Scottish Geographical Magazine* 72, no. 1 (1956), 32–7

Ehrlich, Paul R. *The Population Bomb* (1968), London: Pan Books, 1971

Eliot Hurst, M. 'Establishment geography: or how to be irrelevant in three easy lessons', *Antipode* 5, no. 2 (1973), 40–59

Emery, Fred E. and Trist, Eric L. *Towards a Social Ecology: Contextual Appreciation of the Future in the Present*, London and New York: Plenum Press, 1973

English, Paul Ward, and Mayfield, Robert C. (eds). *Man, Space, and Environment*, New York: Oxford University Press, 1972

Fairchild, Herman LeRoy. *The Geological Society of America 1888–1930: A Chapter in Earth Science History*, New York: Geological Society, 1932

Falconer, William. *Remarks on the Influence of Climate, Situation, Nature of Country, Population, Nature of Food, and Way of Life, on the Disposition and Temper, Manners and Behaviour, Intellects, Laws and Customs, Form of Government, and Religion of Mankind*, London: C. Dilly, 1781

Feibleman, James K. *Foundations of Empiricism*, The Hague: Martinus Nijhoff, 1962

Feynman, Richard. *The Character of Physical Law*, London: B.B.C., 1965

Fischer, Kuno. *Francis Bacon of Verulam. Realistic Philosophy and its Age* (1856), translated by John Oxenford (from: *Franz Baco von Verulam und des Zeitalter der Realphilosophie*, 1856), London: Longman, Brown, Green, Longmans and Roberts, 1857

Forster, George. *Ansichten vom Niederrhein, von Brabant, Flandern, Holland, England und Frankreich im April, Mai und Junius 1790* (Berlin, 1791), ed. Gerhard Steiner, vol. IX (1958) in *Georg Forsters Werke*, 14 vols, Berlin: Akademie-Verlag, 1958–78

Reise um die Welt während den Jahren 1772 bis 1775 (Berlin, 1784), ed. Gerhard Steiner, vol. II (1965) in *Georg Forsters Werke*, 14 vols, Berlin: Akademie-Verlag, 1958–78

A Voyage Round the World in His Britannic Majesty's Sloop, Resolution, commanded by Capt. James Cook, during the Years 1772, 3, 4 and 5 (London, 1777), ed. Robert L. Kahn, vol. I (1968) in *Georg Forsters Werke*, Berlin: Akademie-Verlag, 1958–78

Freeman, T. W. *A Hundred Years of Geography* (1961), revised edn, London: Duckworth, 1971

Fuson, Robert H. *A Geography of Geography: Origins and Development of the Discipline*, Dubuque, Iowa: Brown, 1969

Galilei, Galileo. *Dialogue Concerning the Two Chief World Systems – Ptolemaic and Copernican* (Florence, 1632), translated by Stillman Drake, foreword by Albert Einstein, Berkeley and Los Angeles: University of California Press, 1953

Gallois, L. *Les géographes Allemands de la Renaissance* (Paris, 1890), Amsterdam: Meridian, 1963

Galois, Bob. 'Ideology and the idea of nature: the case of Peter Kropotkin', *Antipode* 8, no. 3 (1976), 1–16

Gendron, Val. *The Dragon Tree: A Life of Alexander von Humboldt*, New York: Longmans, Green and Co., 1961

Gilbert, Edmund W. *British Pioneers of Geography*, Newton Abbot, Devon: David and Charles, 1972

'Geographie is better than Divinity', *Geographical Journal* 128, no. 4 (1962), 494–7

'The idea of the region', *Geography* 45, part 3, no. 208 (1960), 157–75

Gilbert, William. *On the Magnet, Magnetick Bodies also, and on the great Magnet the Earth, a new Physiology, demonstrated by many Arguments and Experiments* (1600), translated into English by Silvanus P. Thompson *et al.* (London, 1900), facsimile, New York: Basic Books, 1958

Gillispie, Charles Coulton. *The Edge of Objectivity. An Essay in the History of Scientific Ideas*, Princeton, New Jersey: Princeton University Press, 1960

Glacken, Clarence J. *Traces on the Rhodian Shore: Nature and Culture in Western Thought from Ancient Times to the End of the Eighteenth Century*, Berkeley: University of California Press, 1967

Gode- von Aesch, Alexander. *Natural Science in German Romanticism* (1941), reprinted, New York: AMS Press, 1966

Goethe, Johann Wolfgang von. *Letters from Goethe*, translated by M. von Herzfeld and C. Melvil Sym, Edinburgh: Edinburgh University Press, 1957

Gray, Ronald. *Goethe: A Critical Introduction*, Cambridge: Cambridge University Press, 1967

Greene, John C. *The Death of Adam: Evolution and Its Impact on Western Thought* (1959), New York: Mentor, 1961

Gregory, Derek. *Ideology, Science and Human Geography*, London: Hutchinson, 1978

Guelke, Leonard. 'An idealist alternative in human geography', *Annals AAG* 64, no. 2 (1974), 193–202

Haggett, Peter. *Geography: A Modern Synthesis*, New York: Harper and Row, 1972; 3rd edn, 1979

Hall, A. Rupert. *From Galileo to Newton 1630–1720*, vol. III in A. Rupert Hall (ed.), *The Rise of Modern Science*, London: Collins, 1963

Hamlyn, D. W. 'Empiricism', in Paul Edwards (ed.), *The Encyclopedia of Philosophy*, vol. I, pp. 499–505, New York: Macmillan, 1967

Harris, C. 'Theory and synthesis in historical geography', *Canadian Geographer* 15, no. 3 (1971), 157–72

Harris, Errol E. *Nature, Mind and Modern Science*, London: George Allen and Unwin, 1954

Hartshorne, Richard. 'The concept of geography as a science of space, from Kant and Humboldt to Hettner', *Annals AAG* 48, no. 2 (1958), 97–108

'"Exceptionalism in geography" re-examined', *Annals AAG* 45, no. 3 (1955), 205–44

'The nature of geography', *Annals AAG* 29, nos 3, 4 (1939), 173–658

Perspective on the Nature of Geography (1959), reprinted, London: John Murray, 1964

Harvey, David. *Explanation in Geography*, London: Edward Arnold, 1969

'Revolutionary and counter revolutionary theory in geography and the problem of ghetto formation', *Antipode* 4, no. 2 (1972), 1–13

Hegel, Georg Wilhelm Friedrich. *Hegel's Lectures on the History of Philosophy*, translated by E. S. Haldane (1892–6), reprinted, 3 vols, London: Routledge and Kegan Paul, 1955

Hegel's Philosophy of Nature: Being Part Two of the Encyclopaedia of the Philosophical Sciences (1830), translated by A. V. Miller, Oxford: Clarendon Press, 1970

Hempel, Carl G. *Philosophy of Natural Science*, Englewood Cliffs, New Jersey: Prentice-Hall, 1966

Herbart, Johann Friedrich. *Outlines of Educational Doctrine* (1835), translated by A. F. Lange, New York: Macmillan, 1901

Herder, Johann Gottfried von. *Outlines of a Philosophy of the History of Man* (1784), translated by T. Churchill (1800), facsimile, New York: Bergman [n.d.]

Herschel, John Frederick William. *Essays from the Edinburgh and Quarterly Reviews, with Addresses and other Pieces*, London: Longman, Brown, Green, Longmans, and Roberts, 1857

Familiar Lectures on Scientific Subjects, London: Strahan, 1867

A Preliminary Discourse on the Study of Natural Philosophy, London: Longman, Rees, Orme, Brown, and Green, 1831

Hipparchus. *The Geographical Fragments of Hipparchus*, ed. D. R. Dicks, London: University of London, Athlone Press, 1960

Hoare, Michael E. 'Johann Reinhold Forster: the neglected "Philosopher" of Cook's second voyage (1772–1775)', *Journal of Pacific History* 2, (1967), 215–24

Hobbes, Thomas. *Leviathan, or the Matter, Forme and Power of a Commonwealth Ecclesiasticall and Civill* (1651), ed. Michael Oakeshott, Oxford: Basil Blackwell, 1955

Hobbes: Selections, ed. F. J. E. Woodbridge. New York: Charles Scribner's Sons, 1930

Hobson, E. W. *The Domain of Natural Science*, Cambridge: Cambridge University Press, 1923

Houston, James M. *A Social Geography of Europe*, London: Duckworth, 1963

Humboldt, Alexander von. *Ansichten der Natur mit wissenschaftlichen Erläuterungen* (1808), 3rd edn, Stuttgart and Tübingen: Cotta, 1849

Asie Centrale. Recherches sur les chaînes de montagnes et la climatologie comparée, 3 vols, Paris: Gide, 1843

(Humboldt, Alexander von, continued)

Aspects of Nature, in different Lands and different Climates; with Scientific Elucidations, translated by Mrs Sabine, 2 vols, London: Longman, Brown, Green and Longmans, and John Murray, 1850

Atlas géographique et physique du royaume de la Nouvelle-Espagne (1812), reprint, Stuttgart: Brockhaus, 1969

Briefe an Cotta, MS, *Handschriftenabteilung des Deutschen Literaturarchivs/Schiller-Nationalmuseum in Marbach am Neckar*, Cotta'sche Handschriftensammlung, Paris 24.1.1805 bis Berlin 11.4.1859

Briefe an Gauss, MS, *Handschriftenabteilung der Universitätsbibliothek zu Göttingen*, MS Gauss: Briefe A, 37 Briefe (1807–54)

Briefe, MS, *Collection, custody Dr Heinz Balmer, Konolfingen, Switzerland*, 15 letters (*c.* 1820–55)

Briefwechsel Alexander von Humboldt's mit Heinrich Berghaus aus den Jahren 1825 bis 1858, 3 vols, Leipzig: Hermann Costenoble, 1863

Briefwechsel und Gespräche Alexander von Humboldt's mit einem jungen Freunde. Aus den Jahren 1848 bis 1856, Berlin: Franz Duncker, 1861

Correspondance d'Alexandre de Humboldt avec François Arago (1809–1853), ed. E.-T. Hamy, Paris: Librairie Orientale et Américaine, 1907

Cosmos: A Sketch of a Physical Description of the Universe, translated by E. C. Otté (1849–58), new edn, with tr. by B. H. Paul (vol. IV) and W. S. Dallas (vol. V), 5 vols, London: George Bell and Sons, 1882–4

Cosmos: Sketch of a Physical Description of the Universe, translated by Mrs Sabine, 3 vols, 8th edn, London: Longman, Brown, Green and Longmans, and John Murray, 1850–52

El Diario Inedito del Humboldt, Bogotá, Colombia: Academia Colombiana de Ciencias Exactas, Fisicas y Naturales, 13, no. 51, 1969

(ed.). *Zur Erinnerung an die Reise des Prinzen Waldemar von Preussen nach Indien in den Jahren 1844–1846*, 2 vols, Berlin: Gedruckt in der Deckerschen Geheimen Ober-Hofbuchdruckerei, 1853

Essai Politique sur le Royaume de la Nouvelle-Espagne; avec un Atlas géographique et physique, 2 vols, Paris: Schoell, 1811

Examen critique de l'histoire de la géographie de Nouveau Continent et des progrès de l'astronomie nautique aux XVe et XVIe siècles (Paris: Gide, 1814–34), facsimile, Amsterdam: Theatrum Orbis Terrarum; New York: De Capo, 1971

Expériences sur le Galvanisme, et en général sur l'imitation des Fibres musculaires et nerveuses, translated from German, with additions by J. F. N. Jadelot, Paris: Fuchs, 1799

Florae Fribergensis Specimen plantas cryptogamicas praesertim subterraneas exhibens . . . Accedunt Aphorismi ex doctrina physiologiae chemicae plantarum, Berlin: Rottmann, 1793

The Fluctuations of Gold, translated by William Maude, New York: Cambridge Encyclopedia Co., 1900

A Geognostical Essay on the Superposition of Rocks in both Hemispheres (1823), translated from the original French, London: Longman, Hurst, Rees, Orme, Brown and Green, 1823

Gesammelte Werke, 12 vols in 6, Stuttgart: J. G. Cotta [n.d.: *c.* 1889]

The Island of Cuba, translated from the Spanish by J. S. Thrasher, London: Sampson Low, Son and Co., 1856

Die Jugendbriefe Alexander von Humboldts 1787–1799, ed. Ilse Jahn and Fritz G. Lange, Berlin: Akademie-Verlag, 1973

Kosmos. Entwurf einer physischen Weltbeschreibung, 5 vols, Stuttgart: Cotta, 1845–62

Kosmos. Entwurf einer physischen Weltbeschreibung. Mit einer biographischen Einleitung von Bernhard von Cotta, 4 vols, Stuttgart: Cotta, 1877

Alexander von Humboldt, Kosmos für die Gegenwart, ed. Hanno Beck, Stuttgart: Brockhaus, 1978

Letters of Alexander von Humboldt to Varnhagen von Ense, from 1827 to 1858, with Extracts from Varnhagen's Diaries and Letters of Varnhagen and others to Humboldt, translated by Friedrich Kapp, New York: Rudd and Carleton, 1860

Letters of Alexander von Humboldt, written between the years 1827 and 1858, to Varnhagen von Ense, Together with Extracts from Varnhagen's Diaries and Letters from Varnhagen and others to Humboldt, authorized translation from the German, London: Trübner and Co., 1860

Lettres américaines d'Alexandre de Humboldt 1798–1807, ed. E.-T. Hamy, Paris: Librairie Orientale et Américaine, 1905

'Mes confessions, à lire et à me renvoyer un jour' (1806), pp. 197–212, in A. Leitzmann, *Georg und Therese Forster und die Brüder Humboldt*, Bonn: Röhrscheid, 1936

Mineralogische Beobachtungen über einige Basalte am Rhein. Mit vorangeschickten, zerstreuten Bemerkungen über den Basalt der ältern und neuern Schriftsteller, Braunschweig: in der Schulbuchhandlung, 1790

Personal Narrative of Travels to the Equinoctial Regions of the New Continent, 1799–1804, by Alexander de Humboldt and Aimé Bonpland, translated by H. M. Williams, 7 vols in 6, London: Longmans, Hurst, Rees, Orme and Brown, 1814–29

Personal Narrative of Travels to the Equinoctial Regions of America during the years 1799–1804, by Alexander von Humboldt and A. Bonpland, translated by Thomasina Ross, 3 vols, London: George Routledge and Sons, 1851

Political Essay on the Kingdom of New Spain. Containing: Researches relative to the Geography of Mexico . . . With Physical Sections and Maps founded on Astronomical Observations, and Trigonometrical and Barometrical Measurements, translated by John Black, 4 vols (London, 1811), reprinted, New York: AMS Press, 1966

Relation historique du Voyage aux Régions équinoxiales du Nouveau Continent Fait en 1799, 1800, 1801, 1802, 1803 et 1804 par Al. de Humboldt et A. Bonpland (Paris, 1814–25), facsimile, 3 vols, Amsterdam: Theatrum Orbis Terrarum; New York: Da Capo, 1973

Relation historique du Voyage aux Régions équinoxiales du Nouveau Continent Fait en 1799, 1800, 1801, 1802, 1803 et 1804 par Al. de Humboldt et A. Bonpland (Paris, 1814–25), ed. Hanno Beck, 2 vols, Stuttgart: Brockhaus, 1970

Researches Concerning the Institutions and Monuments of the Ancient Inhabitants of America, with Descriptions of Some of the Most Striking Scenes in the Cordilleras, translated from the French by Helen Maria Williams, London: Longman, Hurst, Rees, Orme and Brown, Murray, 1814

Versuche über die gereizte Muskel- und Nervenfaser nebst Vermuthungen über den chemischen Process des Lebens in der Thier-und Pflanzenwelt, 2 vols, Posen: Decker, and Berlin: Rottmann, 1797

328   Bibliography

(Humboldt, Alexander von, continued.)

Views of Nature: or Contemplations on the sublime Phenomena of Creation; with Scientific Illustrations, translated by Otté and Bohn, London: Bohn, 1850

Voyage aux régions équinoxiales du Nouveau Continent, fait en 1799, 1800, 1801, 1802, 1803 et 1804, par Al. de Humboldt et A. Bonpland (Paris 1805–34), facsimile, 30 vols, Amsterdam: Theatrum Orbis Terrarum; New York: Da Capo, 1971–3

Vues des Cordillères et monumens des peuples indigènes de l'Amérique, Paris: Schoell, 1810

Humboldt, Wilhelm von. *The Limits of State Action* (1852), translated by J. W. Burrow, Cambridge: Cambridge University Press, 1969

Hume, David. *Hume: Theory of Knowledge: containing: An Enquiry concerning Human Understanding, the Abstract, and selected passages from Book I, A Treatise of Human Nature*, ed. D. C. Yalden-Thomson, Edinburgh: Nelson, 1951

A Treatise of Human Nature: Being An Attempt to introduce the experimental Method of Reasoning into Moral Subjects (London, 1739), reprinted, ed., L. A. Selby-Bigge, Oxford: Clarendon Press, 1949

Huxley, T. H. *Physiography: An Introduction to the Study of Nature* (1877), London: Macmillan and Co., 1882

James, Preston E. *All Possible Worlds. A History of Geographical Ideas*, New York: Odyssey, 1972

'On the origin and persistence of error in geography', *Annals AAG* 57 (1967), 1–24

Jeans, James. *The Mysterious Universe* (1930), 2nd edn with corrections, Cambridge: Cambridge University Press, 1933

Kant, Immanuel. *Kant's Cosmogony, as in his Essay on the Retardation of the Rotation of the Earth, and his Natural History and Theory of the Heavens*, edited and translated by W. Hastie, Glasgow: James MacLehose and Sons, Publishers to the University, 1900

The Critique of Judgement (Berlin, 1790), translated by J. C. Meredith, Oxford: Clarendon Press, 1952

Kant's Critique of Practical Reason and other Works on the Theory of Ethics, translated by T. K. Abbott (1873), 6th edn (1909) reprinted, London: Longman, 1963

Critique of Pure Reason (1st edn (A) 1781, 2nd edn (B) 1787), translated by Norman Kemp Smith, London: Macmillan and Co., 1952

Kant's Inaugural Dissertation and Early Writings on Space, translated by John Handyside, Chicago: Open Court, 1929

Metaphysical Foundations of Natural Science (1786), translated by James Ellington, Indianapolis and New York: Bobbs-Merrill, 1970

Metaphysische Anfangsgründe der Naturwissenschaft, Riga: Johann Friedrich Hartnoch, 1786

Observations on the Feeling of the Beautiful and Sublime (1764), translated by John T. Goldthwait, Berkeley: University of California Press, 1960

On History, ed. L. W. Beck, Indianapolis: Bobbs-Merrill, 1963

Kant: Philosophical Correspondence 1759–99, edited and translated by Arnulf Zweig, Chicago: University of Chicago Press, 1967

Prolegomena, and Metaphysical Foundations of Natural Science, translated by E. B. Bax, London: G. Bell, Bohn's Philosophical Library, 1883

Prolegomena to any future Metaphysics that will be able to present itself as a Science

(1783), translated by P. G. Lucas, Manchester: Manchester University Press, 1953

Immanuel Kant's Sämmtliche Werke, ed. Rosenkranz and Schubert, Leipzig: Voss, 1839

Selected Pre-Critical Writings and Correspondence with Beck, translated by G. B. Kerferd and D. E. Walford, Manchester: Manchester University Press, 1968

Kellner, L. *Alexander von Humboldt*, Oxford: Oxford University Press, 1963

Kepler, Johannes. *Gesammelte Werke*, ed. Max Caspar, Munich: C. H. Beck, 1937

Kirk, William. 'Historical geography and the concept of the behavioural environment', *Indian Geographical Journal*, Silver Jubilee Edition (1952), 152–60.

'Problems of geography', *Geography* 48 (1963), 357–71.

Klencke, P. F. H. and G. Schlesier. *Lives of the Brothers Humboldt, Alexander and William*, translated by Juliette Bauer, London: Ingram, Cooke, 1852

Koestler, Arthur. *The Sleepwalkers: A History of Man's Changing Vision of the Universe* (1959), Harmondsworth, Middlesex: Penguin Books, 1968

Kolakowski, Leszek. *Positivist Philosophy: From Hume to the Vienna Circle* (1966), translated by Norbert Guterman, revised edn, Harmondsworth, Middlesex: Penguin Books, 1972

Kriesel, Karl Marcus. 'Montesquieu: possibilistic political geographer', *Annals AAG* 58, no. 3 (1958), 557–74

Kropotkin, Peter. *Ethics: Origin and Development* (1924), New York: B. Blom, 1968

Mutual Aid: A Factor in Evolution (1902), London: Penguin Press, 1972

'On the teaching of Physiography,' *The Geographical Journal* (London: Royal Geographical Society) 2, July–December (1893), 350–9

Kropotkin's Revolutionary Pamphlets (1927), New York: Dover, 1970

'What Geography ought to be', *Nineteenth Century*, December (1885), 940–56

Krüger, Gerhard, *Alexander von Humboldt. Ein grosser deutscher Naturforscher und Humanist. 14.9.1769–6.5.1859*, Berlin: Deutscher Kulturbund, 1959

Kuhn, Thomas S. *The Structure of Scientific Revolutions*, Chicago: University of Chicago Press, 1962; 2nd edn, 1970

Lakatos, I. and Musgrave, Alan (eds.). *Criticism and the Growth of Knowledge*, Cambridge: Cambridge University Press, 1972

Laplace, Pierre Simon de. *Exposition du Système du Monde* (1796), vol. 6 in *Oeuvres complètes*, 6th edn, Paris: Bachelier, 1835

Laubenberger, Franz. 'Ringmann oder Waldseemüller? Eine kritische Untersuchung über den Urheber des Namens Amerika', *Erdkunde* 13, Heft 3 (1959), 163–79

Lavoisier, Antoine. *Traité élémentaire de chimie. Opuscules physiques et chimiques* (1789), vol. I in *Oeuvres de Lavoisier*, Paris: Imprimerie Impériale, 1864

Leibniz, Gottfried Wilhelm von. *Discourse on Metaphysics, Correspondence with Arnauld, and Monadology*, translated by G. R. Montgomery (1902), Chicago: Open Court, 1916

New Essays concerning Human Understanding, translated by A. G. Langley (1896), 3rd edn, La Salle, Illinois: Open Court, 1949

Leibniz: Selections, ed. Philip P. Wiener, New York: Charles Scribner's Sons, 1951

Leigh, Roger (ed.). *Contemporary Geography: Western Viewpoints*, B.C. Geographical Series no. 12, Occasional Papers in Geography, Vancouver: Tantalus Research Ltd, 1971

Leighly, John, 'Methodologic controversy in nineteenth century German geography', *Annals AAG* 28, no. 4 (1938), 238–58

Leitzmann, Albert, *Georg und Therese Forster und die Brüder Humboldt; Urkunden und Umrisse*, Bonn: Röhrscheid, 1936

Liddell, H. G. and Scott. *An Intermediate Greek-English Lexicon* (1889), reprinted, Oxford: Clarendon Press, 1961

Lind, Paul von. *Immanuel Kant und Alexander von Humboldt. Eine Rechtfertigung Kants und eine historische Richtigstellung*, Erlangen: Junge, 1897

Locke, John. *The Educational Writings of John Locke*, ed. J. L. Axtell, Cambridge: Cambridge University Press, 1968

   An Essay concerning Human Understanding (1690), ed. Alexander Campbell Fraser, 2 vols, Oxford: Clarendon Press, 1894

   Essays on the Law of Nature, ed. W. von Leyden, Oxford: Clarendon Press, 1954

   Some Thoughts concerning Education (1693), in J. L. Axtell (ed.), *The Educational Writings of John Locke*, pp. 109–325, Cambridge: Cambridge University Press, 1968

   The Works of John Locke, 10 vols, London: Tegg, Sharp, 1823

Lovejoy, Arthur O. *The Great Chain of Being. A Study of the History of an Idea* (1936), New York: Harper, 1960

Lowenthal, David (ed.). 'Environmental perception and behavior', *Research Paper* no. 109, University of Chicago, Department of Geography, 1967

   'Geography, experience, and imagination: towards a geographical epistemology', *Annals AAG* 51, no. 3 (1961), 241–60

   'Past time, present place: landscape and memory', *Geographical Review* 65, no. 1 (1975), 1–36

Luc, Jean André de. *An Elementary Treatise on Geology: determining fundamental Points in that Science, and containing an Examination of some Modern Geological Systems, and particularly of the Huttonian Theory of the Earth*, translated by Henry de la Fite, London: F. C. and J. Rivington, 1809

Lyell, Charles. *Principles of Geology, or the Modern Changes of the Earth and its Inhabitants considered as illustrative of Geology* (1830–4), 10th edn, revised, 2 vols, London: John Murray, 1867

Macgillivray, W. *Life, Travels and Researches of Baron Humboldt*, London: Nelson, 1859

McKie, Douglas. *Antoine Lavoisier: Scientist, Economist, Social Reformer*, London: Constable, 1952

Mackinder, Halford J. 'Global geography', *Geography* 28 (1943), 69–71

McLellan, David. *Marx before Marxism* (1970), revised edn, Harmondsworth, Middlesex: Penguin Books, 1971

Macpherson, Anne M. 'The Human Geography of Alexander von Humboldt', unpublished Ph.D. dissertation, University of California, Berkeley, 1972

Malthus, Thomas Robert. *An Essay on Population* (1798), London: Dent [n.d.]

Mannheim, Karl. *Ideology and Utopia. An Introduction to the Sociology of Knowledge* (1929), translated by L. Wirth and E. Shils, New York: Harcourt, Brace, 1936

Markham, Clements R. *Major James Rennell and the Rise of Modern English Geography*, London, Paris and Melbourne: Cassell, 1895

Marsh, George Perkins. *Man and Nature* (1864), ed. David Lowenthal, Cambridge, Mass.: Belknap Press of Harvard University Press, 1965

Marx, Karl. *Grundrisse. Foundations of the Critique of Political Economy (Rough*

Draft) (1939), translated by Martin Nicolaus, Harmondsworth, Middlesex: Penguin Books, 1973

Writings of the Young Marx on Philosophy and Society, translated by L. D. Easton and K. H. Guddat, New York: Anchor Books, Doubleday, 1967

Marx, Karl and Engels, Friedrich. *Briefwechsel*, vol. II, 1854–60 Berlin: Dietz Verlag, 1949

Collected Works, London: Lawrence and Wishart, 1975

Correspondence 1846–95: A Selection, translated by D. Torr (1934), 2nd edn, London: Lawrence and Wishart, 1936

Gesamtausgabe, Berlin: Dietz Verlag, 1976

The Marx–Engels Reader, ed. Robert C. Tucker, New York: Norton, 1972

Mason, Stephen F. *A History of the Sciences* (1953), revised edn, New York: Collier, 1962

Maury, Matthew Fontaine. *The Physical Geography of the Sea, and its Meteorology* (1855), ed. John Leighly, Cambridge, Mass.: Belknap Press of Harvard University Press, 1963

May, Joseph A. *Kant's Concept of Geography and its Relation to Recent Geographical Thought*, Toronto: University of Toronto Press, 1970

Mead, W. R. 'Luminaries of the North. A reappraisal of the achievements and influence of six Scandinavian geographers', *Trans. Inst. Br. Geogr.* 57, (1972), 1–13

Medawar, Peter Brian. *Induction and Intuition in Scientific Thought*, London: Methuen, 1969

Mercer, D. C. and Powell, J. M. *Phenomenology and Related Non-positivistic Viewpoints in the Social Sciences*, Monash Publications in Geography, no. 1, Melbourne: Monash University Department of Geography, 1972

Meyer-Abich, Adolf, and others. *Alexander von Humboldt 1769/1969*, Bonn/Bad Godesberg: Inter Nationes, 1969

Mill, John Stuart, *John Stuart Mill's Philosophy of Scientific Method*, ed. Ernest Nagel, New York: Hafner, 1950

Minguet, Charles. *Alexandre de Humboldt: historien et géographe de l'Amérique espagnole 1799–1804*, Paris: François Maspero, 1969

Minshull, Roger. *The Changing Nature of Geography*, London: Hutchinson University Library, 1970

Moore, Ruth. *The Earth We Live on: The Story of Geological Discovery*, New York: Alfred Knopf, 1970

Münster, Sebastian. *Cosmographiae universalis* (1544), Basle: Heinrichum Petri, 1550

Nelken, Halina, *Alexander von Humboldt: His Portraits and Their Artists: A Documentary Iconography*, Berlin: Reimer, 1980

Newton, Isaac. *Newton's Philosophy of Nature. Selections from his Writings*, ed. H. S. Thayer, Introduction by John Herman Randall Jr, New York: Hafner, 1953

Philosophiae naturalis principia mathematica, London: Streater, 1687

Nidditch, P. H. (ed.). *The Philosophy of Science*, Oxford: Oxford University Press, 1968

Nordenstam, Tore. *Empiricism and the Analytic–Synthetic Distinction*, Oslo: Universitetsforlaget, 1972

Nørreklit, Lennart. *Concepts: Their Nature and Significance for Metaphysics and Epistemology*, Odense: Odense University Press, 1973

Obler, Paul C. and Estrin, Herman A. (eds). *The New Scientist: Essays on the Methods and Values of Modern Science*, New York: Anchor Books, Doubleday, 1962

Odishaw, Hugh (ed.). *Earth in Space*, U.S. Information Service, Voice of America Forum Lectures, 1968

Olsson, Gunnar. 'Some notes on geography and social engineering', *Antipode* 4, no. 1 (1972), 1–22 (special issue: 'Geography, Epistemology, and Social Engineering')

Onions, C. T. (ed.). *The Oxford Dictionary of English Etymology*, Oxford: Clarendon Press, 1966

Paley, William. *Natural Theology*, vol. IV in *Works*, 6 vols, London: Rivington, etc., 1830

Parsons, James J. 'Towards a more humane geography', *Economic Geography* 45, no. 3, 1969

Partington, J. R. *A History of Chemistry*, vol. I, London: Macmillan, 1970

Passmore, J. A. *Ralph Cudworth: An Interpretation*, Cambridge: Cambridge University Press, 1951

Payne, Robert. *Marx*, London: W. H. Allen, 1968

Peddie, R. A. and Waddington, Q. (eds). *The English Catalogue of Books . . . Books issued in the United Kingdom of Great Britain and Ireland 1801–1836* (London: Sampson Low and Marston, for The Publishers' Circular, 1914), reprinted, New York: Kraus Reprint Corp., 1963

Peet, Richard. *Radical Geography: Alternative Viewpoints on Contemporary Social Issues* (1977), London: Methuen, 1978

Peschel, Oskar. *Geschichte der Erdkunde bis auf Alexander von Humboldt und Carl Ritter* (1865), 2nd edn corrected by S. Ruge (Munich: Oldenberg, 1878), reprinted Amsterdam: Meridian, 1961

Pfeiffer, Heinrich (ed.). *Alexander von Humboldt. Werk und Weltgeltung*, ed. Heinrich Pfeiffer for the Alexander von Humboldt-Stiftung, Munich: Piper and Co., 1969

Piaget, Jean. *Genetic Epistemology*, translated by Eleanor Duckworth, New York: Columbia University Press, 1970

Plato. *The Republic*, translated by Paul Shorey (1930–5), Loeb Classical Library, 2 vols, Cambridge, Mass.: Harvard University Press, 1953–6

  Timaeus, Critias, Cleitophon, Menexenus, Epistles, translated by R. G. Bury, Loeb Classical Library, Cambridge, Mass.: Harvard University Press, 1952

Pliny. *The History of the World commonly called The Natural History of C. Plinius Secundus or Pliny*, translated by Philemon Holland (1601), ed. P. Turner, New York: McGraw-Hill, 1964

  Natural History, with an English translation in 10 volumes by H. Rackham, Loeb Classical Library, Cambridge, Mass.: Harvard University Press, 1942

Polanyi, Michael. *Knowing and Being*, ed. Marjorie Grene, Chicago: University of Chicago Press, 1969

  Personal Knowledge (1958), revised edn, 1962; reprinted, New York and Evanston: Harper and Bros, 1964

  The Tacit Dimension, London: Routledge and Kegan Paul, 1967

Poole, W. F. and Fletcher, W. I. *Index to Periodical Literature, 1802–96*, 4 vols, New York: Peter Smith, 1938

Popper, Karl R. *Conjectures and Refutations: The Growth of Scientific Knowledge*, London: Routledge and Kegan Paul, 1963

The Logic of Scientific Discovery (1934), London: Hutchinson, 1959

Objective Knowledge: An Evolutionary Approach, Oxford: Oxford University Press, 1972

The Open Society and its Enemies (1945), 3rd edn, London: Routledge and Kegan Paul, 1957

Power, Henry. *Experimental Philosophy, in three Books: Containing New Experiments: Microscopical, Mercurial, Magnetical. With some Deductions, and Probable Hypotheses, raised from them, in Avouchment and Illustration of the now famous Atomical Hypothesis*, London: Martin and Allestry, 1664

Price, H. H. *Hume's Theory of the External World*, Oxford: Clarendon Press, 1940

Ptolemy. *Geography of Claudius Ptolemy*, translated by E. L. Stevenson, New York: New York Public Library, 1932

Claudii Ptolemaei Geographia, ed. C. F. A. Nobbe (Leipzig 1843–45), facsimile, Hildesheim: Georg Olms, 1966

Claudius Ptolemaeus, Geographia (Strasburg, 1513), facsimile, vol. IV in *Theatrum Orbis Terrarum*, 2nd series, introduction by R. A. Skelton, Amsterdam: 1966

Ray, John. *Further Correspondence of John Ray*, ed. R. W. T. Gunther, London: Ray Society, 1928

Observations Topographical, Moral, and Physiological; made in a Journey through Part of the Low Countries, Germany, Italy, and France, London: Martyn, 1673

The Wisdom of God Manifested in the Works of the Creation. In two Parts. (1691), 12th edn, corrected, London: Rivington, Ward and Richardson, 1759

Relph, Edward. 'An Inquiry into the relations between phenomenology and geography', *Canadian Geographer* 41, no. 3 (1970), 193–201

Rennell, James. *A Treatise on the Comparative Geography of Western Asia*, 2 vols, London: Rivington, 1831

Richards, Paul. 'Kant's geography and mental maps', *Trans. Inst. Br. Geogr.* 61, (1974), 1–16

Ritter, Carl. *Comparative Geography*, translated by William L. Gage, Edinburgh and London: William Blackwood and Sons, 1865

Robinson, Harry. 'Geography in the Dissenting Academies', *Geography* 36 (1951), 179–86

Roslansky, J. D. (ed.). *The Control of Environment*, Amsterdam: North-Holland, 1967

Rotenstreich, Nathan. *Experience and its Systematization: Studies in Kant*, The Hague: Martinus Nijhoff, 1965

Rousseau, Jean-Jacques. *Émile ou de l'éducation* (1762), Paris: Garnier Frères, 1964

Russell, Bertrand. *A Critical Exposition of the Philosophy of Leibniz* (1900), 2nd edn, London: George Allen and Unwin, 1951

Our Knowledge of the External World (1914), New York: Mentor, 1960

Sack, Robert David. 'Geography, geometry, and explantion', *Annals AAG* 62, no. 1 (1972), 61–78

Sauer, Carl O. 'The fourth dimension of geography', *Annals AAG* 64, no. 2 (1974), 189–92

Schaefer, Fred K. 'Exceptionalism in geography: a methodological examination', *Annals AAG* 43, no. 3 (1953), 226–49

Scheffler, Israel. *Science and Subjectivity*, New York: Bobbs-Merrill, 1967

Schmithüsen, Josef. *Geschichte der geographischen Wissenschaft von den ersten*

Anfängen bis zum Ende des 18. Jahrhunderts, Mannheim, Vienna, Zurich: Bibliographisches Institut, 1970

Schopenhauer, Arthur. *The World as Will and Idea,* vol. I, translated by R. B. Haldane and J. Kemp (1883), London: Routledge and Kegan Paul, 1950

Schultze, Joachim H. (ed.). *Alexander von Humboldt: Studien zu seiner universalen Geisteshaltung,* Berlin: Walter de Gruyter, 1959

Schwarzenberg, F. A. *Alexander von Humboldt: or, What may be Accomplished in a Lifetime,* London: Robert Hardwick, 1866

Scurla, Herbert. *Alexander von Humboldt: Ansichten der Natur: Ein Blick in Humboldt's Lebenswerk,* 2nd edn, Berlin: Verlag der Nation, 1962

Sewell, W. R. Derrick, and Foster, Harold D. (eds.). *The Geographer and Society,* Western Geographical Series, vol. I (1970), revised edn, Victoria, B.C.: University of Victoria, 1974

Singer, Charles. *A Short History of Scientific Ideas to 1900* (1959), reprinted, Oxford: Oxford University Press, 1972

Sinnhuber, Karl A. 'Alexander von Humboldt, 1769–1859', *Scottish Geographical Magazine* 75, no. 2 (1959), 89–101

Smith, Adam. *The Early Writings of Adam Smith,* ed. J. R. Lindgren, New York: Kelley, 1967

An Inquiry into the Nature and Causes of the Wealth of Nations (1776), London: George Routledge and Sons, 1893

Somerville, Martha. *Personal Recollections, from early Life to old Age of Mary Somerville. With Selections from her Correspondence,* London: John Murray, 1874

Somerville, Mary. *The Connexion of the Physical Sciences* (1835), 10th edn, revised by Arabella B. Buckley, London: John Murray, 1877

Physical Geography, London: John Murray, 1848

Physical Geography (1848), 2nd edn, revised, 2 vols, London: John Murray, 1849

Spate, O. H. K. 'Muscovite geography: impressions and developments', Handbook of Fifth Meeting of Institute of Australian Geographers, 1–18, Sydney, August 1966

Sprat, Thomas. *History of the Royal Society* (1667), ed. J. I. Cope and H. W. Jones, London: Routledge and Kegan Paul, 1959

Stevens, Henry. *The Humboldt Library. A Catalogue of the Library of Alexander von Humboldt* (London: Henry Stevens, 1863), facsimile, Leipzig: Zentral Antiquariat der Deutschen Demokratischen Republik, 1967

Stoddart, D. R. 'Darwin's impact on geography', *Annals AAG* 56, no. 4 (1966), 683–98

Strabo. *The Geography of Strabo,* translated by H. L. Jones and J. R. S. Sterrett (1917), Loeb Classical Library, 8 vols, reprinted, Cambridge, Mass.: Harvard University Press; London: William Heinemann, 1969

Strahler, Arthur N. *The Earth Sciences,* New York: Harper and Row, 1963

Physical Geography, New York: John Wiley, 1960

Strawson, P. F. *The Bounds of Sense,* London: Methuen, 1966

Taylor, Eva Germaine Rimington. *Late Tudor and Early Stuart Geography, 1583–1650,* London: Methuen, 1934

The Mathematical Practitioners of Hanoverian England, 1714–1840, Cambridge: Cambridge University Press, 1966

Tudor Geography 1485–1583, London: Methuen, 1930

Taylor, Griffith (ed.). *Geography in the Twentieth Century. A Study of Growth,*

Fields, Techniques, Aims and Trends (1951), 3rd edn, London: Methuen, 1960

Teggart, F. J. *Theory of History* (Yale, 1925), reprinted in his *Theory and Processes of History*, Berkeley: University of California Press, 1960

Terra, Helmut de. *Humboldt: The Life and Times of Alexander von Humboldt, 1769–1859*, New York: Knopf, 1955

Tooley, R. V. *Maps and Map-Makers* (1949), 2nd edn, New York: Bonanza, 1961

Townson, Robert. *Tracts and Observations on Natural History*, London: Townson, 1799

Troll, Carl. 'The work of Alexander von Humboldt and Carl Ritter: a centenary address', *Advancement Science* 16, no. 64 (1960)

Tuan, Yi-Fu. *Man and Nature*, Commission on College Geography Resource Paper no. 10, Washington, D.C.: Association of American Geographers, 1971
'Humanistic geography', *Annals AAG* 66, no. 2 (1976), 266–76

Ullman, Edward L. 'Geography as spatial interaction', *Proceedings of the Western Committee on Regional Economic Analysis*, ed. D. Revzan and E. A. Englebert, Berkeley (1954), 1–13

Van Dusen, Robert. *The Literary Ambitions and Achievements of Alexander von Humboldt*, Berne: Herbert Lang, 1971

Van Paassen, C. *The Classical Tradition of Geography*, translated by C. Reith-Aerssens, B. Wevers and R. Symonds, Groningen: Wolters, 1957

Velikovsky, Immanuel. *Earth in Upheaval* (1956), London: Abacus, 1973
Worlds in Collision, London: Victor Gollancz, 1950

Vickers, Geoffrey. *Value Systems and Social Process*, New York: Basic Books, 1968

Wagner, Philip. *The Human Use of the Earth: An Examination of the Interaction between Man and his Environment*, London: Free Press of Glencoe, 1960

Waldseemüller, Martin. *Cosmographiae Introductio* (1507), facsimile, with an English translation: 'Introduction to Cosmography with certain necessary Principles of Geometry and Astronomy, to which are added the Four Voyages of Amerigo Vespucci', by J. Fischer and F. von Wieser, Ann Arbor: University Microfilms, 1966

Walmsley, D. J. 'Positivism and phenomenology in human geography', *Canadian Geographer* 18, no. 2 (1974), 95–107

Walsh, W. H. 'Philosophy and psychology in Kant's *Critique*', *Kant-Studien* 57, (1966), 186–98

Walsh, William. *The Use of Imagination: Educational Thought and the Literary Mind* (1959), Harmondsworth, Middlesex: Penguin Books, 1966

Warntz, William. *Geography Now and Then: Some Notes on the History of Academic Geography in the United States*, New York: American Geographical Society, 1964

Watson, R. A. and Watson, P. J. *Man and Nature. An Anthropological Essay in Human Ecology*, New York: Harcourt, Brace and World, 1969

Whewell, William. *History of the Inductive Sciences, from the earliest to the present Time* (1837), 3rd edn, 3 vols, London: John W. Parker and Son, 1857
Novum Organon Renovatum. Being the second Part of the Philosophy of the Inductive Sciences, 3rd edn, London: John W. Parker and Son, 1858

Whiston, William. *A New Theory of the Earth, from its Original to the Consummation of all Things. Wherein the Creation of the World in Six Days, the Universal Deluge, and the General Conflagration, as laid down in the Holy Scriptures, are shown to be perfectly agreeable to Reason and Philosophy* (1696),

5th edn, *To which is added an Appendix, containing a new Theory of the Deluge*, London: John Whiston, 1737

Whitehead, Alfred North. *The Concept of Nature* (1920), reprinted, Cambridge: Cambridge University Press, 1955

Wilcock, A. A. '"The English Strabo" the geographical publications of John Pinkerton', *Trans. Inst. Br. Geogr.* 61 (1974), 35–45

Wilson, Alan G. 'Theoretical geography: some speculations', *Trans. Inst. Br. Geogr.* 57, (1972), 31–44

Wolf, A. *A History of Science, Technology, and Philosophy in the 16th and 17th Centuries* (1935), 2nd edn, London: George Allen and Unwin, 1950

Wooldridge, S. W. and East, W. Gordon. *The Spirit and Purpose of Geography* (1951), London: Hutchinson's University Library, 1955

Wright, John Kirtland. *Human Nature in Geography: 14 Papers, 1925–1965*, Cambridge, Mass.: Harvard University Press, 1966

Zaunick, Rudolph (ed.). *Humboldt: Kosmische Naturbetrachtung. Sein Werk in Grundriss*, Stuttgart: Kröner, 1958

Zelinsky, Wilbur. 'Beyond the exponentials, the role of geography in the Great Transition', *Economic Geography* 46, 3 (1970), 498–535

'The demigod's dilemma', *Annals AAG* 65, no. 2 (1975), 123–43

Zelinsky, Wilbur, Kosiński, Leszek, and Prothero, R. Mansell (eds). *Geography and a Crowding World. A Symposium on Population Pressures upon Physical and Social Resources in the Developing Lands*, New York: Oxford University Press, 1970

INDEX

Abbot, George, 68
Academy of Sciences, Paris, 102
Achinstein, Peter, 267
Ackerman, Edward A., on ecosystem
  concept, 270–1
Adam, Alexander, 167
Adams, Michael, 167
America, 299 n. 4
  discovery of, 32–3
  Humboldt's research on, 33–4, 221–32,
    234–5, 238, 240–1, 249–52
  named after Amerigo Vespucci, 33–4
  referred to: in geography texts, 68, 70,
    75, 95–6, 98, 150, 156–8, 168–71,
    234–5, 237; by Herder, 217–19
anarchism, 261
Anaxagoras, 259, 286, 288
Anaximander, 87, 192, 282, 296 n. 48
Anschauung, Pestalozzi on, 239
  see also *Weltanschauung*
Apian, Peter (Apianus or Bienewitz), *34*,
  69–70, 72
a priori knowledge, theory of
  in Descartes, 62, 65
  in Kant (categories of), 203
  in Leibniz, 188–90
  in Plato, 20–3
  rejected: by Herbart, 245; by Locke,
    126–8; by Newton, 103
Aquinas, St Thomas, 18, 27, 39
Arcueil, Society of, 195
Aristotle, 15–27 passim
  cosmology of, 15–18, 21–2, 59–60, 68
  on deductive logic, 18, 36–7, 43
  on faculty theory of mind, 19, 24–7,
    39
  on induction (*epagogē*), 35–6
  Metaphysics, 23ff.
  On the Psyche (Soul), 23–7
  Organon, 18
  Physics, 23
  and rational empiricism, 243, 258, 287
atlas, 34
atomic theory
  of Anaxagoras, 286, 288

applied to experience and mind, 128–32,
  178–81
of Democritus, 51, 112, 180
of Empedocles, 23, 180
of Epicurus and Lucretius, 112, 180
Leibniz on, 187–8
Ray on, 112–13
see also materialism

Bacon, Francis, *36–57*, 102, 215–16, 218,
  231, 253
  Advancement of Learning (1605) and
    Novum Organum (1620), 18, 37–56
    passim
  on concepts, 48, 54–6
  and faculty theory of mind, 19, 39ff.
  on geography, 51–7
  on history, 48–50
  on imagination, 48–9, 55–6
  on inductive method, 18, 35, 37–42, 55–6
  objectivism in, 3, 26–7, 41–7, 56–7
  on science, classification of, 47–55
  sense-empiricism of, 3–4, 18, 26–7,
    37–42, 64–5, 253; in Locke, Berkeley,
    Hume, 125–42 passim, 175–90; see also
    under facts
  Sylva Sylvarum and *New Atlantis* (1627),
    57
  on utility of knowledge, 38, 42–54 passim
Bacon, Roger, 31
Barrow, John, 161
Barrows, Harlan, and geography as human
  ecology, 270
Bateson, Gregory, and *cybernetic
  epistemology* and ecology of mind,
  274–5
Bauer, Bruno, 251–3
Beck, Hanno, 210, 250–2, 305 n. 1, 309 nn.
  75 and 79
Bentham, Jeremy, and utilitarianism, 253–4
Berghaus, Heinrich, 240, 242, 249, 254
Bergmann, Torbern, 160
Berkeley, George, 125, *132–4*, 136, 179–80,
  299 n. 10
Berlin, Isaiah, 126, 136, 142–3, 303 n. 7

Berghaus, 240; Forster, 211–12;
Guyot, 171–2; Humboldt, 213–62
passim; Malte-Brun, 235–7; Pinkerton,
233–5; Ratzel, 270; Ritter, 238–9;
Schneider, 159–60; Somerville, 247;
Zeune, 237–8; Zimmerman, 237–8
physical, 101, 118, 157–60, 165, 172, 196,
218, 234–6, 270; as geography of all
nature (*physis*), including man:
Büsching on, 157–9, of Guyot, 171–2,
247; Humboldt on, 231–2, 242, 246–59
passim; Kant on, 159, 203–9, 255; of
Somerville, 247
place of man in, 66–7, 74, 80–5, 206–7
of plants, 8, 213–14, 218–20, 222–5,
231–2; see also biogeography, geognosy
and positivism, 5–7, 10–12, 152, 171–3,
237, 260–75 passim
Rousseau on, 198
and sense-empiricism, 5–12, 51–3, 57–8,
66–8, 79–91 passim, 100–2, 104–6,
122–5, 144–6, 152
of seventeenth century, 66–8, 69–122
passim: Abbott, 68–9; Blome, 96–8;
Bohun, 98–100; Carpenter, 70, 72–5;
Christiani, 81; Clarke, 95–6; Cluverius,
71–2; Comenius, 77; Echard, 99,
144–6; Feind, 91; Gölnitz, 81; Pat
Gordon, 147–8, 153; Heylyn, 70–1;
Keckermann, 69–70; Pemble, 76;
Sanson, 96, 98; Stafford [Prideaux],
76; Varenius, 77–90, 96–7, 276–83·
study of history of, 1, 6, 10–13, 68–9, 87,
91, 124–5, 144, 162, 166–7, 248–51,
282, 290–1 nn. 17, 18, 299–300 nn. 1, 3
and 7, 302 n. 49
theory of: classical–Renaissance models,
18, 27–34; current debates on, 1–3,
5–14, 100–1, 122, 260–75 passim; in
eighteenth century, 122–5, 144–6,
154–61, 165–73, 206–9, 211–14, 218–22;
in nineteenth century, 223–5, 231–42
passim, 246–51, 255–9, 284–9; in
seventeenth century, 51–7, 69–70,
73–107 passim
of twentieth century, 1–3, 5–13, 247–9,
260–75 passim
see also chorography, climate,
cosmography, demography,
Erdbeschreibung, Weltbeschreibung
geology, 116–17, 122, 191, 193, 212–14,
235, 249
Humboldt on, 220–3, 232, 247, 250
geosophy, proposed by Wright, 6
Gilbert, William
Bacon attacks, 40–1
Carpenter on, 72, 74
on earth magnetism, 17, 41

on empirical method, 17
Galileo on, 60
Humboldt on, 40–1
Glacken, Clarence, 121, 290 n. 17, 298 n.
48, 302 n. 41
Goethe, Johann Wolfgang von, 12, 176, 193,
199, 216–17. 225
Gölnitz (Golnitzius), Abraham, 70, 81
Gordon, George, 151
Gordon, Patrick, *Geographical Grammar*
(1693), 147–8, 153–4
Graunt, John, 105, 122
Gregory, Derek, 263
Gregory, J., *Manual of Modern Geography*
(1739), 151
Grosseteste, Robert, 31
Guthrie, William, *New Geographical,
Historical and Commercial Grammar*
(1770), 164–5, 167, 169
Guyot, Arnold, *The Earth and Man* (1849),
171–2, 239, 247

Haeckel, Ernst Heinrich, coins term
ecology, 8, 269
Halley, Edmond, 102
introduces isogonic lines, 105
Hartshorne, Richard, 101, 297, n. 21
Harvey, David, 3, 152
Hegel, Georg Wilhelm Friedrich, 119, 194,
225, 242–5, 253, 255, 258–9, 287–8
Hegelians, 251–3, 255
heliocentric theory, 11, 72, 162
see also Copernicus
Helvétius, Claude-Adrien, 179, 253
Heraclitus, 259, 286–8
Herbart, Johann Friedrich, 245–6
on geography, 245–6
rejects faculty theory of mind, 245
Herbartianism, adapted to
sense-empiricism, 245
Herder, Johann Gottfried von, 159, 194,
216–20, 227, 254
Heron, Robert, 164, 168
Heylyn, Peter, *70–1*, 76, *92–4*, 144, 147–8,
150
Hipparchus, 29, 88
Hippocrates of Cos, *Airs, Waters, and
Places*, 20, 288
history, 9, 32
Bacon on, 48–50
and continuity of knowledge, 9–10, 12,
248–9, 255–9, 263–75 passim
and empirical method, 4–10 passim,
181–3, 185–6, 263–75 passim
etymology, 49
of geography, *see under* geography
Hegel on, 243, 253
Herder on, 217–19

LaVergne, TN USA
19 January 2010
170551LV00003B/34/P